William B. Kays
CONSTRUCTION OF LININGS FOR RESERVOIRS, TANKS,
AND POLLUTION CONTROL FACILITIES

John E. Traister
CONSTRUCTION ELECTRICAL CONTRACTING

William R. Park
CONSTRUCTION BIDDING FOR PROFIT

J. Stewart Stein
CONSTRUCTION GLOSSARY: AN ENCYCLOPEDIC
REFERENCE AND MANUAL

James E. Clyde
CONSTRUCTION INSPECTION: A FIELD GUIDE
TO PRACTICE

Harold J. Rosen and Philip M. Bennett
CONSTRUCTION MATERIALS EVALUATION AND
SELECTION: A SYSTEMATIC APPROACH

C. R. Tumblin
CONSTRUCTION COST ESTIMATES

Harvey V. Debo and Leo Diamant
CONSTRUCTION SUPERINTENDENT'S JOB GUIDE

Oktay Ural, Editor
CONSTRUCTION OF LOWER-COST HOUSING

CONSTRUCTION OF STRUCTURAL STEEL BUILDING FRAMES

CONSTRUCTION OF STRUCTURAL STEEL BUILDING FRAMES

Second Edition

WILLIAM G. RAPP, P.E., S.B.
Consulting Engineer

A Wiley-Interscience Publication

JOHN WILEY & SONS, INC.

New York • Chichester • Brisbane • Toronto

Library of Congress Cataloging in Publication Data

Rapp, William G
   Construction of structural steel building frames.

   (Wiley series of practical construction guides)
   Bibliography: p.
   Includes index.
   1. Building, Iron and steel.   2. Framing (Building)

I. Title.
TH1611.R37   1979   693.7′1   79-19146
ISBN 0-471-05603-0

Printed in the United States of America

10 9 8 7 6 5 4 3 2 1

Series Preface

The Wiley Series of Practical Construction Guides provides the working constructor with up-to-date information that can help to increase the job profit margin. These guidebooks, which are scaled mainly for practice, but include the necessary theory and design, should aid a construction contractor in approaching work problems with more knowledgeable confidence. The guides should be useful also to engineers, architects, planners, specification writers, project managers, superintendents, materials and equipment manufacturers, and, the source of all these callings, instructors and their students.

Construction in the United States alone will reach $250 billion a year in the early 1980s. In all nations, the business of building will continue to grow at a phenomenal rate, because the population proliferation demands new living, working, and recreational facilities. This construction will have to be more substantial, thus demanding a more professional performance from the contractor. Before science and technology had seriously affected the ideas, job plans, financing, and erection of structures, most contractors developed their know-how by field trial-and-error. Wheels, small and large, were constantly being reinvented in all sectors, because there was no interchange of knowledge. The current complexity of construction, even in more rural areas, has revealed a clear need for more proficient, professional methods and tools in both practice and learning.

Because construction is highly competitive, some practical technology is necessarily proprietary, but most practical day-to-day problems are common to the whole construction industry. These are the subjects for the Wiley Practical Construction Guides.

M. D. Morris, P.E.

New York, New York

v

Preface

An economical design must take into account the problems faced by the structural steel erector. Structural steel erection consists in taking various steel members that have been fabricated in a shop from plates, angles, and other rolled shapes and placing them in their correct positions in the field to form a finished steel frame for the walls, floors, and roof of a building or for the permanent nonstructural steel parts of the completed structure. Erection also includes the preliminary steps that are necessary before those members can be set in place and the subsequent operations of aligning, plumbing, and securing them permanently by bolting or welding. Setting up and later dismantling the equipment needed to perform all the operations involved in the production of a finished piece of work are other phases of structural steel erection.

Since the types of structure in which structural steel frames are required are so varied, this book is limited to the work of erecting tier, column-core, and mill buildings, hangars, pier sheds, boiler rooms and power houses, and some miscellaneous structures.

The book is intended for those in management who are directly concerned with structural steel erection, as well as for men actually engaged in erecting the structural steel building frames and for engineers who have had little or no previous experience in that kind of work. It is also intended to be a guide for architects and structural engineers who design and supervise construction, and who need an understanding of steel erection procedures. Students planning to enter the construction field can use this book to good advantage.

For best results there must be complete cooperation among all concerned. The architect and engineer preparing the design, the fabricator of the steel, the general contractor, the builder's office, the field engineers, and the ironworkers in the field are all responsible for achieving a safe, economical and efficient procedure.

The nine chapters of this book deal with different phases of structural-steel erection. Although all are closely interrelated, they are divided specifically into preparation for estimating, actual estimation, servicing and use of tools and equipment, fieldware preparations, selection of field personnel, organizing the job, placement of the structural-steel members with various types of equipment, and all the necessary operations required to complete the work, such as plumbing, fitting, bolting, riveting, and welding.

The assembling (or setting up) and dismantling of equipment is described in detail. Although welds and high-strength bolts are replacing the use of rivets, Appendix E contains sections on rivets and riveting, including procedures other than for a bolted or welded structure, on estimating and erecting a riveted building frame, and on riveting tools for the benefit of those erectors who may be required to use rivets but who are not familiar with the tools and techniques for driving rivets.

Many designers and erectors have had almost no experience with structures erected by guy derricks. Accordingly sections have been devoted to the guy derrick and the procedures necessary for its use. In addition, the need for planning the structure to support such a piece of erecting equipment is explained.

Although a separate chapter is devoted to safety, this subject is discussed throughout the book as normal procedure. Safety must be a part of all planning and actual performance. Accordingly, it is not singled out except in the chapter that summarizes the subject and provides basic facts to help implement safe practices.

A suggested erection safety code is included in Appendix B.

WILLIAM G. RAPP

Larchmont, New York
July, 1979

Contents

4 Safety   89

5 The Erection Scheme   110

CONSTRUCTION OF STRUCTURAL STEEL BUILDING FRAMES

1

Estimating

1.1 Preparation

Requests for bids may come to a steel erector from a fabricator who has no erection facilities but who desires to submit a single price for furnishing, fabricating, and erecting a steel structure. An owner may desire prices for furnishing and fabricating, separate from erection, to compare various bids for each and thus secure the best combination of the various prices. Occasionally there is a "public letting" where the request for bids has been through publication, advertising the basic requirements. Specifications are then either furnished for a fee, often returnable on relinquishing the specifications and drawings, or they may be available on request. The successful bidder retains his set; all other bidders' sets are subject to recall. In such an advertisement requesting bids, the details will be stated if separate bids are desired for furnishing and fabricating, for erection only, or for a combination of the two.

The request for a bid can also come from an architect or a structural engineer. Often only one erector will be asked to quote when the erector has established an excellent reputation for low prices, efficient and safe work, speedy erection, and dependability to meet contract dates and obligations. Occasionally the cost of erection will exceed the cost of furnishing and fabricating the steel, in which case the erector will probably submit a combined bid after securing a price from a fabricator for his portion of the contract. In any of these situations the steel erector must be prepared to estimate the erection cost.

To be prepared, an erector should have prior records of all features of the work involved. The best records are those based on past experience of the firm or past experience of the individuals preparing the estimate. Two identical structures are rarely erected exactly alike, or for the same costs, or at identical rates of progress or production. Accordingly, in addition to having records of costs, progress, and production, the estimator must be able to judge from past experience what to expect in preparing a new estimate.

To provide records of value for future estimates, the field force should be told what information will be needed on a project; what records to keep in order to furnish worthwhile details to the estimator; and what reports must be prepared to aid not only the estimator but also the office engineer responsible for the safe, efficient, economical operation in the field.

Assuming that a contract has been secured and a complete analysis has been made, the instructions to the field will spell out what information to assemble. This information should be such that the office can set up a systematic list of field production results according to the type of structure, kinds of equipment used, location area, the labor that was available, and individual operations.

For example, the records should show the cost, time, and size of gang required to unload, assemble, set up, reeve, jump or move, dismantle, and load out individual pieces of equipment by types. These types will cover the various kinds of derricks and travelers, including any type of derrick mounted on the traveler platform; they will be divided into low travelers and tower travelers; crawler, truck, tower, and other types of cranes; jinniwinks; "Chicago booms"; jigger sticks; gin-poles; dutchman; handline. The records should provide similar information on unloading from the various forms of steel transportation and on each of the different types of equipment used for sorting, distributing, and raising. They must be separated into type of structure such as low, medium, and tall tier buildings; lightweight apartment houses; mill buildings; church, theater, or convention-hall types; hangars; and all the various types of steel-framed structures the erector expects to bid on. The cost of passing and relaying steel from one rig to another and from an unloading point to the erecting equipment is needed.

A system should be set up whereby the records for each type of structure are kept separately, with the kind of equipment used being so distinguished that it can be spotted quickly in preparing a new estimate for a structure of similar type that can be erected by similar equipment. When there has been no prior experience with equipment selected to erect a structure being estimated or on prior experience on that type of structure, judgment must be used in deciding on production and size of gangs, i.e., tons or pieces per day per gang, or—as it is usually expressed—per gang-day. The cost per gang-day can be readily figured from the labor rates for the area where the work will be done, after deciding on the size of gangs.

The cost per ton or per piece can be calculated from this and from a reasonable production rate.

A check of estimated cost based on production by tons per day should be made against production by pieces per day. Too often a structure will have many pieces per ton so that if the estimate is based on what seems a reasonable tonnage to expect per day per piece of equipment, it may turn out to be equivalent to a figuratively astronomical number of pieces per day. By the same token, a reasonable tonnage erected per day may involve so few pieces that production will be estimated too far below what can really be expected. The estimate will then be so far out of line that the erector will not be competitive and thus will not be awarded the contract.

In addition to cost, progress, and production by type of structure and by kind of erecting equipment, records must be secured from actual fieldwork of such operations as bolting and welding. Records should be kept separately for machine (unfinished) bolts, high-strength bolts, turned bolts (which often require reaming of the holes in addition to installing the bolts), centerless-ground bolts, body-bound bolts, ribbed bolts, and any other special bolts. The records must show clearly if the bolters performed the fitting up or if a special fitting-up gang preceded the bolters. This will be reflected in the cost but probably not in the progress or elapsed time.

The records of welding should be assembled from the feet of different sizes and types of welds and for positions of welding, or preferably, from the weight of electrodes deposited, taking into account the weight of wasted electrode stubs and lost or spoiled electrodes. Production using ordinary electrodes (E60XX series) should be kept apart from that using alloy or high-strength electrodes.

Additional information should be developed in the field for use in the office by the estimator, covering any other operations performed such as plumbing columns, aligning portions of the structure, laying, picking up, and moving planks, installing and tightening bracing and sag or tie rods, installing crane rails, drilling holes, removing rivets in old structures by cutting out or burning, and unloading, assembling, moving, dismantling, and removing falsework.

In addition to all such records derived from field reports, overhead items and costs other than direct field labor must be assembled. Even if the erector's equipment will normally move by truck from his toolhouse, costs should be on hand for renting various sizes of trucks and for running the erector's own trucks. These latter costs must include not only the actual on-the-road expense, but also maintenance, repairs, depreciation, insurance, and replacement costs based on a time-life or mileage-life basis or a combination of the two.

Costs of transporting key men should be available, including any living expenses that may be incurred. The availability should be determined, and the costs developed, of renting erecting equipment and tools. This should be kept up to date. A cost-accounting system should be set up for the erector's own equipment and tools. The cost of maintaining a toolhouse must be prepared, taking into account not only the running cost of the labor involved, but also rent, taxes, insurance, depreciation, and replacement of the structure as well as major equipment and tools. This must all be assembled and a unit cost decided on, such as a yearly cost divided by anticipated tonnage to be erected, to give an average cost per ton for inclusion in the estimate. The yearly cost can be divided by the estimated total field payroll to give an average cost per dollar of estimated field labor. An additional cost per dollar, or cost per ton, of field labor should be calculated for the main office labor, expense, and overhead items.

A good erection organization office force will include, as part of management, a man to act as manager of erection with assistants as needed; a chief engineer or chief draftsman, with engineers and draftsmen to design and detail equipment, to prepare erection schemes, and to work on engineering problems; estimators familiar with steel erection and with the equipment owned by the company. In addition there should be a staff to handle estimating, engineering, contracting, legal work, purchasing, insurance, record-keeping, and managing the operation of the entire organization. The cost of this organization is usually estimated as a percentage of the field payroll.

The field office organization should be tied in to the main office through a resident engineer, field engineer, or junior engineer, and timekeeper. The cost for these men must be included in the estimate, either as a percentage of field payroll or as an estimated cost per ton, or by calculating the individual cost of each man who will be involved in the particular project being estimated.

Costs of all other phases of the work other than direct field labor should be assembled in anticipation of need in estimating a new job. These costs may include plank or timber; painting the structural steel; railroad or barge charges for shipping (if the equipment is to be sent by rail or water); cost of electrical installation of hoists, compressors, welding machines, etc. (if such equipment will be electrically powered), or cost of gasoline or diesel fuels (if the equipment is so driven). Other incidental costs such as electrodes, manila rope, wire rope, rope blocks, timber blocking, steel or wood falsework, scaffolding, small tools, shanties, and field offices should be available. With experience, they can sometimes be estimated as a percentage of labor cost. The overhead items are usually more than half of the total cost. By using percentages for these considerably smaller costs, the error will normally be minor and the time saved by not having to estimate the cost of each individual small item will expedite the making of the estimate. On some complicated erection schemes, the cost of purchasing, renting, or building such items as falsework or special equipment can be appreciable and must be recognized.

Insurance rates should be secured in advance to cover the many possible requirements other than the usual workmen's compensation rates (which depend on location and type of work), public liability and property damage insurance (which may depend on surrounding areas and type of structure), and the various forms of automobile insurance for erector-owned vehicles.

With all of these records of costs, production, and progress on hand, the estimator must still have some knowledge of the various ways a job can be erected; of how it should be erected for safety, efficiency, and economy; and of what equipment is available, both owned and rented, since it is sometimes better to rent equipment locally near the job site than to ship one's own to the job. He must be able to decide the probable adequacy and availability of men in the particular location. He must have an understanding of the abilities of the erector's superintendents and key men and the type of work each is best suited for. Above all he must have the courage to envision new methods of erection to achieve an erection estimate low enough to get the contract, high enough to leave a profit at the end, and yet based on a safe erection scheme.

When work will probably be done in a particular location, an attempt should be made in advance to secure copies of all applicable local ordinances and state and federal regulations, as well as standard specifications that may affect the work, such as those issued by the U.S. Army Corps of Engineers, AIA (American Institute of Architects), AISC (American Institute of Steel Construction), and AWS (American Welding Society). Information should be developed on license and permit requirements and on safety construction codes and standards that may apply—either local, state, or federal—for example, NSC (National Safety Council), ANSI (American National Standards Institute, formerly ASA—American Standards Association), and OSHA (Occupational Safety and Health Administration) *Safety and Health Regulations for Construction.* The above deal with work in the United States. Similar standards must be dealt with when working in foreign countries that have established similar safety regulations.

It is also advisable to keep posted on current labor rates and fringe benefits in areas where work will be estimated.

1.2   *Site Visit*

After a request has been received to bid on a proposed structure or the erector has been asked to bid on the structure, the first step is a thorough study of the proposed specifications, drawings, and form of contract. Immediately after a preliminary review of these documents, in order to get a good idea of the extent and type of work involved, the estimator or a competent erector or engineer should visit the site to determine conditions that may affect the erection method and all work involved, and that will affect the estimate. This first, preliminary review should develop a familiarity with the high spots of the job, including probable heavy pieces, probable awkward pieces or assemblies (either as fabricated or as assembled at the site before being lifted into place), unusual size or shape of pieces, pieces with probable poor lateral stability that may require special treatment in hauling from an unloading point or in lifting into place at the site, or even when erected but not yet tied in with bracing or other lateral support.

Other features in the specifications and drawings may require investigation at the site before pricing is done. These would include the area of the structure, its height, time requirements, unusual features such as providing and maintaining facilities for the owner's or customer's inspectors or engineers, housing, and facilities for radiographic equipment. The type of structure and size may suggest a logical type of equipment and erection method to use. The site visit must then determine the feasibility of using such equipment and method. It is also important to check on possible labor restrictions such as the kind of labor required, requirements of citizenship, local residence, etc., since the site visit will be used to develop availability of the restricted labor in the area.

Upon visiting the site, not only must the particulars of the specifications and drawings be investigated, but a survey must be made of possible obstructions or hindrances to delivery, unloading, erection, and use of equipment; and an investigation must be made as to the cost of their removal or elimination, or the additional expense that may be incurred in overcoming them. The streets between the site and the point where the fabricator will deliver the steel must be checked. Any inadequate bridges or street paving on the probable route should be noted for load prohibitions. Overhead wires (power, light, telephone, or telegraph) that might affect the permissible height of the load must be noted. These wires can often be raised or removed, but the cost for this should be included in the estimate, unless provided for in the specifications as an obligation of the owner or general contractor.

The probable delivery point by boat or barge to a dock, on cars to a railroad yard, or by truck to the site must be determined. Unloading facilities from railroad cars at a yard, or from a railroad lighter to trucks at a dock, may be costly. Their use is often at the erector's expense. Unloading facilities, if away from the site, may affect the estimated cost. A local check may develop restrictions on the erector's portion of the hauling obligation. Some cities restrict delivery to particular hours of the day, to particular days of the week, to delivery on the side of the street nearest the site at limited times. All of these must be checked to decide how they may affect the steel erection cost.

A possible sorting yard to expedite erection, or a storage yard in which to preassemble trusses, etc., can be found at this time, and rental rates determined for inclusion in the estimate. Permits and licenses may be needed to operate trucks or equipment and the fees for these should be obtained. Traffic conditions that may affect the operation should be observed. Any protection that the erector may be responsible for can often be determined at this time, such as protection of roofs of adjoining buildings and protection of pedestrian or vehicular traffic.

Ground conditions must be observed because they may affect the type of equipment that can be used, or they may indicate the need of supplying mats, timbers, or planks to permit setting up or moving equipment. If foundation contractors are already at the site, their equipment could be available and would reduce the cost of setting up the erector's rigs. Probable interference by others, and of others by the erector, can sometimes be determined at this time. Space that will be available for field offices and tool or change shanties for the men should be checked. If none is available, nearby rooms or stores may be needed, and so rental rates should be secured. Types of anchorage that can be installed for equipment, and site conditions for use of cranes, for setting up equipment, and for possible falsework requirements should all be investigated. Any preliminary work done on permanent foundations should be observed since it may limit the anchorages that can still be placed for erection equipment and may give a clue to the time when erection of steel must be started.

The local labor situation should be investigated for available skilled men: whether nonunion men can be used or if union men must be hired, and whether the erector can bring his own men in or if he must hire locally. Gossip of local troubles, labor or otherwise, should be investigated, how they were settled or remedied, how they might affect the erector; what labor rates will probably be demanded by the men and possible increases or fringe benefits. If the site is in a busy city area, a check should develop areas for the men to park their cars. In an out-of-the-way area, housing may be needed for the men, and the details should be worked out.

Weather conditions in the area should be investigated, since adverse weather can delay operations and increase costs. The supervisory force and some key men are usually paid straight time whether the job works or not, so that adverse weather can cause considerable expense for such "lost time" wages. If the area has experienced considerable rainfall at the time of the year when the job will probably be under way, this will affect the lost time payments to these men as well as delay the completion of the job. At the same time, a knowledge of probable weather delays due to ice, snow, rain, hurricanes, or tornadoes will aid the estimator in deciding on how much time to allow for completion.

If the contract time is limited and delays are liable to be frequent it may mean estimating the use of additional equipment and men to expedite the work during the period when work can proceed unhampered. This in turn will affect the estimated cost of shipping equipment to the job, since extra costs for working overtime may have to be included.

1.3   *Estimate*

Having secured all information from the site visit that could affect the erection contract, the estimator is ready to proceed with the actual estimate. If possible, the "take-off" customarily made by the fabricator to estimate the cost of furnishing and fabricating the structural steel should be obtained. If this is not done, the estimator, or men skilled in this operation, must go over the design drawings in great detail, "taking off" the number and sizes of pieces to erect, totaling the weight of these pieces, checking the individual weights of heavy pieces, and estimating the number and type of bolts or welds as shown on the design drawing. If the quantities are not shown, the probable quantities must be estimated. He is now ready to decide on the probable erection method to arrive at a cost.

He will next decide on the type, capacity, and number of pieces of erecting equipment that will be required; the size and number of hoists, compressors, welding generators, transformers, or rectifiers. He will estimate the tonnage and number of pieces to be erected by power and by handline, separately. If by crane, by derrick, by traveler, or by a combination of different types of equipment, he will break down the tonnage and pieces since the rate of progress varies with the kind of equipment used.

As a check, he should find the pieces per ton or tons per piece for the powered equipment, and pieces per ton or pounds per piece for the material estimated to be set by hand. Extra heavy pieces and knocked-down trusses should be segregated because they may mislead an estimator when he decides on production if the weight and number of such pieces are left in the totals. He should decide on the production in tons per gang-day and check this against pieces per gang-day for each type of equipment, as explained previously. For handline work the production is best estimated in fractions of tons, or pounds, per man-day.

Columns should be analyzed for field splices and floors per tier. Probable moves in a mill building type of structure should be decided upon. Trusses should be studied to see if they will be shipped as complete units or partially assembled; or if they should be shipped knocked down and assembled at the site to be picked as complete trusses; or if they should be erected in place piece by piece on falsework because of the capacity of equipment, or local conditions, or other limiting requirements.

Girders should be similarly examined to determine if they should be shipped complete or in sections to be assembled at the job because of shipping limitations. If they must be assembled in place on falsework because they would be too heavy to pick in one piece by the equipment selected for erecting the rest of the steel, additional money must be included in the estimate for falsework, for assembly, and for permanently fastening the spliced sections in place aloft. If they appear to be laterally unstable, money must be added for bracing to make them safe to handle and to leave in place until permanent bracing can be erected to connect to them. If special hitches or special slings will be needed to handle any of these girders, the cost must be taken into account.

Possible shop assemblies to reduce work in the field are analyzed, bearing in mind the limitations of fabricating, shipping, and hauling, as well as the capacity of the erector's equipment and possible additional costs to the fabricator that the erector may have to pay for.

After a production decision has been made concerning unloading, sorting, distributing, and raising the steel itself by the various pieces of equipment or by handline; and concerning the production of bolts, welds, and the like to be installed per day; and an estimate is made of the tons and pieces in the structure, the estimator is almost ready to make his final estimate of costs. He must decide on the size of the various gangs and the labor rate to be paid, from which he can figure the total field labor cost of each type of gang.

For example, a guy derrick gang usually consists of a foreman or pusher, a hoist operator, six structural ironworkers, and an apprentice or helper covering several gangs to bring drinking water and erection bolts and do odd jobs. A crane gang is composed of a foreman or pusher, crane operator, oiler, and four or five ironworkers.

The estimator should have enough familiarity with steel erection to know the makeup of the various other kinds of gangs, so that by figuring the cost per day of each gang and the production he can expect per day from each, he can arrive at unit costs and days required for each operation.

For instance, estimating the number of a particular type of bolt that a two-man gang can install and tighten in a day, and dividing this into the cost of the two men plus part of their foreman and an apprentice feeding them bolts, will give the estimator a cost per bolt. This can be multiplied by the total estimated number of that type of bolt for a total cost. It will also enable him to estimate the number of two-man gangs needed to complete the bolting within a day or two after completion of actual erection of the steel.

In the same way, dividing the tons to be erected per day per rig into the total estimated tonnage, and comparing this time with the contract allowed time, the estimator will have an immediate clue to the number of rigs he must use. From this he can calculate the shipping and delivery costs depending on the number of pieces of equipment required to complete the job in time.

By being familiar with the work, the estimator will know what supervision will be required for the men actually doing the work. This will include superintendents plus assistant superintendents on a large job; general foremen if there are a number of foremen in charge of many different gangs; possibly a "floor boss" on a tier building with four or more erecting rigs working simultaneously; a resident engineer, field engineers, or junior engineers, some of whom may handle the necessary surveying instead of hiring surveyors for intermittent needs; timekeepers and clerks.

1.4   Forms

It is highly advisable to develop a standard form for preparing estimates. With spaces marked for all the possible items, there is less chance that the estimator will overlook anything. For example, there can be spaces for number of tons and pieces by different types of equipment, and for handwork. There should be spaces marked for number and types of bolt (machine bolts, high-strength, turned, body-bound) and their diameters; spaces for welds by type and size (butt, fillet, overhead, horizontal, and flat) and to show the total length of weld or weight of electrodes to be deposited for each. There should be spaces for quantities and diameters of sag rods, tierods, bracing rods, etc. In entering these items on this form, a good estimator will check the drawings for any unusually difficult connections that may affect production and the specifications for any unusually costly restrictions.

The weight and number of grillages, base plates, slabs, and setting or leveling plates should be listed since these are erected at different rates than the structural steel above them. Unloading, setting in approximate position, and then bringing them to final exact position and correct elevation by means of shims or other leveling devices will give a production in pieces per gang-day or tons per gang-day, much less than that which can be expected for the columns, beams, girders, etc., of the structure they will support.

The ideal estimating forms will list all probable operations to act as a tickler to prevent overlooking any items of work, supervision, or overhead. This listing should first cover all the items of labor and then all the items of overhead and materials or services, purchases, and any parts of the work that may be subcontracted.

With the portion covering field labor there should be spaces to note the number of job days estimated, as distinguished from days for an individual operation. For example, if the job should involve 1200 tons and the estimator is planning on an erecting rig that erects 40 tons per gang-day, this would be 30 operation days. But if he is figuring on shipping and using two rigs, this is only 15 job days. In determining the total time by adding up the total job days for each operation, he must guard against overlapping. If this same job involves high-strength bolting, and with the number of bolting gangs the estimator is assuming, the bolting will take 16 days, probably 13 or 14 of these 16 days will coincide with 13 or 14 of the erecting days. So in adding, he must use only 2 or 3 days in addition to the 15 erecting days in order to include both erecting and bolting.

After determining the total days required to do the work, he must then decide on how much lost time may be encountered. As stated before, certain key men such as the foreman and crane and hoist operators are usually paid straight time. These are in addition to the supervisory force of superintendent, engineers, timekeepers, etc. They are paid on holidays and bad-weather days when the rest of the field force does not or cannot work. The cost of this so-called lost time is part of the estimate. In summer, in a normal climate, the job may encounter an average of perhaps half a day, or even no days, per week of bad weather that prevents work. In winter, at the same locale, this may be one or more days lost per week. The time of year when the work will be done—if the climate is expected to be very hot or very cold, very windy, very dry, or extremely rainy, snowy, or icy; if there are local or national holidays to be recognized—all of these seasonal factors must be considered and a percentage for lost time added to the time and estimated cost of the actual performance. The cost of the supervisory force depends on this total time and is figured accordingly.

The form should list items in a definite sequence, based somewhat on a chronological pattern of probable operations on the job; for example, on a building estimate, production could be estimated and the labor costs priced in the following suggested order.

1.5   Labor

Unload equipment (derricks, cranes, hoists, compressors, welding machines, small tools, toolhouses, shanties, offices).

Build barricades, fences, sidewalk bridge, etc. (if in contract).

Set up equipment (derricks, cranes, compressors, welding machines, etc.; office, toolhouses, shanties, storehouses, etc.).

Dismantle equipment, load out equipment and tools. (This is an example in which dismantling and loading out are not in chronological order, but are placed here as a reminder to the estimator to be sure to include a cost; otherwise, this is easily overlooked. A good estimator can take a percentage of the items of unloading and setting up equipment, based on experience, without going through the details of the cost of dismantling each piece of equipment, loading out, etc.)

Derrick jumps, moves; crane or traveler moves; climbing crane jumps.

Relay derrick or crane.

Set anchor bolts, screed angles, shims, forms for grout, grout, engineering, surveying.

Unload and set grillages, base plates, slabs (loose or on columns), setting or leveling plates, wall-bearing plates. (Use tons per gang-day and check that against pieces per gang-day—if unreasonable, reevaluate production.)

Unload steel (possible storage yard, sorting yard, reload for delivery to site, unload from cars, trucks, carriers, barges, boats; at yard, at site).

Distribute steel (sort, possible relay hoisting, passing).

Assemble trusses, girders, subassemblies.

Erection (by derrick, crane, traveler, pole, combination, hand, etc.; estimate separately heavy pieces, trusses in one piece or knocked down, heavy girders, bulky or special pieces, relay, balance of steel).

Hand erection (pieces per man-day or per gang-day; check against tons or pounds per man-day, special equipment needed).

Planking (decide on own, rented, or purchased; time for laying, picking up, unloading, loading out; special for "open panel" construction).

Plumbing, aligning (columns, girts, spandrels, etc.; excessive aligning and guying; check if columns are liable to be self-aligning or will need guying).

Fitting up (for erection, bolting, welding; estimate probable percent of fitting-up bolts needed against total holes; use bolts per man-day).

Bolting (machine, turned, high-strength, ribbed, body-bound; estimate number of bolts per man-day installed and tightened).

Welding (size, length, type of welds, welding machines or transformers or rectifiers, number of points, feet or pounds of electrodes per man-day, cost of supervision, helpers, machine operator).

Lintels (unload, sort, distribute, erect, align, adjust; store loose lintels; possible return trips after completion of steelwork for adjustment of hung lintels).

Handrail (erect, align, fasten; feet per man-day or per gang-day).

Crane rails (erect, align, fasten; feet per gang-day).

Old structure (alter, connect to, knock out rivets, cut welds, drill holes, remove bolts, ream old holes, burn: square inches, number of points, location; fabricate, caulk, weld, falsework or shoring, method of dismantling and reerecting).

Falsework (unload, frame, erect, dismantle, remove, ship out or scrap, possible preassembly before contract time, salvage: sell locally or ship away).

Shoring.

Track (unload, lay, take up, load out; feet per gang-day).

Clean-up.

Cost of probable lost time.

Overtime (in areas where men are scarce it is often necessary to plan on working overtime as an inducement for men to come to the area; where not working overtime the job will be extended into bad-weather time of the year involving snow, ice, hurricanes, etc. Also, on a penalty contract, it is often better to work overtime and pay the extra amount of double time or time-and-a-half, as the case may be, than not to work overtime and instead pay penalty time because of not completing the work in contract time; whichever is cheaper can be used unless other considerations dictate to the contrary. For example, working overtime may increase production beyond the ability of the fabricator or carrier to produce or deliver the material fast enough).

Miscellaneous and contingency items.

Supervision (depends on actual erection time to unload, distribute, erect, and complete all other work; deduct overlapping time, add lost time, and then use for supervisory time and cost).

The form should then list all the items, other than actual field labor, that must be included in a proper estimate of the total cost of the work; this will be included in the steel erection contract. This is usually considered as overhead in contrast to direct labor.

1.6   Overhead

Field expense (aim at a percentage of estimated labor based on past experience; include expendable items such as gasoline or diesel fuels, oxygen, acetylene, grease and oil; incidentals such as hardware, telephone, paper cups, stationery; rental or purchase of office equipment and supplies such as adding machines, typewriters, computers, and water cooler).

Transportation of tools and equipment (secure freight rates, hauling rates; estimate the number of truckloads or carloads, both ways, of equipment, tools, supplies, etc. based on weight and size; demurrage on equipment, on structural steel, on materials; incidental trucking, miscellaneous trips to and from and around the job site; hauling steel, usually figured as a price per ton by a subcontractor hauler or erector's own trucks; include cost of unloading at delivery point if not done by erector's own forces; delivery of cranes and rented equipment; estimate number of loads at hourly or daily rate or price per ton).

Travel expense (for men, for supervisory force, preliminary trips of engineers, estimators, office men, trips of office personnel after job starts).

Tools (preparation, loading, unloading, of tools and equipment in toolhouse; fabrication of special equipment; an amount should be decided upon as a cost per ton or cost per dollar of field payroll, to cover toolhouse maintenance cost not directly chargeable to the job, such as depreciation and replacement of tools due to normal use and wear of such things as chucks, drills, reamers, sockets, drift pins, erection bolts, slings, wire rope, manila rope, electric cables, safety goggles, welders' shields or helmets, protective welding glass, electrode holders; also repairs to equipment and tools. The purchase of new equipment, original or replacement, should be provided for in this cost, such as for hoists, compressors, derricks, cranes, welding generators, transformers, rectifiers, and similar equipment and tools or supplies. See Sec. 3.5).

Rental (rented tools and equipment: cranes, trucks, derricks, compressors, welding machines, hoists; compare cost of preparing in toolhouse and shipping, against direct purchase or rental at site locally).

New material (wire rope, manila rope, etc. if not included in items of field-expense percentage or toolhouse charge suggested above).

Electricity (installation, removal; use of electricity should be under item of field expense).

Paint and painting (if in erection contract).

Plank (ship own, rent locally, or purchase with delivery directly to job site; compare costs including preparation, loading, and later unloading at toolhouse; if usable on several jobs, split cost of new planks between the jobs by using a "holding account," which, while slightly complicated, will give truer estimating costs; or else assume a definite percentage of the job payroll as a charge for equipment, tools, plank, etc. plus current cost of material not returned to the toolhouse: lost, stolen, or scrapped, to be charged directly against the final cost of the job).

Engineering (estimate number of drawings and cost per drawing for preliminary work, erection scheme drawings, details, special drawings or sketches; office engineering supervision, design to "heavy-up" permanent structures for erection loads, and special equipment; at-site engineering if not in field labor cost).

Special equipment (special tools, "Chicago boom" seats, jumping beams, "open panel" beams, special planks, special cables and turnbuckles; special slings, locomotive dinkeys, scaffolds, other special equipment or supplies; shims, screed angles, base plate and grillage leveling devices).

Falsework (engineering, material, shipping, delivery, return; fittings; salvage value to be deducted).

Marine equipment (boats, barges, launches).

Welding electrodes (unless included in material furnished by the fabricator).

Railroad charges (flagmen, work train, demurrage, switching charges, track, shifting; take into account the number of shifts by the railroad per day or days of the week, as it might affect time of completion due to insufficient delivery speed).

Insurance (rates are usually a percentage based on the estimated field payroll or cost per $100 of payroll; as a minimum, this item should include workmen's compensation, public liability, property damage, and automotive liability and property damage insurance, depending on the circumstances anticipated to be found if the contract is awarded to the erector. If the specifications or proposed contract call for it, additional insurance that may be desired or required should be included, which might cover fire and extended coverage, railroad, owner's, or contractor's protective liability and property damage, builder's risk, marine, marine cargo, contractor's contractual public liability, and property damage insurance).

Office overhead (percentage to be added for cost of running main office).

Profit (decide on percent profit on labor, tools and equipment, overhead items; consider competition, need of work to keep organization intact; check allowable profit if governmental agency is involved in the contract).

1.7   Exclusions

It is advisable to exclude items that can be done more economically by the owner, customer, or general contractor, or that would probably be a duplication of work that has to be done by other subcontractors as part of their normal operations. Accordingly, it is important in presenting a bid (when it is not a formal bid that prohibits any conditions or exclusions being stated) to specify clearly in the proposal what the erector will do, and what is to be done for him by others at no cost or at a specified cost; in other words, what is included in, and what is excluded from, his bid. He should state the time he expects to take, his anticipated starting date, and if escalation will be permitted, the labor rate estimated to be paid to the structural ironworkers. In addition it is advisable to tie in the starting of erection and time of completion with the requirements of the site, foundations, and footings, and with no delay or interference by others, in order to complete in the time stated. This is especially important when wall-bearing steel is involved.

Too often the erector will give a time for completion of the structural steel erection and then, through no fault of his own but because of the failure of others at the site to do their work as expected, he is unable to start on time, continue uninterruptedly, or finish on time. This may be because footings or piers are not ready when he is ordered or required to start, because the foundation contractor has not completed enough piers as rapidly as the steel erection speed requires, or because of interference by other trades. In some cities, on a tier building, local ordinances prohibit the steel erector from erecting more than a stipulated number of floors ahead of completion of permanent floors below. If the floor contractor falls behind, the steel erector must then legally stop work until enough floor arches are poured, or decking or forms placed, to reduce the number of floors of open steel-work to the legal maximum. This can result in additional expense not estimated.

It should be expected that certain things will be done by others either before or after the steel erector starts to work. By listing these items as they apply to the particular contract being estimated, it becomes clear what the erector has included in his price, and these exclusions should appear in the proposal or bid. If it was intended that the erector would not include them, no harm is done by stating them. If, on the other hand, it was expected of the steel erector, a compromise may have to be reached; or the erector will be asked to give a price to cover such items. This price can be compared with what a general contractor or other subcontractor might want for the same operations, and a decision can then be made by the owner as to who is to do the work in question. The worst that can happen is that the customer may demand that the erector include the work in his proposal price, and then the erector must decide if he will absorb the extra cost, stand on his rights to demand his estimated additional cost, do the additional work on a cost-plus basis, compromise his price, or relinquish the contract.

Items that a steel erector might reasonably assume will be taken care of by others (at no cost to him), or that could probably be done by others at a lower cost, or that might involve special trips are listed below. The minimum number applicable to the contract being bid should be given in the proposal. In addition, they should be stated in the simplest possible terms. If, on the other hand, the specifications or drawings, or the invitation to bid states that any of these conditions will be provided by others, they should not be mentioned in the proposal.

Removal of wires, obstructions.

Furnishing of water, electrical outlets, compressed air (if in a plant), space to operate and work, clear site, suitable ground conditions for equipment, toilet facilities.

Installing anchor bolts, derrick anchors, tie-downs, door frames, loose lintels, loose angles.

Setting of wall-bearing material, leveling or setting plates.

Grouting of anchor bolts, equipment anchors, grillages, slabs, base plates, wall-bearing plates.

Protection of traffic: pedestrian, highway, vehicular, railroad.

Protection of surrounding structures, machinery, running ropes (for erector's equipment), temporary guys.

Erection of material left down for customer's convenience.

Erection of small pieces not connecting to the main structure.

Shoring; no cross-lot bracing interference with steelwork.

Cutting for others, holes for others, exposing old steel for alterations or dismantling, exposing old connections for new steel, demolition of masonry, concrete, steel (in connection with work on old structure).

Site access for cranes, trucks, derricks, equipment, men.

Watchman service.

Concrete cleaning before painting of steel (if painting is to be done by the steel erector).

One floor of bearing walls ready at one time (specify number of return trips estimated).

Floor planking required beyond a reasonable amount to be furnished by the erector (sometimes needed because the concrete subcontractor fails to keep up with the steel erector's progress).

If it is evident from the design drawings or the specifications that the steel erector's work will not be done in one continuous operation but will require one or more return trips, it is advisable to specify how many total trips are included in the estimate. Often the steel erector must set up a portion of the structure, for example, a column-core building, and another subcontractor then completes certain work before the erector can proceed further. The erector may have assumed that the other contractor will complete each step of his work before the customer requires the steel erection to be resumed, whereas the other contractor may only partially complete his portion of the necessary intermediate work before the erector must come back. In this case the erector may have to leave and return several times, which was not included in his estimated cost. Equipment must be tied up or extra trucking must be used to haul it back and forth on additional trips. Furthermore, there are extra costs in starting and stopping work, in getting an organization reassembled and back to work, in getting the job under way again, and in supervision wasted in the cleanup of work done at the end of each trip so that additional trips not estimated can turn out to be quite costly. Therefore, it is wise to specify if the proposal is based on one continuous operation, or how many trips are assumed in the estimate and the proposal. If the customer then insists on additional trips, the erector is in a far better position to request or demand additional recompense.

2

Office Procedure

2.1   *Preliminary Studies*

The engineer in charge of preparing the erection scheme should, if possible, be the one who will be responsible later for the proper performance of the work in the field. He should be thoroughly familiar with the conditions in and about the site of the contract before starting to prepare the erection scheme. He should note field conditions that might affect the method of erection. These should be called to the field supervisory force's attention and should be included in the instructions to the field, or shown on erection schemes or special drawings.

The effect of surrounding structures on the steel erection procedure should be observed. If the site is hemmed in, especially in the case of a very narrow lot, the choice of erecting equipment may be limited. In addition, the superintendent may have to be warned to protect adjoining structures from possible dropped bolts, tools, or other equipment. Protective material must then be included in the job tool list.

Any trolley, telephone, telegraph, or power wires that may endanger normal operations should be noted. A safe procedure must be set up or arrangements made for their removal, deenergizing, realignment, or the installation of protective coatings. Poles supporting these wires may interfere with delivery or raising of the steel, and may have to be moved or removed. All this should be specified in the instructions to the field in order for it to be checked to make sure it has been done before work is actually started.

Railroad tracks may interfere with delivery or erection, and details of safe working and cooperation with the railroad company should be arranged well in advance. Similarly, vehicular or pedestrian traffic could

20

present hazards. Conferences may be necessary with the local police or with the general contractor to provide necessary protective measures.

Information should be developed and given to the field regarding delivery facilities and arrangements. If railroad delivery is involved, the delivery points for carload and less-carload (L.C.L.) shipments, and time of shifts or days when train movements can be expected must be given. If steel delivery is to be by truck, the proper person to contact should be determined to permit the field to have control of the steel delivery.

The location of the job office, men's shanties, toolhouse shanty, storage space, hoist placement, working space, and any other critical items are usually determined by a conference with the general contractor's or the owner's representative at the site. This should then be shown on the erection-scheme drawings or in the written field instructions. The site visit may indicate the need for a definite sequence and direction of erection. Poor ground conditions may indicate the need for corduroying an access road or laying planks over part of the working area, or the need for timber mats for the equipment to operate on.

Any restrictions on the use of gasoline, diesel fuel, steam, or electricity, or on operations such as hoisting from the street, storing material or placing equipment on the sidewalks or highway, or any police or fire department requirements should all be developed. When these affect the work, the field should be informed. This will enable the superintendent to avoid violating such restrictions or requirements. An investigation should be made of license and permit needs, and if these are required, they should be secured and included with the material prepared for use in the field.

A conference with the general contractor, owner, or customer at the site will develop the schedule of other trades with which the steel schedule must be meshed. Any anticipated delays to steel erection that are caused by other trades can be ascertained at that time. By means of advance warning, the superintendent will be able to keep his progress and his working force under better control, thus avoiding unnecessary costs due to premature tie-ups or unexpected delays. At this time, too, arrangements can usually be made for other subcontractors to set any needed anchors, tie-downs, anchor bolts, wall-bearing plates, or small steel members to be embedded in the foundations. The time when these will be needed and their delivery points should also be worked out so that the needed material will be shipped correctly and on time.

At such a conference arrangements can often be made to have piers brought up to finished elevations, or to have the foundation subcontractor set leveling or setting plates to grade. (Just before the job is to start, it is highly advisable to send engineers or surveyors to the site to check elevations of piers, slabs, etc. as prepared by the foundation contractor. The locations of any anchor bolts or anchorages that have been set in advance should be checked. If there are any discrepancies that must be corrected, the steel contract will not be delayed if the corrections are made before erection actually starts.) Scheduling of foundations should be discussed to be sure that there will be enough piers ahead and a fast enough schedule maintained for the balance, so that the steel erector will not be stopped by insufficient footings.

Details of access ramps and removal of obstructions should be determined. If anchorages are to be embedded in the foundations for the erector's equipment, drawings will be required in time for the foundation contractor to know where and how to place them: their angle, depth, direction, etc. (The toolhouse should be informed when these anchorages are wanted at the site so that there will be no delay in their delivery in time to be installed.)

If the cooperation of the general contractor can be secured, blank sketches of the column locations should be provided for him to indicate at regular intervals what progress is being made on the foundations, piers, or footings. In this way the engineer can see if the foundations will be ready for steel erection to start on the scheduled date, without having to visit the site repeatedly to observe the progress. These sketch sheets should have convenient spaces at each of the column locations, or a code agreed upon in advance, to show the necessary details of piles driven, footings prepared, concrete poured, leveling plates set, final elevations of piers, and locations of anchor bolts that have been checked.

As part of the preparatory office procedure, it is advisable to confer with the fabricator to develop and agree on backcharges, if any, for errors in the steel as furnished, whether due to shop or drawing-room errors. This may be merely a rate per hour for the different classes of men involved in corrections, such as foremen, structural ironworkers, apprentices, and engineers, based on the wage rates plus overhead and a reasonable profit. Unit prices for individual operations can be set up, such as drilling holes, installing bolts, making welds, burning or cutting steel, and excessive reaming.

Normally, before any corrective work is done, the fabricator should be notified so that he can check the error and possibly specify the manner of correction. For example, there may be a question of responsibility for reaming, a moderate amount being usually expected. (See AISC *Code of Standard Practice for Steel Buildings and Bridges,* section on Erection.)

As part of the preliminary advance office work, subcontracts should be prepared for any work the erector will not do with his own forces but that is included in his contract. When applicable, such subcontracts should caution the subcontractors that they must check the steelwork when their work involves the elevation and location of the steel members. The tolerance in rolling structural shapes, in fabrication, and in erection may cause a variation in the elevation or location of the steel members. Overrun or underrun of columns varies and can cause beams between to be slightly out of level. (Where discrepancies are known to exist, it may be advisable to notify the customer's or owner's subcontractors to avoid later back-charges from them based on claims that the steelwork was incorrectly placed.)

On a power-house contract, grating and railings may or may not be included in the steel erector's contract. If such material is in his contract, and for one reason or another the steel erector sublets its erection, then the subcontractor must be made aware of the need for checking the location of the steel members to which he must connect in order to have a satisfactory installation. It may even become necessary to alter some of the material to conform to the steel location and elevation.

2.2 Checklists

A checklist that includes all possible items to be taken care of well in advance of the time a job is to start should be used. Such a list will depend on the type of structure and the method of erection to be used. A master list, disregarding all inapplicable items, can be used. The arrangement should be such that a check mark of some kind, or a date, can be inserted in a column opposite each item to indicate that action has been started. Another mark or date in a separate column can indicate that the item has been completely arranged for, or a mark can even indicate that an item is not required for the particular contract. A separate column can be used for items involving an outside agency or a subcontractor. An example of a possible checklist follows:

1. Prepare erection scheme drawing.
2. Set up crane areas or divisions for each crane or derrick divisions for each derrick for shipping and sorting.
3. Decide on column tier divisions, location of shop, and field splices.
4. Determine hitches needed on slabs and grillages, columns, girders, trusses, and eccentric pieces.
5. Material to be shop-assembled for shipment, such as knee braces, connection angles, plates for welding, and splice plates.

6. Shims or filler plates for connections.
7. Derrick beam connections, crane-supporting beam connections, connections to be "heavied-up," shores, and use of extra members for supporting equipment.
8. Pins for hitches.
9. Fitting-up bolts, drift pins, and extra long fitting-up or erection bolts.
10. Layout of derricks, location of cranes, sizes, and capacities.
11. Layout of anchors (if derricks or guy-derrick cranes will be used) to be embedded in footings, in walls, and in rock.
12. Maximum loads to be handled: weight, size, possible job preassembly.
13. Picking delivered loads from basement or "hole."
14. Preliminary erection schedule.
15. Shipping schedule sent to shop.
16. Necessary screed angles, shims, wedges, leveling devices for setting grillages and/or slabs.
17. Derricks, cranes, blocks, and fittings; availability, capacity, and schedule.
18. Hoisting engines: availability, capacity, condition, capacity of drums, and schedule.
19. Check of selected crane-engine or hoisting-engine lead-line pull, number of parts required in falls to handle loads, in topping lift.
20. Wire running ropes: condition, length, new or used.
21. Sufficient slings to handle light and heavy loads, unusual-size loads.
22. Major equipment and small tools (other than already listed); prepare job tool list; send to toolhouse.
23. Plank for working floor, for extra floors, for skids, and for scaffold planks.
24. Need for relay derricks if setbacks will interfere.
25. Need for material platform on setbacks, reinforcement of permanent structure to carry extra load, capacity of platform, and material to construct platform.
26. Investigation of electric power source if needed, availability, ac, dc, voltage, amperage required.
27. Contract for hauling steel, equipment, and tools.
28. Contract for paint and painting (if part of erection contract).
29. Print of grillage sent to surveyor to establish lines for settling grillages later.

30. Check of available derrick or climbing crane, jumping beams, strength, and need of shores.

31. Layout of electric power lines for hoists, welding equipment, compressors, hand tools, signal system, and lighting.

32. Contracts for supplies, such as coal, oil, grease, gaosline, diesel fuel, oxygen, acetylene, propane, wire rope, manila rope, antifreeze, waste, planks, and timbers.

33. Arrangements for medical setup: first aid, minor treatment, major treatment, ambulance, stretchers, doctors (general and specialists such as eye, heart).

34. Insurance arranged, certificates secured: workmen's compensation, public liability, property damage, fire and extended coverage, theft, railroad protective, contractor's protective, owner's protective, builder's risk, marine, marine cargo, automotive, special, etc.

35. Provisions for moving hoists out of "hole" or cellar, basement; ramps for cranes.

36. Provisions for loading out at end of job: hoists, cranes, derricks, equipment such as compressors and welding machines.

37. Record of weights, areas, volumes of tiers, by derricks, by cranes, and by totals.

38. Instructions to the field: superintendent, field engineers, and timekeepers.

39. Drawings assembled for field use: details, erection diagrams, special sheets, grillage plan, location of anchors, and erection scheme drawings.

40. Detailed lists of bolts and welds.

41. Quantities of machine bolts, high-strength bolts, welds, etc. broken down into categories by tiers and totals.

42. Breakdown of pieces and tons, by derricks and tiers, cranes and areas, and totals.

43. Electrical power installation ordered (if needed).

44. If electric power ordered, a space should be provided as a check that it has been ordered discontinued on completion of the contract.

45. Derricks, cranes, and equipment removal on completion, arrangements where to ship for storage, refurbishing, or reuse.

46. Check need of special scaffolds, floats, needle beams, and planks.

47. Check of legal requirements, permissive operations in area.

48. Securing of licenses, permits, and bonds.

49. Abstract of specifications for field use, contract requirements.
50. Selection of field supervisory force: superintendent, resident engineer, field engineers, junior engineers, timekeepers, clerks, surveyors, and safety engineer or safety observer.

If Required by Specifications or Contract

51. Secure approval of paint, painter, payroll form, payment by check or cash, erection sequence, erection method, and welding procedures.
52. Check requirements of citizenship, residence, and union or nonunion status.
53. Arrangements for customer's or owner's engineering office, telephone, light, heat, drinking water, cleaning of office, and other maintenance.

Additional Items for Open-Panel Construction

54. Wire-rope floor-support cables (two sets), turnbuckles, and end fastenings.
55. Temporary supporting beams instead of wire ropes.
56. Special planks for floors.
57. Special skids for landing steel: timber, steel.
58. Possible use of permanent material borrowed for temporary supports.
59. Details of framing to use borrowed steel.
60. Warning of necessary precautions in field instructions.
61. Arrangements for safety meeting briefing.

Additional Items for Column-Core Construction

62. Derrick beams.
63. Hitches on columns, check of stability for lifting from horizontal.
64. Location of column splices (as many splices should be shop-assembled as can be handled safely with the field equipment).
65. Column guy anchors.
66. Timber or steel struts and braces.
67. Bolts for struts and braces.
68. Guys: horizontal and vertical.
69. Arrangements for removal of guys.

Additional Items for Cross-Lot Bracing Jobs

70. Check interference with erection of columns, beams, etc.
71. Need of tower falsework for derrick.
72. Need of tower falsework for hoisting engine.

In checking each item it must be kept in mind that some items require information to be sent to the fabricator's drawing room or to his chief engineer. The toolhouse is involved in some of the items. Instructions to the field superintendent and others in the field supervisory force must include information on many of the items; some involve the customer or the general contractor. Additional columns for specific purposes may aid in making sure everyone involved has been informed.

2.3   *Reports and Records*

Costs and records must be kept by the timekeeper or engineer on the job in sufficient detail to aid in future estimating as well as to permit checking actual performance and costs against those anticipated. In this way any production falling below a reasonable amount can be noted promptly and possible remedial action can be taken in time to prevent unnecessary excessive costs. It also permits comparison with the achievements of other, similar work and gives a basis for scheduling and estimating future contracts. In order to be sure that the items that the estimator will need or that the office engineer in charge of the particular contract will want are reported, the engineer responsible for the contract should refer to a master list. From this master list he can select the particular items needed by him and by the estimator for which costs and production should be kept in the field for reporting regularly. This should be noted in the instructions to the field.

When a job is completed all field unit costs such as cost per piece for raising, for bolts, welds, rods, etc., as well as costs per ton, should be translated into costs at a wage rate of $1.00 per hour by dividing all such actual unit costs by the average wage rate. This is an easy form in which to use costs, because it is then a simple matter to multiply such costs by a new wage rate in estimating a similar job to be done at a different rate. It also is an easy way to compare jobs done at different wage rates, since the costs will all be on the same basis. In addition, these figures of work done at a theoretical rate of $1.00 per hour will automatically be the man-hours per unit of work and can be used in estimating and scheduling for determining the total time required for the various operations.

Many firms use different standard cost and production report forms listing basic items common to the type of structure for which a particular report form is used. Such standard reports could cover high tier building structures on separate forms from mill buildings, or derrick-erected structures on separate forms from those for structures erected by cranes. Production report forms should provide spaces for reporting unloading, sorting, distributing, and raising of steel by individual rigs, by totals, and by tons and pieces per gang-day. These should be shown separately for material erected by each different type of powered equipment and for steel erected by hand.

The report should provide spaces for bolts by totals and by pieces per man-day or per gang-day, breaking down the production into types where radically different production is expected; for example, machine bolts are listed separately from high-strength bolts or from turned bolts. Space is needed for tabulating welding by length and by types of welding, or pounds of electrodes used or deposited (the difference is the unburned stubs or wasted rods), and the individual footage or pounds per man-day.

On derrick or climbing crane jobs the production report should show the time taken for erecting each area or tier before jumping. The actual time consumed in jumping should be recorded separately, since this is invaluable information. The actual setting of light material can vary appreciably from erecting heavy material, but the time of jumping will probably be the same.

The effect of accidents on production and costs should be investigated and used in promoting the safety program. For the cost report it is usually more convenient for the field to keep records in a somewhat chronological or a sequential order.

A checklist for the engineer in charge to outline what costs and production the timekeeper should report is described below; the list should be used only as a guide to make sure he has not missed any items of real value. Individual costs that are of insufficient value if reported separately should be lumped under group headings, as indicated. The less detail the timekeeper must record, the better the information is likely to be. By following a chronological or standard pattern sequence for all jobs, one can more easily use the reports and records for cost estimating, future scheduling, and checking job progress and costs. Making a comparison with other jobs will also be simplified.

Example of Groups of Items to Be Reported

1. Unload equipment and tools, load out equipment and tools. Tools and office: unload, set up, dismantle, and load out.
2. Unload steel at yard, load trucks at yard. Unload to storage, load out from storage.
3. Ramp, corduroy road, mats, install delivery tracks.
4. Grillages, base plates, slabs, setting shims, screed angles, leveling, aligning, and engineering.
5. Derrick, crane, traveler, pole, etc.: unload, load out, assemble, set up, reeve, and dismantle.
6. Unload steel at site, distribute, and sort; erect by derrick, crane, traveler, or pole.
7. Assemble trusses, girders; preerection assemblies.
8. Derrick jumps, climbing crane jumps, extend fixed tower crane. Move derrick, crane, traveler, or pole.
9. Pass steel; relay hoisting.
10. Material buggy, material tracks, temporary tracks, railroad work.
11. Hand erection.
12. Planking, extra planks, decking, protective covering.
13. Plumbing-up, jacking.
14. Fit up for erection.
15. Bolting: machine, high-strength, turned, ribbed, body-bound, special.
16. Fit up for welding; welding; test welders; qualify welding procedure; qualify welders.
17. Rods: bracing, sag, and tie.
18. Railing, crane rails: erect, align, and fasten.
19. Lintels, canopies, expansion dams: erect, align, and fasten.
20. Grating: erect, fasten.
21. Machinery, tanks: erect, set in position.
22. Joists: unload, distribute, erect, bolt, and weld. Bridging angles: erect, bolt, and weld.
23. Falsework: unload, frame, assemble, erect, remove, and load out. Piles: unload, drive, pull, and load out.

24. Drill; ream.
25. Burn; cut.
26. Cut out rivets in old structure, welds.
27. Work on old structure; dismantle old steel, reerect old steel; shore old structure; alterations.
28. Spot: bolts, welds.
29. Painting.
30. Safety: meetings, briefings, individual indoctrination. Safety engineer; safety observer.
31. Miscellaneous: clean-up, snow removal, straighten steel, physical examinations, etc.
32. Nonproductive time; overtime.
33. Lost time (due to weather, holidays, tie-ups, etc.); reporting time; waiting time (for steel delivery, delays by others).
34. Superintendent; general foreman; engineers; timekeepers; clerks; watchmen.
35. Repairs.
36. Shop errors and drawing room errors (to be backcharged to fabricator); foundation errors (possible backcharge to customer).

2.4   Instructions to the Field

Instructions to the field should include details of what production and cost records are to be kept and reported regularly, as well as any special details particular to the job. Attention should be called to special notes on erection diagrams or on detail drawings and to any special fieldwork alterations that must be done before certain material can be erected. Any cautions on the erection scheme drawings should be emphasized. Attention should be called to having the load falls vertical when lifting, thereby avoiding side pulls that could wreck a boom or tip a crane. The field should be warned to check ground conditions for stability of equipment, for safe loading, for safe hauling, and for possible unforeseen hazards.

Work should be planned, but the engineer who prepares the erection scheme must be ready to change the plan if necessary or advisable. When feasible, the superintendent in charge in the field should be included in the advance planning. If the superintendent has a reasonable, economical, efficient plan that is as good as or better than the one set up, adopting his plan may well be worthwhile if this can be done easily and without excessive extra costs. The superintendent in turn will then do his utmost to make his suggested change be successful. He will work harder to try to prove it is a better plan and will have no alibi, which might be the case if he had to execute a plan he disapproved of. After the scheme is finally prepared and the erection scheme drawings are made, there should be an admonition to the field to follow the erection scheme implicitly and to check any deviation with the engineer who prepared the scheme or with any other responsible person. If someone discovers a better scheme that could be used on other work, he should be encouraged to report it.

If any assemblies of material should be made before erection of such material, this should be stated in detail, unless it is covered on the erection diagram or on the erection scheme drawings. When dismantling or demolition is involved, the sequence must be given, and any precautions necessary for the safety of the structure as removed and of the portion remaining must be clearly and explicitly stated. When the work will be of open-panel construction, the necessary precautions should be detailed and a note should advise of the need to brief the men in advance.

If two rigs will be operating jointly to erect girders, trusses, or other heavy or limber material, the instructions should be clear on the procedure to be followed. When important, the points at which the two rigs are to hook on to the piece must be shown. When this is not given on the erection scheme drawings, additional information should be provided to indicate where equipment is to be set up. In the case of cross-lot bracing, details should indicate if any steel must be left down temporarily or if there will be any interference from the bracing. Any unusual sequence to be followed must be given.

In addition, a list should be furnished of available medical facilities such as first aid, minor and major medical treatment of injuries, doctors, specialists, hospital, ambulance, and stretcher location. Details should be given of any arrangements already made for banking facilities for payrolls, cashing checks, securing petty cash for field purchases, local credit for purchase of incidental supplies, for electric power installation and later discontinuance, telephone service, etc. Copies of contracts made for hauling, trucking, painting, and other subcontracted items should be included, as well as information on supplies arranged for in advance such as paint, fuel, oxygen, acetylene, coal, electrodes.

When painting is involved, it should be specified whether to paint the entire structure, portions of the structure only, connections omitted in the shop because of welding or high-strength bolting, spotting of bolts or welds. If no painting is to be done in the field, this should be stated specifically in order to avoid costs for work done but not included in the erection contract.

Inspection requirements should be stated, giving details of who is to inspect and who is to be notified, and when "outside" inspection is to be used for such operations as steel erection, welding, high-strength bolting, and painting.

Frequently, if the superintendent is given the production per day as estimated and thus expected of him—broken down into the various operations such as raising, bolting, and welding—he will achieve even more than asked for. In most cases the superintendent will try to show he can do better than the estimator figured; also, by having such a series of goals, he will have a definite production estimate to beat for his own satisfaction as well as possible unexpected profit for his employer.

The shipping schedule should be given together with information as to whom to notify in case the schedule must be revised. The erection schedule should be stated, broken down into dates for completion by tiers or areas, and, where important, broken down further into dates for completion of unloading, raising, bolting, welding, clean-up, etc. At least one copy of the list of tools furnished to the job should be given, and, where advisable, copies not only for the superintendent, but also for the resident or field engineer, and the timekeeper. Any notices legally required to be posted must be furnished.

Even though it may be standard, an outline of the safety program should be added, emphasizing the issuance of safety codes to the men or the use of safety hats, goggles, flash goggles; stating when the goggles are to be worn, when safety meetings or briefings are to be held, how often, and who is to attend; what safety posters are to be displayed; and what safety literature is to be distributed.

A supply of erection diagrams, erection scheme drawings, bolt and welding lists, and any special drawings should be sent with the instructions to the field to meet the needs of the particular job—usually based on the number of foremen who will use these drawings. One complete set of all drawings, including the detail drawings from which the steel is fabricated, should be kept in the job office for reference.

A complete set of supplies should be prepared, including such things as payroll sheets, checks (if payment of wages will be by check), cost report forms, production report forms, stationery, accident report forms, and any additional forms required for the proper operation of the job. These could include authorization for medical treatment, for ordering supplies, and for securing necessary temporary permits. Other forms could be included for recording such information as carloads or truckloads delivered or shipped away, or any standard forms used by the office to tabulate information from the job, to check invoices, to compare information from other jobs, and for permanent records.

When the starting date is in sight and field conditions have been verified to be sure the site will be ready for uninterrupted erection, certain preparations should be made by the office. The shipping schedule should be confirmed. Steel should be ordered shipped so that the first material required will arrive on time to be used when needed and will not be late, which could cause unnecessary expense to the field force awaiting its arrival. Correspondingly, if it arrives too early, demurrage on railroad cars, barges, or trucks could result; or there might be no unloading equipment or men on hand ready to receive it and release the carriers. Sufficient time should be allowed for the steel in transit to meet the field readiness.

Arrangements should be made for copies of the shipping notices to be sent to the field office for checking the shipments and for use on the job, and for copies to be sent to the main office for the records. If the customer requires copies, this should be the obligation of the fabricator-shipper. Details of which carrier will deliver and to what point should be verified. Any arrangements for payment of railroad demurrage, lighterage fees, crane unloading charges, or heavy lift charges should be made and the field informed in the instructions.

The instructions to the superintendent should clearly inform him whom he is to contact for subsequent shipments, since it is better for him to control them than for someone in the main office to try to do so. Any unforeseen delays requiring temporary suspension or delay in shipments can be reported much more quickly to the shipper. The superintendent should also be told what arrangements have been made for release of emptied carriers.

In addition to ordering the first steel shipments, the office should contact the toolhouse to release the tools and equipment in time for delivery to the site when needed. The details of release, carrier, delivery point, etc. should be given to the superintendent, or else he should be permitted to make his own arrangements with the toolhouse for the delivery of the tools and equipment. If equipment is being rented or purchased, the release to ship should be made at the proper time. Contracts for the use of rented equipment (as well as those for purchase) should be clear as to type, size, capacity, supplementary equipment to be furnished, time and location of delivery, and—in the case of rental—the probable extent of the rental period. A rental price per day, per week, and per month should be established, with an agreed rate for nonuse on holidays or idle time due to weather or causes beyond the control of the erector. Details should indicate whether or not the vendor will be responsible for transportation to and from the site, and for the necessary liability and other insurance in transit and while in use at the site. If the vendor disclaims responsibility, the necessary insurance to protect the erector should be arranged for in ample time before delivery.

With the job starting date determined and verified, supplies and services that will be required should be ordered or shipped, such as fuel for equipment, planks or equipment being purchased and not sent with the erection material from the toolhouse, oil, grease, coal, electric power, telephone, mail delivery, bank deposit to cover payrolls, water for drinking and for cooling equipment, rental of office typewriter, check writer, adding machines, etc., if they are not furnished from the toolhouse.

2.5   *Field Supervisory Personnel*

The office should have selected the superintendent and notified him to report in time to start the job. This time should be far enough in advance of actual operations so that he can check conditions, confer with the customer's and general contractor's representatives at the site, and arrange for men to report for work on the starting date. The superintendent will be in charge of labor, while the resident engineer will be in charge of engineering and contacts with the customer that do not involve the superintendent. This leaves the latter free to execute the work more efficiently.

The superintendent chosen should be one experienced in the type of work to be done. An efficient superintendent should be a trustworthy, steady, sober man who is able to get the most from his men. He should be able to keep peace with other trades, be presentable to the owner's and/or customer's representatives, be able to read and interpret drawings, and be able to foresee the proper time for additional equipment, steel, and men. He should have the ability to plan work for the men, have the right number of men, be able to judge men in order to pick the right men for the different positions on the job—the right men for foremen, for general foreman, for assistant superintendents, and for "floor boss." Preferably, he should be a man who has come up from the ranks, from apprentice or helper to ironworker to foreman to superintendent.

If the job is large or complicated, the resident engineer selected to be in charge of engineering at the site should be familiar with the particular job, with the erection scheme, and with the need of engineering service. Preferably he should be the man who prepared the preliminary work in the office, which would have included the erection scheme and the shipping and erection schedules. He should also have been the one making the necessary arrangements with the customer and with the fabricator's drawing room for erection requirements. He should be a man with experience as a field engineer on a similar type of erection, preferably an engineering graduate. He should have at least an engineering background if he is not an engineering graduate. He should have the personality to deal with the superintendent, the foremen, and the men. He should be able to make decisions when unforeseen erection problems arise, or when errors in the drawings or fabrication, or errors due to improper erection are discovered. He must be able to determine errors by comparing drawings against the actual steel, and should be able to satisfy the owner's and the customer's representatives that work is progressing satisfactorily. The ideal man will have an even temperament, a cool head, and be firm (not easily swayed)—with an understanding of the need for safe, economical, efficient results. He should have the ability to make sure that erection is proceeding in accordance with plans, specifications, and the erection scheme.

When the job is small and simple enough to be handled by the engineer in the office, a field engineer will often be sufficient to handle the engineering work in the field. He will not usually be as experienced as a resident engineer, but he should be a man open to instructions so that he can gain experience for more complicated jobs in the future. He should have the attributes of a resident engineer, with the ability to learn either under a resident engineer's supervision on a large job, or under the engineer working out of the office. It is also preferable to use an engineering graduate for the position of field engineer.

The man selected to act as timekeeper will be in charge of keeping the records, making the reports, and handling the field paperwork. He should have enough intelligence and experience to be able to keep cost and production records, to make intelligible reports, and to do the usual routine work of checking the men working to determine what the various gangs are doing. He must be able to prepare payrolls and calculate the necessary taxes and deductions (unless the payrolls are calculated in the main office) and to arrange for purchases of incidental supplies locally. He should have the ability to perform the usual clerical work required to furnish the superintendent with the necessary records and costs promptly, and to keep costs current.

A man should be selected as safety engineer to cover only a single job, if advisable, due to its size or hazards. Otherwise one man can often cover several jobs, visiting each as often and for as long a time as needed to achieve safe operating procedures. He will be in charge of the safety program on the job. The man selected should have a knowledge of safe erection methods, be personable, and be able to educate the men. He must be able to "sell" safety to the workers, be able to analyze accidents and accident reports to prevent future accidents, and should have had erection experience in the field as engineer, superintendent, foreman, or even as a structural ironworker. He should assist the superintendent in holding safety meetings and briefings whenever possible. He should report unsafe conditions to the superintendent for speedy correction and also to the engineer in the office, since occasionally conditions arise which the superintendent cannot correct without additional help from the office. He should be fully conversant with the requirements of OSHA (Occupational Safety and Health Administration) regulations on work in the United States and with similar legal codes and standards where applicable in foreign countries.

The superintendent, the resident engineer or field engineer, and the office representative should all watch the actual erection progress, including setting up equipment and unloading steel, assembly of trusses or girders, and preassemblies of groups of pieces before raising, plumbing and fitting, bolting, welding, and clean-up. This will help them see that the original schedule is maintained or improved and that costs are being kept down. They must watch for and endeavor to prevent delays due to weather or hauling, railroad or water delivery, erection troubles, equipment breakdowns or repairs, and interference by others.

The office should avoid specifying particular men to be employed in the field labor force. When this is not done, friction will often develop between such designated persons and the superintendent. The superintendent should be responsible for selecting and hiring his direct assistants, such as general foreman and foremen, so that he can have no alibi for unsatisfactory results, as often happens when men selected by the office are used in such positions. Similarly, the superintendent should hire the men who actually perform the work such as the structural ironworkers (often called "bridgemen" even on building work), boilermakers (if platework is involved), hoist and crane operators, oilers, firemen (if coal-fired steam equipment is used), ironworker apprentices, boilermaker helpers, and any other skilled or unskilled workers such as electricians, carpenters, truckdrivers, and laborers.

The size of the gangs hired by the superintendent should be watched by the office to make sure he keeps the number of men to the minimum required for safe, economical, expeditious work, but still enough to keep the work progressing efficiently and up to schedule.

Regardless of the duties of the various members of the field supervisory force, there should still be an office representative responsible for the work. At his vantage point of being divorced from actual physical contact with the work being done, he can often see things that need correction better than the men too close to the work to realize what can be improved. In addition, there are many times when help from the office will keep a job running smoothly.

In cases where local requirements do not permit the erection of the steelwork to proceed too many floors above the forming or pouring of the permanent floors on a tier building, an office man can often secure quicker results from the customer, or the general contractor through the owner, by calling attention to the failure of the floor contractor to keep within the maximum number of open floors permitted legally. Similarly, where insufficient progress is being made with the foundations so that a tie-up of erection is imminent, the office man can often get better action to have the foundation work expedited than is possible for the superintendent or the engineer at the site. In other words, once the job starts, cooperation between the field and the office must be maintained right up to the end of the erector's contract.

3

Erection Tools and Equipment

3.1 Toolhouse

There should be an adequately maintained, properly organized, and efficiently run toolhouse. Space should be provided for servicing and storing small tools and equipment such as compressors, welding machines, and hoists under cover to protect them from the elements. An area should be provided for planks, timbers, and blocking to be covered if the equipment is not otherwise protected from rain, snow, and ice. Cranes, trucks, and similar machines are not usually stored economically under cover, but there should be some protection against theft or vandalism.

The layout of the stored items should be arranged for ease in preparing shipments and servicing the equipment. Some thought must be given to the flow of equipment returned from a job so that small tools can be routed through a checking and servicing area to their storage bins or areas. Adequate overhead or equivalent crane facilities should be provided to handle the unloading of returned material or to load out equipment to be sent to a job. A railroad spur leading into the toolhouse is ideal where rail shipments are likely to be made.

Adequate paved roadway connections to a highway are necessary for truck deliveries, preferably directly into the covered areas of the toolhouse facility. Some outside, uncovered area is generally economical because not all equipment needs overhead protection. Anything so stored should be inspected frequently and, where required, cleaned and painted with a protective coating.

Stored planks should be piled so that a load for a railroad car or a truck can be picked as a unit, with sufficient separators where needed to prevent dry rot or other damage to the wood. Such separators can be

1 × 2-in. strips to permit air circulation. A good arrangement would permit picking the planks four wide and about fifteen high to give good loading in railroad cars or on trucks, and for reasonable handling at the erection site. If the planks are not stored under cover, heavy roofing paper, properly held down against the wind, will usually provide enough protection against damage by rain or snow.

Machines such as hoists, compressors, and the like should have lifting lugs installed so that when picked at those points the machine will not be damaged and will hang reasonably level for handling. Where slings or other devices are to be used instead of lifting lugs, the points where they should be placed or attached should be clearly marked or indicated. Hoists that may have drums in place or removed, or that may have one or two or even three drums in place under different requirements, should have the proper lifting lugs or lifting points clearly marked for the various conditions. A hoist with all drums removed to lighten its weight in handling will require lifting at a different point than with some or all of the drums in place. Similarly, with booms and masts for derricks or cranes, the lifting points will vary depending on how much of the boom (or a mast for a derrick or tower crane) is assembled for shipment or for later assembling and handling at the site.

Where it is inadvisable or inexpedient to mark lifting points, the center of gravity of individual pieces and of assemblies should be marked. On masts, booms, falsework, or other items that will have protective coatings, a small bead can be welded, about $\frac{1}{16}$ in. high, in addition to marking the lifting points and/or the centers of gravity with a distinctive paint mark. When welding these beads, points of concentrated stress should be avoided and the beads should be made only under controlled, approved procedures, especially on alloy or high-strength steels. Then, when such members are wire brushed, sand blasted, or otherwise cleaned, and the paint is thus removed or obliterated through usage, the bead will permit speedy remarking after the pieces have been repainted. The locations of the centers of gravity should be added to the detail drawings of the pieces for future use. Locating these points can be done by trial and error in the toolhouse when fabricating, when loading, or when storing, or they can be found by mathematical calculations.

3.2   Toolhouse Personnel

The toolhouse superintendent should have men available in his working force who are able to weld, burn (cut) with oxyacetylene or similar type of torches, and do minor blacksmithing repairs. Some of his men should be able to repair and service pneumatically or electrically driven equipment as well as internal combustion machines.

The best type of man to be superintendent of such a toolhouse is one who has had experience in the field as a superintendent or foreman, as well as having actually performed the various phases of erection (raising), bolting, welding, etc. so that he can know what the men in the field will want and how the equipment will be used. He should know how to service the tools and equipment. He should have had experience in supervising men and thus be able to assign his working force to the best advantage.

It is also advisable to have a good clerk to keep records of tools, equipment, and supplies, so that at any time the superintendent can readily determine what is on hand and available for use. Material shipped to the jobs should be deducted from the totals owned; then, as equipment and tools are returned, they should be tabulated and added to the inventory.

3.3   Job Tool List

A standard form of job tool list should be used to permit the toolhouse force to select material easily. The list of tools and equipment for a job should be prepared by a man with sufficient field experience to be able to visualize all operations to be performed and thus be able to list and order just enough adequate equipment. Too little could delay operations, and too much could lead to careless use in the field with the possible loss of tools. Excess equipment costs more to prepare and load, and later to unload and store, and adds extra costs for the field force to handle at the site. There will also be extra transportation costs.

The tools selected for a job should be based on the erection scheme; on the quantities of pieces, tons, bolts, welds; on the sizes and weights of awkward or heavy pieces; on the number of gangs raising, fitting, bolting, welding, plumbing up, and performing handline operations; and on the number of men who will be in the working force. The time allowed for the completion of the work must also be considered, for this may affect the number of derricks, cranes, welding machines, compressors, etc. that should be on hand.

A well-planned blank tool list can be printed for repeated use. The arrangement of such a standard list can be alphabetical or based on groups such as pneumatic equipment, electrical, wire rope and manila rope, and blocks and sheaves. All tools that could possibly be used for any sort of erection scheme should be included in the list, even though the erector does not own all the items. When worthwhile, tools not owned can be purchased or probably rented. With such a list, the man preparing the tool list—preferably the office engineer responsible for the job—is reminded of things he might otherwise overlook.

In the case of purchased equipment, the manufacturer or his distributor will frequently send an operator or a representative to the site to make sure the equipment is used properly and to explain to the erector's operator exactly how it should be operated. A representative from the toolhouse should be on hand at that time to become familiar with the machine so that it can be properly serviced or repaired if necessary either at a future job or on its arrival at the toolhouse on completion of use at the job. He can also ask questions that only the manufacturer might be able to answer. When equipment is not owned, a decision must be made on purchase, rental, or use of an alternate erection scheme that will utilize available equipment instead of purchasing or renting.

3.4   Special Sheets

In addition to the standard job tool list, additional sheets can be used. These are arranged so that all fittings and equipment for a particular type of derrick or crane can be listed in advance on a special sheet. The man preparing the list for a job merely indicates the number of derricks or cranes of the particular type; the length of mast, boom, and jib as the case may be; and the diameter and length of running rope required.

Such a sheet would include a list of the diameter, number, and size of splice bolts; all necessary sheaves, pins, links, overhauling weights, wire rope blocks, hoisting spreaders, and sorting spreaders; street slings; signal system complete with bells, or lights and bell cord, or electric cables and signal box; etc.

Other special sheets itemize the tools needed for one bolting gang or other similar gangs. The number of different gangs is then stated, without the need to list all the different items wanted. Possible exceptions are noted on these sheets to cover the cases where more than one gang is to have tools furnished. For example, one pneumatically equipped bolting gang may require a certain size manifold and valves at the compressor, whereas more than one gang would need not two manifolds, but only one with more outlets.

A sample sheet for a guy derrick would assemble and list various items, with quantities and sizes for the particular size and capacity derrick, as follows:

Guy-Derrick Checklist

Blocks.   Wire rope: fiddle, gate, load, topping lift, whip; manila: single, double, snatch.

Bolts.   Splice, washers.

Boom.

Boom shoe.

Bullstick (or bullwheel).

Footblock.

Guys.   Foot, jumping, mast, engine anchors, footblock anchors, clips.

Jumping beams.   Steel, timber.

Manila lines.   Falls, tagline, and hook.

Mast.

Overhauling weight.   Light, heavy.

Pins.

Shackles.

Sheaves.

Slings.   Column anchor, column-setting hickey, eye-and-eye setting, hoisting spreaders, sorting spreaders, unloading, "street slings."

Spider.

Turnbuckles, spools, or thimbles.

Wire rope.   Running for load, boom, whip (runner).

3.5   Charging Costs (See Also Sec. 1.6)

An accounting system should be developed so that a charge can be made against a job and taken into account in the original estimate as an overhead item. This charge can be made to credit "holding accounts" for new or replacement equipment or to pay for servicing tools and maintaining the toolhouse. This maintenance cost must include not only the wages paid to the toolhouse working force but also the rental of the building; if owned, the interest and amortization if not fully paid for; taxes; insurance; and all the usual running expenses such as electricity, water, and heating.

A decision should be made if the job is to be charged for material lost or damaged on that particular job, or if a general charge against all the jobs is to be made, high enough to cover the average cost of lost, damaged, or completely used equipment as well as the cost of repairs due to normal wear and tear through usage. These charges are most easily assessed as a percentage of the field payroll total, on the basis of tonnage erected, or by any other method that divides the total cost fairly among all the jobs. The actual costs of picking out and loading equipment and tools for a particular job, the cost of transporting them to and from the site, and the later cost of unloading, servicing, and placing back in storage should be assessed directly against the total cost of the job.

The jobs can be charged an hourly or daily rate for the use of machines such as cranes, derricks, hoists, compressors, welding machines, trucks, barges, boats, or launches. This charge can then be credited to the different holding accounts. A similar set of rates to which profit has been added can be used to establish reasonable charges for the use of such equipment by the customer, owner, or other subcontractors. This is especially helpful when equipment is not being used by the erector since it provides a lucrative revenue instead of having the machines lie idle.

When such rental to others is made, it is advisable to have a signed agreement clearly stating the conditions of use, the rental rates, and, most important, the acceptance of liability for loss or damage to the machine as well as liability for injury to persons or property in connection with the use of the machine by those who rent it. It is well to have such an agreement prepared by the erector's legal advisers. A standard form for such a rental agreement can then be printed; then one merely needs to fill in the details at the time of renting.

3.6   Processing Job Tool Lists

When the job tool list is received, the toolhouse superintendent should check his equipment for availability at the time it will be needed. If some is in use on another job but is expected to be returned in time to be serviced and made ready for the new job, this must be checked later to make sure no delay on the current job will prevent the equipment from being ready when needed. The condition of the tools ordered should be checked to see if they need repairing, lubrication, cleaning, and painting, and to be sure everything is in safe working order. The schedule of work ahead in the toolhouse should be watched to make sure there will be enough time to have the particular equipment and tools ready for delivery when needed.

The method of delivery should be decided on, and rented trucks, the erector's own trucks, railroad cars, or—in the case of water delivery—barges, etc. should all be ready for loading at the right time. The proper description should be decided on to obtain the lowest freight rate if by rail or boat. The required fitting-up bolts should be lined up for "running down" and lubricating, sorted for shipment, and then boxed, kegged, or bagged. (When fitting-up bolts other than permanent bolts are to be used, it is well worth the extra cost to purchase cone-pointed, heat-treated bolts because of their extra life and strength, but a special effort must be made at the site to prevent loss of such high-cost bolts.)

Planks should be examined for safe condition. (They must never be painted with anything that can hide cracks or other defects.) Blocks should be inspected to be sure the sheaves are in proper condition, bearings lubricated, and keeper bolts in place with the holes for the keeper bolts in the ends of the pins located so that the nuts can be turned. (Keeper bolts are preferable to cotter pins because the latter can break too easily or even work out of place.) Required derricks, crane booms, and similar equipment should be checked, and where their condition is questionable, paint should be removed for visual, magnaflux, or radiographic inspection. Repairs or replacement of parts should be made in time, checking booms and masts for evidence of blows, abrasions, dents, bent or torn pieces, cracks and breaks (especially at pin holes, bolt holes, or rivet holes in equipment originally riveted), and defects in welded areas. If repairs are to be made by welding, it should be done under a controlled, approved procedure, especially if the material is high-strength or alloy steel. Bent pieces should be straightened cold or heated by controlled, approved methods.

Bearing surfaces and pins should be examined at points of moving parts. If gouged, grooved, or too badly worn for safe operation, they should be replaced or built up and refinished. All lubricating and greasing points should have old, hardened lubricant or grease and dirt removed and fresh lubricant or grease applied, making sure it flows freely and reaches the points needing it. Sheaves in blocks should turn freely and not wobble. If they do wobble they may need new pins, bushings, or bearings. Grooves in bushings should not be excessive because they can cut into the pins. Wire rope and manila lines should be in good condition for the intended use, and not too badly frayed, dried out, or otherwise damaged so that they are unsafe for use.

At least one complete list of tools shipped to the job should be furnished to the field superintendent and one should be kept at the toolhouse. On completion of the work, the field superintendent can then use this list to check equipment and tools returned to the toolhouse. If anything is missing, it will be a simple matter for him to look around the site for it. In addition, any losses in transit can be determined by comparing the list of material shipped back to the toolhouse with the check that should be made when the cars, trucks, etc. are unloaded there, so that a claim can be made against the carrier.

The field superintendent can use a form to good advantage to note the condition of any specific items requiring repairs, overhaul, or other attention. This is largely advisable for major equipment such as compressors, cranes, derricks, engines or hoists, welding machines, tractors, electric tools, or pneumatic tools. Such a report can enable the toolhouse superintendent to overhaul and properly repair equipment more intelligently than if he had no knowledge as to what material might be faulty, damaged, worn, or otherwise in need of attention.

3.7   Erection Tools

A complete checklist of the usual tools and equipment that might be used in connection with the erection of structural steel building frames would include the following: for convenience, it has been tabulated alphabetically. The selection of items for a standard tool list will depend on the needs of the erector, on the type of structures he expects to erect, and should be arranged to suit his particular needs. When the item is normally found in everyday life, no explanation or description is given. When it is unusual, or distinctive to steel erection, it will be described or illustrated enough to identify it or to explain its usefulness and need on the job.

List of Tools

Adze.

Anchors.   Hairpin (Fig. 3.7.1); split-end eye (Fig. 3.7.2). A supply should be kept on hand in various sizes or capacities so that they can be shipped to the job site in ample time for installation. Frequently the erection contract is let after the foundation work has begun. The anchors must then be rushed to the site in order to embed hairpin anchors in the forms before concrete is poured, or to have the holes drilled in rock for the split-end eye anchors while the foundation contractor still has his equipment available.

Auger.   Ship (Fig. 3.7.3); for drilling holes in timber.

Automobile.   Passenger; truck, pickup; truck, heavy-duty; Jeep. Licenses should be checked to ensure legal use of the cars in the area in

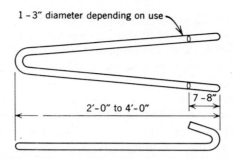

FIG. 3.7.1.   Hairpin anchor (for concrete).

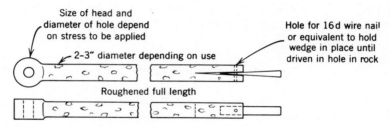

FIG. 3.7.2.   Split-end eye anchor (for rock).

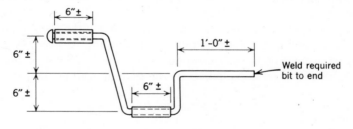

FIG. 3.7.3.   Ship auger.

which they will be used. In some states truck cranes require license plates, and these should be secured if they will be needed to transport the crane through any intervening states or for use at the job site.

Axe.

Backing-out punch.   Hand (Fig. 3.7.4). After rivets that must be removed have one head cut off by a buster, cutter, or burning torch, the backing-out punch is used to drive the beheaded shank out of its hole. It is also used to remove bolts, which are tight in their holes, after the nuts have been removed.

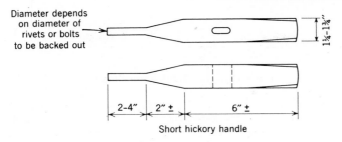

FIG. 3.7.4   Hand backing-out punch (known as "B. and O.").

Balance beam (Fig. 3.7.5).   A device used when lifting with one piece of equipment but where the piece is unstable when picked at its center because it is long or limber, or where bridled slings cannot be used safely. It can also be inverted for picking with two pieces of equipment if the piece to be lifted can be picked at its center of gravity (Figs. 5.25.5 and 5.25.8).

Barrels.

Bars.   Chisel (Fig. 3.7.6) heavy, light; connecting, heel (Fig. 3.7.7); flat (straight) (Fig. 3.7.8); claw (Fig. 3.7.9); crow (Fig. 3.7.10); pinch (Fig. 3.7.11). The chisel bar has one end for cutting, and the other end or

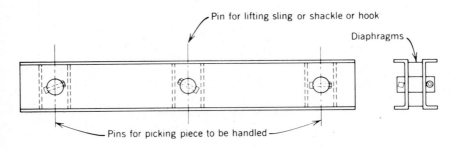

FIG. 3.7.5.   Balance beam (typical–schematic). Can be reversed for two pieces of equipment to lift piece at center.

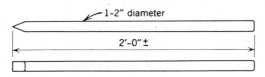

FIG. 3.7.6.   Chisel bar.

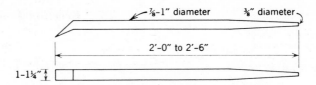

FIG. 3.7.7.   Heel connecting bar.

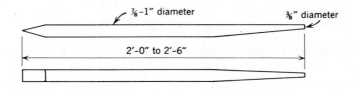

FIG. 3.7.8.   Flat (straight) connecting bar.

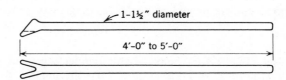

FIG. 3.7.9.   Claw bar.

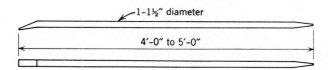

FIG. 3.7.10.   Crow bar.

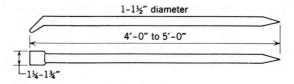

FIG. 3.7.11.   Pinch bar.

48

head is designed to be struck by a maul or other form of beater. The claw bar and the crow bar have their gripping end at a slight angle to the rest of the bar to provide leverage for prying. The connecting bar point is used to help align corresponding holes in the members being connected; the flat end is used to guide a piece into place.

Basket. Bolt (Fig. 3.7.12); a metal basket for bolts, washers, and small hand tools that permits good housekeeping as well as providing a safe means for pulling up such material on a handline. Its dimensions should prevent it from tipping when lifted by the handle.

Bells. Manual (trip gong) signaling sets, cord, pulleys (swivel); electric signaling sets, lights, signal box, wire cable. These are used chiefly for derrick operations and where the hoist operator cannot see the signalman directly or clearly. In the case of the manual system, a spring-operated clapper is used to strike the bell when the bell cord is pulled, lifting the striking clapper to a point where it is released so that the spring can cause it to strike the gong. A set will depend on what is needed to operate the boom, load, runner, bullwheel, etc. The bells should be of different tones for each of the various operations.

The bell cord for the manual system should be a heavy clothesline or light, flexible wire rope. Pulleys are used to enable the signalman to "pull the bells" in various directions from the point where the cord goes down to the hoist. They are mounted on a temporary wooden or metal stand that is moved up to the working floor after each jump on a tier building, feeding the bell cords through the swivel pulleys on the stand, down to the hoist.

In the electrical system the lights can be of different colors, or if they are to be mounted above their respective operating drums, they can all be white. If electrically operated bells are used instead of lights, they should be of different tones as in the case of the manually operated system. The signal box should be waterproof with pushbuttons to actuate the electric bells or lights. Normally there is a strap so that the box can hang from the

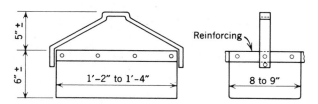

FIG. 3.7.12.   Bolt basket. Main material: 16-gauge hot-rolled annealed sheets with machinery guard perforations.

signalman's neck. The electrical system should have a safety device incorporated to indicate to the hoist operator when the power has failed or if the wires have been cut or become disconnected.

Blacksmith's anvil.   Blower; forge; kit; vise.

Blocks.   Manila: single sheave (Fig. 3.7.13), double sheave, triple sheave, multiple sheave; with hook, shackle; snatch (Fig. 3.7.14); gate (Fig. 3.7.15). Wire rope: single sheave (Fig. 3.7.16), double sheave, triple sheave, quadruple sheave, multiple sheave; with hook, shackle; fiddle (Fig. 3.7.17); gate; snatch; tandem (Fig. 3.7.18); weighted; spare sheaves.

Bolts.   Fitting; fitting washers; falsework; falsework washers; splice for boom, mast, tower, jib, sills, legs, gin-pole, bullwheel, etc. If the permanent bolts on the structure will be high-strength bolts, these can be used in place of special fitting bolts. Falsework bolts usually have two threaded ends for the nuts instead of a head at one end and threads for a nut at the other end.

Bolt bag.   Shoulder; belt. Bolt bags of canvas can be secured with provision for slipping the connector's belt through loops; or a long-shoulder-strap-type can be used. The latter is preferable since it enables the man to remove the bolt bag more readily when he must perform operations other than connecting or bolting; for example, jumping a rig or handling planks.

Boring machine (for wood).   Pneumatic; electric; bits; chuck. These machines vary in the manner of inserting the bits. Those requiring a chuck must have one that will fit the configurations of the ends of the bits as well as the machine itself.

Brace.   Carpenter's; bits.

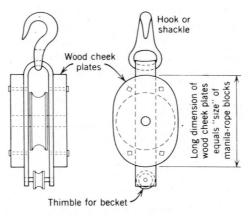

FIG. 3.7.13.   Single-sheave manila-rope block.

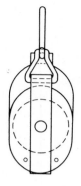

Hook or bail must be rotated 90° to engage lug on loop holder to prevent accidental disengagement from hook or bail in working position.

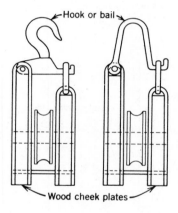

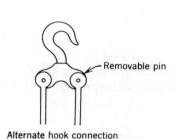

FIG. 3.7.14.   Manila-rope snatch block (wire-rope snatch block similar except for cheek plates).

Branding iron.   For identifying planks and timbers.

Broom.

Brushes.   Paint; wire; wire cup, scratch, wheel. Wire cup, scratch, and wheel brushes are used in powered tools such that, by rotating the brush, rust, paint, and other foreign matter can be cleaned from the steel.

Bucket.   Paint; water. A container with a spigot and a supply of disposable paper cups is more sanitary than a water bucket and common dipper, and is required by law in many places, as well as by OSHA in the United States.

Burning torch (cutting torch).   Box; gauges, oxygen, acetylene; regu-

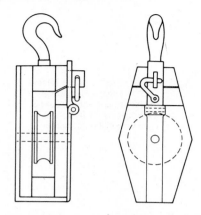

FIG. 3.7.15.   Manila-rope gate block.

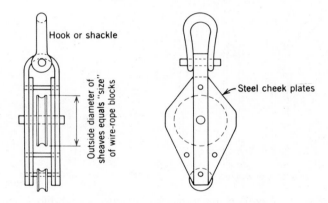

FIG. 3.7.16.   Single-sheave wire-rope block.

lators; hose, oxygen, acetylene, combined; hose couplings; hose mender; lighter; tank key; cutting tips; wrenches for torch, for fittings.

Buster.   Hand (Fig. 3.7.19). Useful for cutting off heads of rivets to be removed.

Can.   Safety gasoline. For filling gasoline tanks on equipment. All such tanks should have a fire-safety strainer in the filling opening. A warning should be painted on the tanks prohibiting the removal of such strainers when filling the tanks.

Chipping hammer.   See *Hammer.*

Chisel.   Hand; bits (for powered tools).

Removable filler to permit
reeving upper block

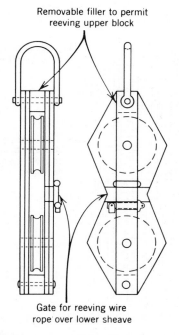

Gate for reeving wire
rope over lower sheave

FIG. 3.7.17. Fiddle block (banjo block).

Coffing-hoist. A patented lever-hoist-type device for pulling pieces together, so that a little effort exerts a greater force.

Come-along. A device to clamp on a wire rope so that it grips and permits pulling the rope by a set of falls or other means.

Compressor. Diesel; gasoline; electric; steam (rarely used). The tool list should specify the number and capacity required; also the manifold outlets and air receiver, if needed. The capacity of the compressor should be more than just enough to operate the pneumatic tools to be used. With a larger volume than needed, there will be fewer outlet pressure fluctuations; and with an air receiver in the line, the compressor will cut in and out less frequently and will "labor" less to meet the compressed-air demand, all of which can increase the useful life of the machine. Furthermore, with an excess air capacity, the compressed air in the receiver or tank on the machine will cool by remaining in the tank or receiver. This, in turn, releases moisture and gives a better supply of air and better performance of the pneumatic tools.

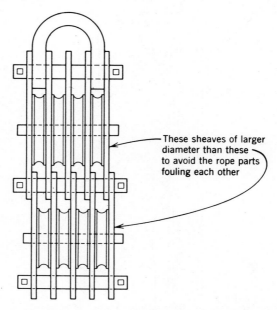

These sheaves of larger diameter than these to avoid the rope parts fouling each other

FIG. 3.7.18.   Tandem block (for 16-part falls).

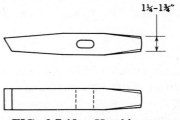

1¼–1¾"

FIG. 3.7.19.   Hand buster.

Crab (*winch*).   Single purchase (single acting) (Fig. 3.7.20); double purchase (double acting) (Fig. 3.7.21). Used for wire-rope single-part or several-part falls or tackle, where a small force on the handles will produce a greater force on the wire rope lead line being wound on the drum. The device can be clamped or bolted on a gin-pole, a dutchman, or a column for use when erection is to be by hand instead of by powered equipment.

Cranes.   Crawler: diesel, gasoline, electric, steam (rarely used); boom; jib; spare treads. Truck: diesel, gasoline; boom; jib; spare tires. Wagon: diesel, gasoline; boom; jib; spare tires. Locomotive: diesel, gasoline, steam;

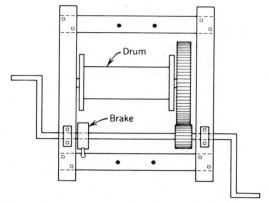

FIG. 3.7.20. Hand crab or winch—single purchase.

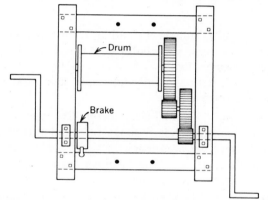

FIG. 3.7.21. Hand crab or winch—double purchase.

boom; jib. Cranemobile: diesel, gasoline; boom; jib; spare tires. Tower: diesel, gasoline, electric; boom; jib; mast or tower; spare tires. Guy-derrick: mast, boom, jib, guys.

Cups.   Paper. (See comments under *Buckets.*)

Cutters.   Hand; cross, diamond point, side (Fig. 3.7.22); straight (Fig. 3.7.23); bits for powered equipment. Used for cutting off material, the shape depending on the configuration of the material to be removed. The cutting end can be round nose, flat, cape, caulking, diamond point, etc.

Derricks.   A-frame: front legs, back leg, front sill, back sill, links, pins. Guy: boom, mast, footblock, boom shoe, bullstick, bullwheel, spider, links, pins. Jinniwink: front legs, sill, back leg, links, pins. Stiffleg: boom,

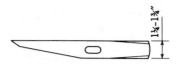

FIG. 3.7.22.   Side cutter.

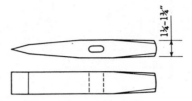

FIG. 3.7.23.   Straight cutter (short hickory handle).

mast, jib, back legs, sills, bullwheel, footblock, pins, counterweights. Traveler: boom, mast, jib, legs, sills, bullwheel, frame, platform. Dutchman: mast, sill, braces, fittings. Gallows frame: side legs, cross-piece, fittings, chain hoist. (For descriptions of derricks and their uses, see Chapters 7 and 8 on derricks and travelers.)

Dinkey locomotive.   A small-size, locomotive-type machine, usually four-wheeled, running on standard-gauge rails and driven by an internal combustion or battery-powered engine. Used for pulling railroad-type equipment where railroad switching facilities are not readily available or would be uneconomical for the usually intermittent service required.

Dogs.   Beam (Fig. 3.7.24); girder (Fig. 3.7.25). Beam dogs are slipped over the beam top flange and grip as long as there is a lifting strain on the purchase ring to which the two scissor-like arms are attached. The points of girder dogs are hooked well in toward the web under the top flange of a girder being lifted. Where possible, the slits in the jaws of the dogs are placed over stiffeners or at bolt heads to help prevent the girder from sliding sideways through the dogs. Both types are made in various sizes depending on the capacity required. The girder dogs are often forgings or castings. A small piece of wood should be placed between the points and the steel surface to grip and help prevent movement of the girder laterally in the dogs.

In addition, it is advisable to install screw clamps ("C" clamps) on each side of the dogs, secured to the member being lifted, to help prevent the piece from sliding through the dogs.

Dolly, riveting.   Jam; striking. (See Appendix E.)

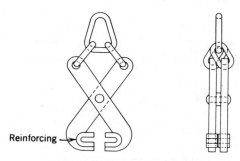

FIG. 3.7.24.   Beam dogs.

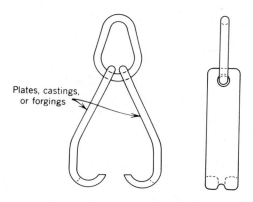

FIG. 3.7.25.   Girder dogs (thickness and size depend on weight to be lifted).

Dolly, timber (Fig. 3.7.26).   Used for rolling timber, pipe, steel, etc. over reasonably even surfaces.

Drift pins (Fig. 3.7.27).   These are made in various diameters, the barrel being the same diameter as the holes in the steel to be connected and pinned. The point is driven through the various plies of steel being assembled in a connection and helps drift the various plies into alignment. The quantity ordered in each diameter should be based on a percentage of the number of holes of the respective diameters to be bolted or the number of holes provided for erection purposes in the case of welded connections. It will also depend on the number of raising gangs, men fitting up, and bolting gangs.

Drills.   Center-spindle (Fig. 3.7.28); side-spindle; close-quarter (Fig. 3.7.29); electric, pneumatic; hand ratchet. The center-spindle drill is used where the work is accessible, requiring two men to hold the machine and

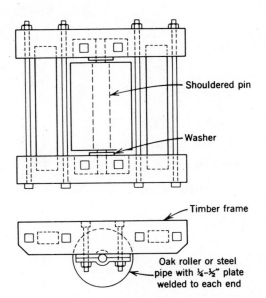

Shouldered pin

Washer

Timber frame

Oak roller or steel
pipe with ¼-½″ plate
welded to each end

FIG. 3.7.26. Timber dolly.

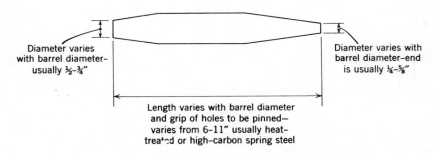

Diameter varies
with barrel diameter—
usually ½-¾″

Diameter varies with
barrel diameter—end
is usually ¼-⅝″

Length varies with barrel diameter
and grip of holes to be pinned—
varies from 6-11″ usually heat-
trea⁺ᵉd or high-carbon spring steel

FIG. 3.7.27. Drift pin.

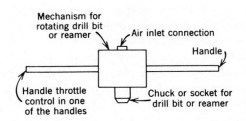

Mechanism for
rotating drill bit
or reamer

Air inlet connection

Handle

Handle throttle
control in one
of the handles

Chuck or socket for
drill bit or reamer

FIG. 3.7.28. Center-spindle drill.

58

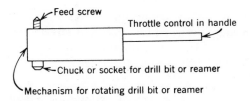

FIG. 3.7.29.   Close-quarter drill.

the bit in place safely. The side-spindle and close-quarter drills are used where one man can handle the machine safely; the latter is used for working in close quarters where there would not be enough space for the other types. The hand ratchet drill turns the drill bit by operating the handle back and forth, the ratchet working against a toothed gear to turn the drill bit.

Drill bits.   For steel; for wood; twist (Fig. 3.7.30); "bridge" reamers for steel: rose, straight (Fig. 3.7.31), tapered (Fig. 3.7.32); "flute" reamers for steel: spiral (Fig. 3.7.33), straight (Fig. 3.7.34), tapered, twist, etc.; drill and reamer bit chucks; Use-em-up sockets.

Drums.   Oil; water.

Electric cable.   Connectors; switches; transformers; wires. (Ground-fault circuit interrupters are required by some regulations for use with electric-powered hand tools.)

Emery wheels.   Powered; hand; spare wheel.

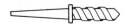

FIG. 3.7.30.   Twist drill bit.

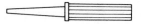

FIG. 3.7.31.   Straight bridge reamer.

FIG. 3.7.32.   Tapered bridge reamer.

FIG. 3.7.33.   Spiral flute reamer.

FIG. 3.7.34.   Straight flute reamer.

Falsework.   Steel; timber; connecting plates (scabs); washers. False-work material is usually designed for specific needs and situations, and the material is then fabricated at the toolhouse or the job site. The bill of materials should include all fittings to be furnished.

Files.   Hand.

Fire extinguishers.   Class or type of extinguishers depends on the hazards to be anticipated. The proper type of hand extinguishers should be furnished in the cabs of all cranes and trucks, and at all locations of internal combustion and electrical equipment.

Floats.   Plank (Fig. 3.7.35; plywood (Fig. 3.7.36). Some erectors prefer plywood floats, while others like plank floats with spaces between the boards. Both types are used for hanging from the structure to permit bolters, welders, fitters, etc. to work at points not readily accessible from the steel itself.

Gin-pole.   Steel, wood; shoe; splice.

Goggles, safety.   All-purpose type: clear, shade; burner's type: clear

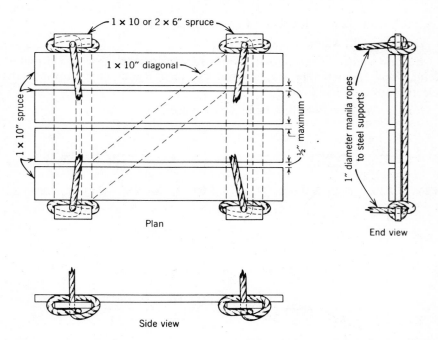

FIG. 3.7.35.   Plank float; 1 × 10s preferably bolted to supports with carriage or elevator bolts, or nailed with nails cinched underneath.

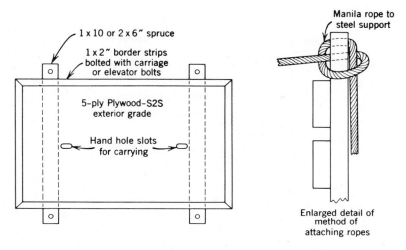

FIG. 3.7.36.   Plywood float. Ropes attached as for plank float.

cover, shade; cover-all type: clear, shade; cup type: clear, shade; spectacle type: clear, shade, with side shields, without side shields; welder's flash: shade.

Grease.

Grease gun.   Dot; Zerk; Alemite; pump.

Grinders.   Electric; pneumatic; hand; spare wheel.

Grindstone.   Hand.

Hammer, hand.   Chipping (Fig. 3.7.37); claw; maul (Fig. 3.7.59); peening (Fig. 3.7.38); sledge; hand (Fig. 3.7.39).

Hammer, powered.   Chipping: electric, pneumatic; rivet: regular, close-quarter, electric, pneumatic (Fig. 3.7.40); spare plungers. Chipping hammers are similar to rivet hammers except that they are designed to ac-

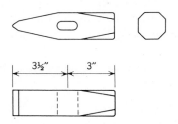

FIG. 3.7.37.   Hand chipping hammer.

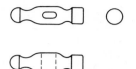

FIG. 3.7.38.   Peening hammer.

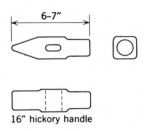

FIG. 3.7.39.   Hand hammer.

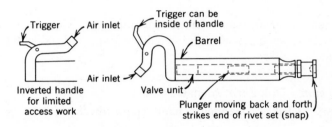

FIG. 3.7.40.   Pneumatic rivet hammer.

tuate a smaller shank chisel or chipping bit than the larger-diameter rivet set or snap. The close-quarter rivet hammer usually has an inverted handle to reduce its overall length. The rivet hammer (see Appendix E) is often used to drive drift pins through holes in a connection, if difficulty is encountered in trying to drive the pins with a maul or other hand tool in "fairing" holes in a connection. Rivets are occasionally used for specific situations.

Handles.   Adze; axe; hand backing-out punch; buster; chipper; cutter; hammer; maul; pick; sledge.

Hatchet.

Hats.   See *Safety hats.*

Hickey.   Beam (Fig. 3.7.41); column-setting (Fig. 3.7.42). The beam

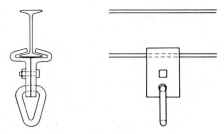

FIG. 3.7.41.   Beam hickey.

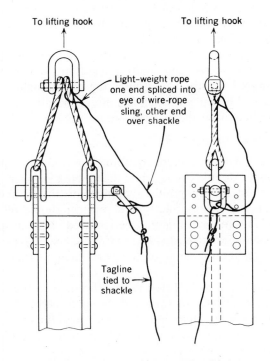

FIG. 3.7.42.   Column-setting hickey.

hickey is clamped on the bottom flange of a beam to hold a set of falls or tackle, usually for handling light loads as required when a detail gang erects small material that has been left out by the raising gangs.

The column-setting hickey is used to rotate a column from its horizontal, delivered position to its vertical position for erecting. In addition, it eliminates the need for a connector to climb up a column, after it is erected in place, to remove a setting sling. It also permits the column to hang more nearly vertical than if a sling were used to erect it. This, in turn, makes the erection of the column easier, safer, and quicker.

Holes 2 to 2½ in. or more in diameter, depending on the weight to be lifted and the diameter of the lifting pin, must be provided in the splice plates at the top end of the column. A pair of shackles is slipped over the plates, a long steel pin is then inserted through the shackles and the holes in the two splice plates. The two shackles in turn are secured by slings to a single shackle above. The end of the pin is kept from falling later by a lightweight sling from a small shackle at the end of the pin to the upper eye of one of the two main lifting slings. Or this protective sling can be secured directly to the shackle on the lifting hook.

The tagline is secured to the small shackle at the end of the pin. After the column has been erected and safely pinned and bolted in place, the lifting hook is lowered slightly, and a quick tug on the tagline, at as flat an angle as the area will permit, will pull the pin clear. This releases the lifting shackles and disconnects the hickey from the column.

Hoist, hand.   See *Crab* (*winch*).

Hoist, powered.   Diesel; electric; gasoline; pneumatic; steam; tractor; Tugger; single drum, double drum, triple drum, etc.; swinger: separate, to connect to main hoist. In listing hoist requirements, the number, type, and horsepower, number of drums, and type of swinger should all be specified according to the lead-line pull that will be required. This, in turn, depends on the speed of the wire rope, on the maximum quantity of rope on the drum when full, on the diameter of the wire running rope, and on the reeving of the falls or tackle.

The height the load is to be lifted affects the time of the lift; the more parts in the falls, the slower the time and the more cable there will be on the drum. But then the lead-line pull required is less than with fewer parts for the same load. If a hoist has air controls, the field should be warned to watch the pressures to avoid overloads on the controls.

The frictions are usually adjusted by regulating the pressure at the control valves, normally using 40 to 50 psi for heavy work and 20 to 25 psi for light work. The brake adjustment should be such that the brakes will hold the maximum load with the foot pedal half way down. The drum dogs, if in proper working order, should make full contact with the dog-ring teeth when set and should give full release when withdrawn.

The Tugger hoist is a small, single-drum device that can be clamped on a column or beam and is usually pneumatically powered. It is adapted to the use of small-diameter wire rope for setting small pieces, where a larger hoist would be unnecessary and uneconomical. A tractor hoist is a single-drum hoist mounted on the rear of a tractor for operating light-capacity falls where the tractor's mobility permits the hoist to be used in several locations.

Hooks. Cant (Fig. 3.7.43); timber-carrying (Fig. 3.7.44); ladder (adjustable) (Fig. 3.7.45); safety (Fig. 3.7.46); scaffold; sorting (Fig. 3.7.47); tagline (Fig. 3.7.48); timber (Fig. 3.7.49). A cant hook is used for rolling and skidding heavy timbers, and a timber-carrying hook permits two men to carry timbers from place to place. An adjustable ladder hook permits hanging a short wooden ladder from the top of a girder or deep beam or truss so that connectors, bolters, fitters, etc. can reach the bottom flange and splices easily and safely. The safety hook has a latch or catch to prevent a sling on the hook from coming off until the latch or catch is specifically lifted or depressed (as the case may be).

The sorting hook has a point that can be slipped into end connection holes on beams, channels, etc., as well as a bent portion designed to grip the end of the web of a beam against the underside of the top flange. The tagline hook is slipped into the end of a load being handled, or an individual piece being erected, to help guide the load clear of obstructions, or to guide a piece to the connectors waiting for it above. The timber hook is used to lift timber by powered equipment, the points digging into the wood to hold it fast.

Hose. Air: ½-in. leader, ¾-in., 2-in.; steam; water; couplings; gaskets; valves: Cleco, Quick-as-a-wink, Thor, etc. An air hose must be checked carefully to make sure it is in safe condition and more than adequate to withstand the pressure to be applied.

Hose mender. Mending wire.

Impact wrench. Electric; pneumatic; accumulator; sockets. An impact wrench rotates a socket for bolting by means of a series of rotating im-

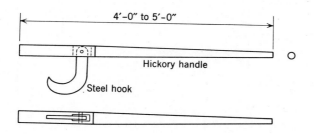

FIG. 3.7.43. Cant hook.

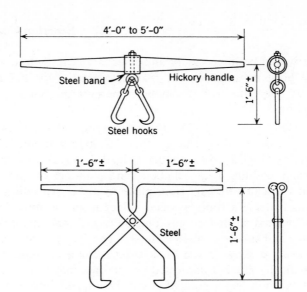

FIG. 3.7.44. Timber-carrying hooks.

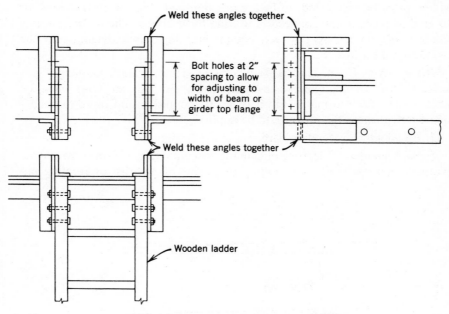

FIG. 3.7.45. Adjustable ladder hook.

66

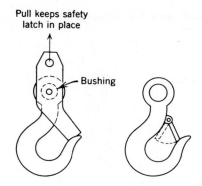

FIG. 3.7.46. Safety hooks.

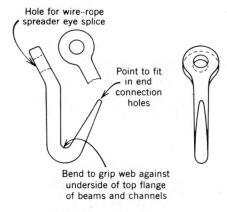

FIG. 3.7.47. Sorting hook.

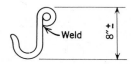

FIG. 3.7.48. Tagline hook.

67

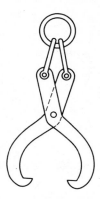

FIG. 3.7.49.   Timber hook.

pacting blows. Its use in tightening high-strength bolts is advisable since it permits a controlled tightening of such bolts.

Jacks.   Bridge (Fig. 3.7.50); hydraulic, fittings, gauges, tubing; journal or bottle (Fig. 3.7.51); push-and-pull (Fig. 3.7.52). See also *Steamboat ratchet;* screw (Fig. 3.7.53); track (Fig. 3.7.54); wedge (Fig. 3.7.55); wedge handle (Fig. 3.7.56).

Jumping beams.   Steel; wood.

Keel.   Marking crayon.

Ladders.   Straight: steel, wood; extension; hooked.

Lanterns.   Red; clear; spare globes.

Launch.

Level, spirit.

Life preservers.   Ring; vest.

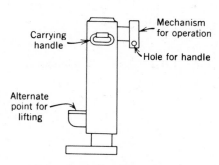

FIG. 3.7.50.   Bridge jack.

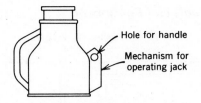

FIG. 3.7.51. Journal or bottle jack.

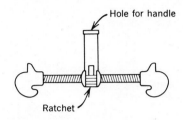

FIG. 3.7.52. Push-and-pull jack.

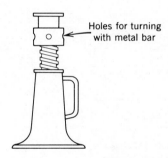

FIG. 3.7.53. Screw jack.

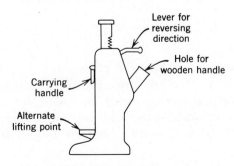

FIG. 3.7.54. Track jack.

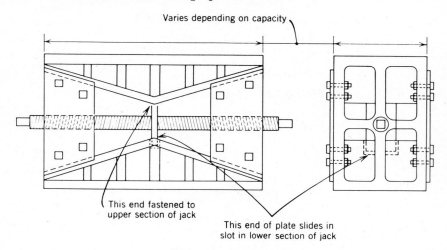

FIG. 3.7.55. Wedge jack.

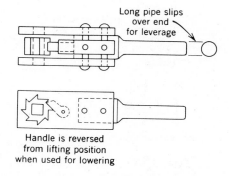

FIG. 3.7.56. Wedge-jack handle.

Locks and keys.

Mats. Timber (Figs. 3.7.57 and 3.7.58).

Maul (Fig. 3.7.59). Used for driving drift pins and bull pins into connection holes, for hitting steel members into place, for straightening bent material, and wherever force is required that can be achieved by a beating action. The commonly used erector's maul has an 8-lb. head and a 30-in. handle.

Oil. Pneumatic hammer; cylinder; engine; etc.

Oil can. Straight; squirt.

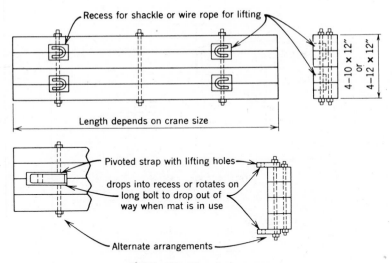

FIG. 3.7.57.   Large mat.

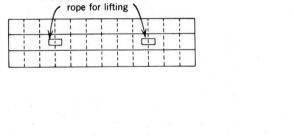

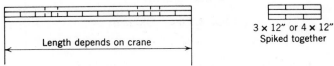

FIG. 3.7.58.   Small mat.

Old man (Fig. 3.7.60).   The base is clamped or bolted to the work where a hole is to be drilled. The arm is adjusted to the length of the drilling machine and bit, with the feed screw in its retracted position. As the drill bites into the material, the feed screw on the machine is forced out against the arm of the old man until the hole has been drilled through.

Overhauling weights.   Light; heavy; one-piece (Fig. 3.7.61); assem-

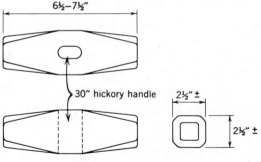

FIG. 3.7.59. Maul.

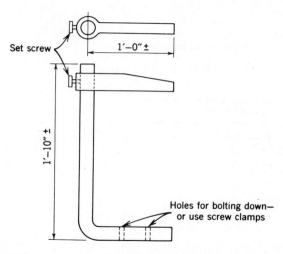

FIG. 3.7.60. Old man.

bled (Fig. 3.7.62). The overhauling weight, or "ball," is used to help overhaul the main or auxiliary load falls to help bring the load block and lifting hook down after a load has been lifted and the hook has been freed. The weight must overcome the friction of the various sheaves in the blocks as well as any idler or lead sheaves over which the rope runs. Overhauling balls or weights are usually 500 or 1000 lb. or heavier, depending on the need.

When a derrick is in use on a tall building, in order for the load block to be lowered (overhauled) to the ground for picking a load, a heavy weight is needed to overcome the weight of the lead line between the hoist

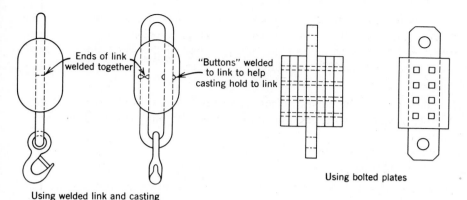

FIG. 3.7.61. Overhauling weight: one-piece (overhauling ball or "headache pill").

FIG. 3.7.62. Overhauling weight: assembled (overhauling ball).

below and the top of the boom, through the foot of the mast, plus the friction of the various sheaves. The weight of the lead line may be great enough to pull the load blocks together if the overhauling weight is insufficient for its purpose. When a climbing crane or a mobile crane with an excessively long boom is used, the same trouble may occur since the ball must bring the load block and lifting hook back down to the ground after the load lifted to the maximum height of the boom has been landed at the upper level.

Too often, when the ball is of insufficient weight, the men are forced to pull the lead line upward in order for the various parts of the falls to overhaul and permit the load block to be lowered. In this case they are doing the work that the overhauling ball should be doing.

Overhauling balls may be solid castings for forgings, shaped like a round ball or elongated, with a welded or forged link cast inside, leaving a loop at the top and at the bottom protruding from the ball. The link transmits the load through the ball to the lifting hook; the metal of the ball surrounding the link serves only as a weight to help overhaul the lifting falls. A hook may be installed in the lower loop before casting. The type without the hook incorporated is preferable because there are times when slings or sorting pendants should be shackled directly to the lower loop. Another type has heavy weights bolted together over a long plate with holes at the top and bottom to which to shackle the lower load block and the lifting hook, respectively.

When there are many parts in the load falls, a weighted block can be used as the lower load-falls block to reduce the size and weight of the overhauling ball. These are merely wire rope blocks with heavy cheek plates bolted to each side of the block instead of the usual thinner cheek plates found on the ordinary block.

Pick.

Pile driver. Pneumatic; electric; steam; internal combustion; gravity; vibratory; sonic; hydraulic-ram type; leads; hammer; points; ring; hook; extractor.

Piles. Steel; wood.

Pins. Bull (Fig. 3.7.63); drift (see *Drift* pins). The bull pin is used by connectors to help drift heavy material into place by driving the bull pin through corresponding connection holes. Drift pins are then driven into other open holes and left there while the bull pin is removed for further use.

Pipe. For air, steam, water; bushings; couplings; elbows; fittings: nipples; plugs; reducers; tees; tongs; valves; vise.

Pipe cutter. Stock and dies.

Plumbing-up cord. Wire; hooks ("crow's feet") (Fig. 3.7.64); fittings; plumb-bob or weight; plates (Fig. 3.7.65).

Pumps. Diesel; electric; gasoline; pneumatic; steam; hand; for water; for hydraulic jacks; for gasoline.

Punches. Center (Fig. 3.7.66); screw (Fig. 3.7.67); punch and dies. The center punch is used to form a small indentation to aid in starting a drill bit in the correct location to drill a hole and also to mark centerlines and other location points on steel. The hand-operated screw punch is for punching holes in reasonably thin material where it would be uneconomical to set up a powered drill or to use a hand ratchet drill.

Railroad dolly car. A small, four-wheeled platform with railroad wheels on standard gauge, to roll on railroad tracks. It is used for transporting small material, steel beams, etc., without the need of loading on railroad equipment. The car can usually be pushed by hand or pulled by means of a set of falls or a wire rope from powered equipment.

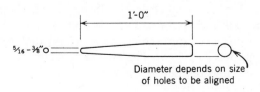

FIG. 3.7.63.   Bull pin.

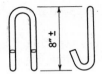

FIG. 3.7.64. Plumbing-up hook ("crow's foot").

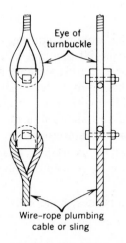

FIG. 3.7.65. Plumbing-up plates.

FIG. 3.7.66. Center punch.

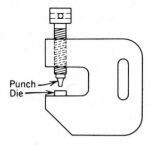

FIG. 3.7.67. Screw punch.

Railroad rails. Frogs; spikes; splice bars; splice bolts; hook bolts; washers; switch plates; ties; track gauge; truck.

Reamers. See *Drill bits.*

Receiver, air. A check must be made that the receiver furnished will comply with the legal requirements of the state and/or city in which it will be used.

Respirator. If any burning, welding, cutting, heating, etc. is to be done in an enclosed space or on material that, when heated, will produce toxic fumes, a respirator should be provided suitable for protection against the fumes produced. This may be a simple face mask with the proper filters and/or cartridges being supplied with it, or an air-line respirator. The air-line feeder should be located so that no injurious fumes will be drawn in with the air being supplied to the respirator.

Ring, purchase (Fig. 3.7.68). A loop so shaped and designed to permit lifting at one point, two or more bridled slings connected to the piece being lifted.

Rope. Manila: derrick gantlines; hand falls; float lines; scaffold lines; slings; spool lines (on hoist); tagline.

Rope, wire. See *Wire rope.*

Rowboat. Oars; anchor.

Safety belt and line. Safety belts should have quick-release buckles.

Safety hats (hard hats). With brim (Fig. 3.7.69); without front brim (for welder's shield) (Fig. 3.7.70); extra bands: leather, leatherette, plastic, sponge rubber; winter liners.

Saws. Cross-cut: two-man, one-man; hand; hack: frame, blades.

Scaffolds. Boatswain's chair; floats (see *Floats*); needle beams; ship. Needle-beam scaffolds consist of two 4 × 6-in. dressed timbers of Sitka spruce (or the equivalent in strength and weight) about 26 ft. long, held up by ropes near the ends and at the center. Short scaffold planks, usually 2 × 9 or 2 × 10 in. and 10 to 16 ft. long, are laid across the 4 × 6-in. timbers. The planks have drop bolts or cleats at their ends to keep them in place.

Scale box. Steel (Fig. 3.7.71), wood, or metal for handling kegs of bolts, and other small material.

Where men are to be lifted to work that is not easily accessible by ladders, or where climbing to a high level is too fatiguing, a form of scale box, sometimes known as a "suspended work platform," should be used. Preferably this should have a railing approximately 42 in. high, a toeboard about 4 in. high, and a mid rail, all on three of the sides, with a removable

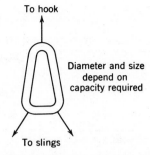

FIG. 3.7.68. Purchase ring.

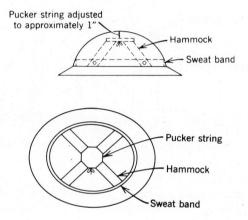

FIG. 3.7.69. Safety hat (hard hat): with brim.

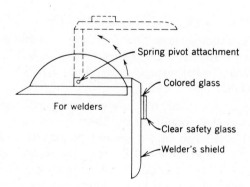

FIG. 3.7.70. Safety hat (hard hat): without front brim.

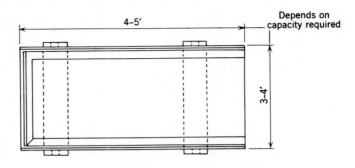

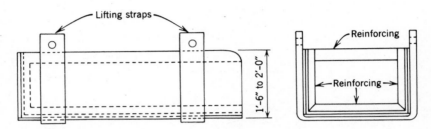

FIG. 3.7.71.   Steel scale box (for material and heavy equipment only).

guard or railing on the fourth side for access. This platform should be used only for men and their tools or equipment, and the scale box only for materials and heavy equipment.

Screw clamp ("C" clamp).   Structural (Fig. 3.7.72); chain; hook.

Screw driver.

Shackles.   Pin: screw (Fig. 3.7.73), standard (Fig. 3.7.74).

Shanties.   Office; toolhouse or "men's shanty"; combination office and toolhouse or "men's shanty"; knock-down (Fig. 3.7.75); portable one-piece (Fig. 3.7.76); trailer: single-axle (Fig. 3.7.77), two-axle (Fig. 3.7.78); large, small.

Shims.   A supply of shims for use in setting grillages, base plates, slabs, and screed angles should be on hand to be shipped promptly if needed at the job site. The supply should be replenished subsequently. These are usually 3 × 3 in. or 4 × 4 in., and ⅛, ¼, and ½ in. thick. For jobs where leveling of base material is critical, a supply of very thin shimming material should also be on hand. Screed angles are customarily ordered specifically for individual grillages since their length depends on the

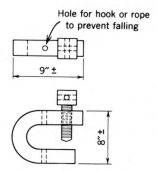

FIG. 3.7.72. Screw clamp ("C" clamp).

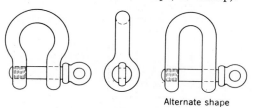

Alternate shape

FIG. 3.7.73. Screw pin shackles. Diameters of shackle and pin depend on capacity required.

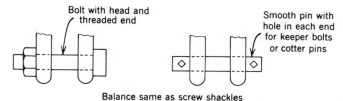

Bolt with head and threaded end

Smooth pin with hole in each end for keeper bolts or cotter pins

Balance same as screw shackles

FIG. 3.7.74. Standard pin shackles.

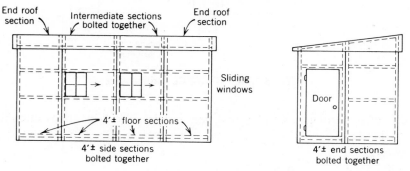

End roof section

Intermediate sections bolted together

End roof section

Sliding windows

Door

4'± floor sections

4'± side sections bolted together

4'± end sections bolted together

FIG. 3.7.75 Knock-down shanty.

79

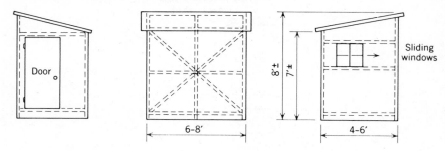

FIG. 3.7.76.   One-piece shanty.

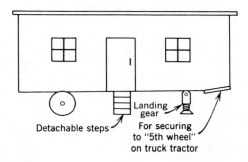

FIG. 3.7.77.   Single-axle trailer office.

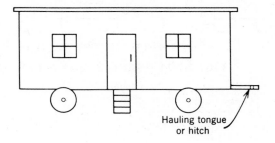

FIG. 3.7.78.   Two-axle trailer office.

width of the grillage assembly and may require 3×5-in. shims on which to be leveled.

Shovel.

Signal bells.   See *Bells.*

Signal system, voice.   Earphones; loud speaker; transmitter; wire.

Sledge.   See *Hammer, hand.*

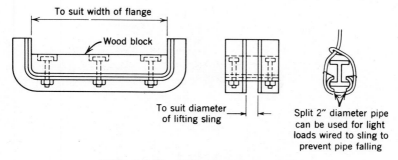

FIG. 3.7.79.   Sling protector (saddle).

Sling protector (Fig. 3.7.79).   For use on bottom flange of heavy gird-ers to prevent the corners of the flange from cutting or damaging the sling. The size depends on the width of the flange and the diameter of the wire rope sling to be used.

Spools (Fig. 3.7.80).   These are designed to protect a wire rope guy, etc., where it is attached to a pin or bolt. The sizes vary with the diameter of the wire rope to be used. They are usually used where a guy is bent over an attachment and a temporary loop is made in the guy by means of wire rope clips. When an eye at the end of a single-part guy is secured to a pin or bolt, a thimble is preferable, with the thimble installed at the time of splicing the eye so that it will not drop out (Fig. 3.7.81).

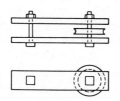

FIG. 3.7.80.   Spool and plates for attaching guys to turnbuckle eye.

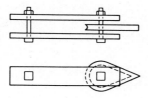

FIG. 3.7.81.   Thimble and plates for attaching guys to turnbuckle eye.

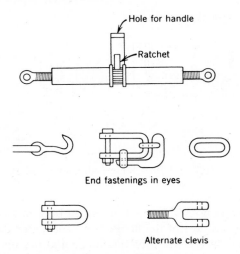

FIG. 3.7.82. Steamboat ratchet.

Steamboat ratchet (Fig. 3.7.82).   Used for pulling two pieces together when a stronger force is required than could be achieved by means of a turnbuckle. The ends are pulled together by operating the handle and ratchet back and forth. A push-and-pull jack is similar except that it not only pulls pieces together but can also push them apart.

Stock and dies.   Bolt; pipe.

Tank, water.

Tape.   Steel; cloth. Measurements are usually made in feet broken down into sixteenths or eighths of an inch. Occasionally a tape is required that will read in tenths and hundredths of a foot.

Tarpaulins.   For protecting equipment from the weather or from falling dirt, sparks, scale, slag, etc. It is advisable to treat tarpaulins to make them fire and mildew resistant.

Thimbles (Fig. 3.7.81).   For guys or special slings. See *Spools.*

Timber.   Blocking; buggy; falsework; floats (see *Floats*); mats: large, small (see *Mats*); needle beams (see *Scaffolds*); floor planks; scaffold planks; skids; shores. Planks for working floors and extra floors are usually 2 × 12 in. by 22 to 24 ft. of rough, select Douglas fir, structural grade, or the equivalent in strength and weight. Where necessary, because of the span on which they will be used, 3 × 12 in. are sometimes required for the loads to be carried, but they should be reduced in length if possible to make them safer for two or three men to handle. Timber should not be

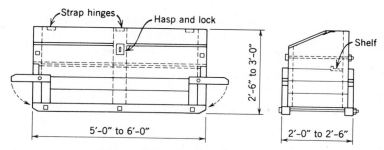

FIG. 3.7.83. Wooden toolbox.

painted since this may hide possible defects. Clinching strips driven into the two ends of planks usually help prevent splitting.

Skids for landing loads of steel or individual pieces when distributing are usually 4 × 4 in.-timbers from the cars or trucks in which steel is shipped, or they can be two or three 2 × 12 in.-planks laid one on top of the other, two or three high, using planks that have become unsafe for use as floor planks or scaffold planks. The quantity ordered depends on the area being worked and the number of raising and other gangs operating simultaneously.

Timber dolly (Fig. 3.7.26). Rollers; wheels. (See also *Dolly, timber.*)

Toolboxes. Large (Fig. 3.7.83), small; crane; derrick; engine; superintendent's.

Toolhouses. See *Shanties.*

Tractor.

Traveler, frame for. A-frame derrick, guy derrick, stiffleg derrick.

Turnbuckles (Fig. 3.7.84). For guy derrick: boom guys, footblock guys, mast guys; hoisting engine anchor guys; for gin-pole guys; for

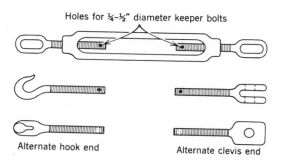

FIG. 3.7.84. Turnbuckles. Size and ends depend on use and capacity required.

plumbing-up guys; eye-and-eye ends; eye-and-clevis ends; clevis-and-clevis ends.

Vise.   Bench; blacksmith's; pipe.

Washers.   See *Bolts.*

Waste.

Water carrier.   See *Buckets:* water.

Wedges.   Steel (Fig. 3.7.85); wood (Fig. 3.7.86).

Welding.   Clamps; electrode holder; ground wire; lead wire; hammer; helmet; helmet glass: plain (cover), shade; shield.

Welding machines.   Skid-mounted; wheel-mounted; diesel, electric, gasoline; rectifier; transformer.

Whistle.

Wire-rope clamps (Fig. 3.7.87).

Wire-rope clips (Fig. 3.7.88).   For guy derrick: boom jumping guys, mast guys; gin-pole guys; traveler guys; load-line becket; topping-lift (boom-falls) becket; whip-line (runner) becket; plumbing-up guys.

Wire-rope guys.   Jumping; guy derrick boom; gin-pole; guy-derrick mast; plumbing-up.

Wire-rope lashing.

Wire-rope, running.   For air hoist; crab or winch; crane boom line, load line, whip line; derrick boom line, load line, whip line. Wire rope should be kept well lubricated and checked frequently for wear, broken wires, corrosion, deterioration, etc. and removed from service promptly when inadequate or unsafe for the strength and service required. A crite-

FIG. 3.7.85.   Steel wedge.   **FIG. 3.7.86.**   Wooden wedge.

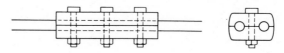

FIG. 3.7.87.   Wire-rope clamp.

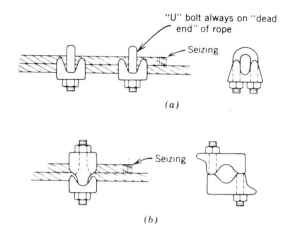

FIG. 3.7.88.   Wire-rope clips: (**a**) "Crosby" type; (**b**) "Laughlin" type ("fist-grip" type).

rion should be established to determine when used running rope should be discarded, based on reduction in diameter due to wear or chafing, or on the number of broken wires in any one lay of any one strand, kinks, birdcaging, etc. Discarded running rope can usually be used for lashing or for plumbing-up guys where the reduced strength is still more than enough for the purpose.

Wire-rope slings.   One part (Fig. 3.7.89a); plaited or braided (Fig. 3.7.89b); column anchors; column-setting hickey; engine anchors; eye-and-eye: erecting, unloading; "street slings." Length, diameter, and type should be specified in ordering slings, based on the size and weight of the material to be handled. Wire rope slings should be discarded when they are

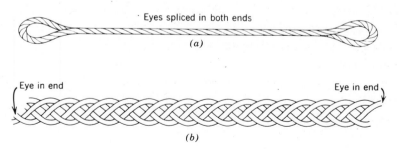

FIG. 3.7.89.   Wire-rope slings: (**a**) single-part eye-and-eye; (**b**) endless braided.

so badly kinked or damaged that they can no longer adequately grip the piece being lifted, or when broken wires make them unsafe for the load and dangerous for the men to handle. Spliced ends should be trimmed carefully for safe handling.

"Street slings" are eye-and-eye slings that are long enough to be choked around a load being unloaded from a car or truck, by threading the eye at one end through the eye at the other end. The free-eye end is then hooked on the hoisting spreader hooks.

Wire-rope spreaders.   Hoisting, eye and heavy lifting hook; sorting, eye and light sorting hook; spare hooks. Some erectors splice the hooks directly into the eye at one end of the pair of spreader slings; other erectors use eye-and-eye slings and shackle the hooks into the eyes.

Wrenches.   Box (Fig. 3.7.90); "Crescent"; key (Fig. 3.7.91); monkey; open-end spud (Fig. 3.7.92); socket; Stillson.

Equipment for the field office usually includes some or all of the followings items:

Field Office Equipment

Adding machine.
Bulletin board.
Check writer.
Computor.
Files.
First-aid cabinet.
First-aid supplies.
Level: builder's; carpenter's; engineer's (surveying).
Level rod (surveying).
Office cabinet; chairs; desk; drawing file rack.
Plumb bob (surveyor's).
Stove: large, small; coal, oil, electric, steam radiator, propane.
Stretcher.
Transit (surveying).
Typewriter.
Safety codes.
Safety posters.

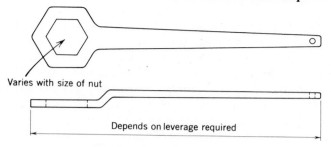

Varies with size of nut

Depends on leverage required

FIG. 3.7.90.   Box wrench.

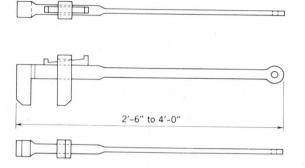

2'-6" to 4'-0"

FIG. 3.7.91.   Key wrench.

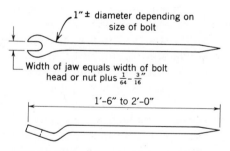

1" ± diameter depending on size of bolt

Width of jaw equals width of bolt head or nut plus $\frac{1}{64}-\frac{3}{16}''$

1'-6" to 2'-0"

FIG. 3.7.92.   Open-end spud wrench.

3.8   Records and Reports

At the end of the job, after the tools and equipment have been returned to the toolhouse and checked, the toolhouse superintendent should prepare a report of tools and equipment shipped but not returned (lost, stolen, mislaid, discarded) as well as what items he has scrapped because they are badly damaged, worn out, or unsafe for reuse. A value should be placed on these items to give an indication of how well the field superintendent has used and cared for the tools and equipment furnished him.

The toolhouse records should include a job identification number or mark, the date the tools were ordered, where they were shipped from, and the date, the consignment, mode of transportation, car numbers (if by rail), truck numbers (if by common carrier trucker), or own trucks' identifying marks. By using a standard columnar form, a running record can be maintained, listing opposite each item, in individual columns, the number of each item ordered, shipped, returned, missing, scrapped, and the unit cost, as well as the total cost of the lost and scrapped tools.

A permanent report for future use in ordering tools and equipment should be prepared for estimating purposes and cost accounting. This should give a brief description of the job and type of erection, such as crane, derrick, traveler; high building; low building; apartment house; mill building. The total tons and pieces erected should be given, as well as the average number of pieces per ton, broken down into erection by powered equipment and by hand; the total number of bolts by types (machine bolts, high-strength bolts, and average diameters); welds by sizes, length, type, weight of electrodes furnished, and probable weight deposited.

The cost of lost and scrapped tools should be given as totals, as a percent of the total payroll, and as a cost per ton of steel erected. (These last two figures are important for use in future cost estimating.) The information should list major equipment used by type, size, and capacity, with important details such as length of boom, mast, tower, jib. The starting and completion dates should be stated as well as the amount of time lost because of bad weather, holidays, work stoppages, or other causes.

For future reference, the names of the supervisory personnel (superintendent, resident or field engineers, foremen, and timekeepers) should be included.

These records can assist a toolhouse superintendent in operating his toolhouse efficiently. The report is invaluable to the estimator, to the engineer lining up future jobs, to the cost accountant, and to all others involved in seeing that jobs are run efficiently and economically. A well-run toolhouse, supplying adequate, proper, and safe equipment at a reasonable cost, can contribute significantly to the safe, economical, efficient, and speedy erection in the field.

4
Safety

4.1 Management

For a structural steel job to operate with maximum safety, certain fundamental requirements must be recognized and made effective. Everyone involved should be aware that the safety drive must be cooperative and continuous.

Management must be convinced that safe operations produce the most efficient, economical results. Management must be willing to approve expenditures for safety devices, equipment, and tools. Proposed erection schemes and erection methods must be analyzed for safety. There must be a real conviction, forcibly expressed, that efforts to have a safe operation are considered essential and worthwhile. Not only must the idea of safety be inculcated throughout the planning stages in the erection scheme, in the toolhouse, and in the actual execution of the work, but the field supervisory force must be assured of management's backing in all their efforts to establish safe conditions and safe workmanship. Management should endeavor to put enthusiasm and interest into the safety program. Safety must be emphasized as well as production since safety is really a part of all operations.

4.2 Supervision

Supervision must be "sold" on safety, must always be alert to possible unsafe conditions, and must be able to distinguish between safe and unsafe workmanship. The entire supervisory force must be trained to convince the workers that safe work is required. Newly hired men must be carefully watched and, when necessary, safe working methods explained or demonstrated to them. The men should be observed until they show that they are safe workers.

Since many erecting firms operate over a large area of the country, they use only their own superintendents and sometimes their own foremen. The men who will actually do the work are hired locally. In hiring such men, the superintendent should investigate and try to make sure that they are skilled in the particular duties they are to perform: that connectors are experienced; that hoist and crane operators are familiar with the types of hoists or cranes to be used; that toolroom men are skilled in the proper methods of caring for tools, sharpening drills, and repairing pneumatic tools. On a large project it is sometimes effective to use a skilled master mechanic to supervise the repair and maintenance of equipment and tools. Unsafe equipment can lead to serious and costly accidents.

Some erectors employ only union men, in which case the superintendent can go directly to the union hall in the area to check the records of available men. With experience in an area a superintendent will soon learn who are safe, experienced, skilled workers, and who are careless and unsafe. In going from one job to another in the same general area, entire gangs can often be taken along: a raising gang to erect and one to bolt, as well as individual men such as hoist and crane operators and others. But unless the superintendent or the foremen actually know the men hired, it is important for them to observe the men carefully as the work is started. A good supervisor can quickly spot an inexperienced, unskilled, and often unsafe worker. Helpers or apprentices must be instructed so that they work safely and efficiently as they learn to become skilled workers.

When a new man is hired, if arrangements can be made to have a local doctor give him a basic physical examination, it may prevent a serious accident later. A crane or hoist operator subject to a heart attack could cause a catastrophic accident. A man with a hernia could become crippled for life if he happened to be assigned to a task involving heavy lifting. Although the man may not know of his condition before the examination, he might be able to have it corrected and then be fit again for any work. A man with defective vision could misjudge his footing when walking on a beam and possibly fall. If he is informed of his defect at the time of the physical examination, he might be able to correct his vision with prescription glasses. Thus, he will be safer, but will also be aided in his normal living. Safe methods must be continuously enforced.

4.3   Men

The men actually doing the work must be made safety-conscious. They should be trained in safe methods of operating and the safe use of tools and equipment. They should be indoctrinated with the benefits of wearing safety hats (hard hats) and shown how to fit them properly for maximum protection. They should be convinced of the need to wear safety goggles when performing eye-hazardous duties. They should be shown the correct way to fit and wear the goggles.

The men should be informed that they are expected to cooperate with supervision and management to keep the job safe and to protect themselves, their fellow workers, and other trades working in the area from accidents and injuries. They should be encouraged to ask their foremen when they have any questions about the proper, safe way to perform any part of their work. Their attendance at safety meetings should be strictly enforced. They should be induced to use protective equipment properly, to study any safety codes available, to observe safety posters and to follow the guidelines normally depicted on such posters.

In the United States (and in foreign countries where similar regulations are in effect) the requirements of OSHA *Safety and Health Regulations for Construction* must be strictly observed. Penalties imposed on the employer can be severe. It is the employer's obligation to provide a safe and healthy place for the workers, but it is then imperative that the conduct of the men be watched constantly since it is their obligation (without penalty) to "comply with all occupational safety and health standards, rules, regulations, and orders issued under the Act, that apply to [their] actions and conduct on the job." Such regulations include the obligatory use of safety hats and, where needed, safety lines, belts, and goggles.

4.4   Implementation

To implement management's obligations, forethought should be given to providing a safe erection scheme and safe working conditions and surroundings. Equipment and tools should be inspected regularly, kept in safe condition, and be of adequate strength. Unsafe material such as broken planks and damaged or worn out manila and wire ropes should be discarded promptly.

A safety incentive plan involving monetary considerations for the supervisory force, or even the award of pins, certificates, or prizes to individuals for accident-free records, will give tangible proof that the employer wants the men as well as the supervisory force to work safely. While men's lives cannot be measured in dollars and cents, an incentive paid to the supervisory force, based on the job-accident record will aid materially in reducing accidents, increasing production, improving costs, and reducing insurance premiums.

The records of accidents should be kept and constantly analyzed; they should list frequency (number of accidents per million man-hours worked), severity (number of days lost by injured men due to accidents, per million man-hours worked), and costs. Superintendents should be informed not only about their own safety records, but about the records of other superintendents working for the same employer. Comparison with others can often be the spark needed to improve a man's efforts.

Safety records should be tied in to production to help convince the supervisory force that "safety pays." Safe surroundings, equipment, and methods will generally attract a better type of skilled worker which, in turn, will result in better accident records, lower insurance costs, and lower erection costs.

One of the best ways that management can help the superintendent is to appoint a safety engineer (Sec. 2.5). He must work with the superintendent for best results, aiding him in watching for unsafe conditions, seeing that safety literature to be distributed reaches the men, making sure that safety posters are displayed for best results, keeping accident records, seeing that safety devices are used properly, and acting as an extra pair of eyes for the superintendent.

When a safety engineer is not on hand at all times, a few men can be appointed as a safety committee to watch for unsafe conditions that the superintendent or foreman might miss. Such situations can then be caught promptly enough to prevent accidents. It is helpful to change the personnel of such a committee weekly, making sure that the men on it do not shirk their regular duties. On a job covering a large area, even with or without a safety engineer on hand, it may be advisable to appoint one man as a safety observer, permitting him to devote his full time to covering the entire job (see Sec. 6.21).

The superintendent should enforce the correct use of equipment and tools, and the use of ample sizes and strengths of manila and wire ropes, planks, turnbuckles, shackles, etc. An accident can result from improper use, or the use of improper size and strength, of tools and equipment. The superintendent should inform his foremen and men about the plan of work as it unfolds in the course of the operation. This will help him secure their cooperation. Mutual problems should be discussed freely with his foremen and engineers.

Safety meetings must be held regularly and in addition as often as needed. At these meetings, the superintendent, with the help of the safety engineer, must convince the men that safety is a requirement for continued employment. Special safety meetings are advisable with only the superintendent and his foremen present. Gang meetings, sometimes called "stand-up" or "toolbox" meetings, with just the foreman and his individual gang attending, are advisable. They are held where the work is being done. Any unsafe action can be pointed out immediately instead of waiting for a general meeting. Some erectors have evening dinner meetings at which the men can really relax and are more likely to discuss otherwise unnoticed conditions that might cause accidents. Although this practice may seem costly, it often pays real dividends in fostering a more cooperative spirit toward both safety and production on the entire job.

A briefing should be held for all the men involved, before any special, unusual, or especially hazardous operation. An explanation can be given then, in detail, of what is to be done and how it is to be done. Any pitfalls that might be encountered can be pointed out, making sure the men understand what each one is to do to coordinate every man's work into the plan as a whole. Safety depends on teamwork and cooperation.

Safety posters applicable to the particular operation should be displayed conspicuously, changed frequently, made attractive, and protected from the weather or other damage. Safety subjects should be covered as fully as possible by such posters. For example, a poster could describe the safe working distance from electric wires or crane contact rails, precautions to avoid heat exhaustion or sunstroke, the danger of using broomed or mushroomed heads on percussion tools, or similar hazardous conditions.

Any requirements pertaining to safety in the contract or specifications, in any applicable building codes, in any laws whether municipal, state, or federal should be abstracted and included in the instructions to the field. Various codes and laws are currently becoming stricter in requiring safer surroundings in which to work, safer equipment with which to work, and safer methods to be followed. Labor unions are generally cooperating with the safety movement by stressing safety in their apprentice training courses, teaching the proper and safe way to use tools and equipment, and instilling in these men the need to protect themselves and their fellow workers. Accidents mean suffering and loss of income to the injured man, expense to the employer, and delays to the project that will affect the owner. In fact, everybody loses something from an accident, whether it is a minor or a major one.

Mechanically, equipment has had more and more safety features built in over the past few years. Manufacturers have conferred with users and have analyzed the use of their equipment to learn where improvements can eliminate hazards. More attention is being paid to the strength of parts for their use in a machine. Manuals are prepared to inform users of the proper and safest method of operating, lubricating, and maintaining equipment, in the interests of longer, safer life. But safety features built into tools and equipment are only as good as the safe practices of the men using them. The misuse or abuse of any tools or equipment should be guarded against.

The job tool list will include all safety equipment that may be of value on the job. Personal safety equipment for each man should be provided, such as hard hats, goggles, safety belts with proper size ropes, and—if there is a water hazard—life preservers and vests. Safe (size and strength) manila lines and wire rope, slings, shackles, pins, ladders, etc. should be used. In the case of water operations, boats and/or launches should be on hand.

Medical facilities should have been arranged well in advance so that in case of an accident the field office will know immediately what doctor, specialist, or hospital to use; where first aid can be secured; what ambulance service is available. When posting notices of medical service or insurance coverage is a legal requirement, such notices should be secured before the job starts. Safety posters, safety codes, and safety literature to be distributed should be on hand when needed (see Appendix B).

Any posters required by federal or state regulations should be secured in advance so that the requirements for posting can be followed, because sometimes there are penalties in some instances for not posting such notices, which may be intended to notify workers not only about safety, but also about nondiscrimination, wage payments, and laws intended to protect the men.

4.5   Aids to Safety

Shanties or trailers to protect the men against rain, sleet, or snow may help prevent illness, which could cause a man to work unsafely, or could even prevent a man from being absent from work because of illness, leaving a gang shorthanded and unsafe. A place of refuge from the weather will enable the men to change into dry clothing. Frequently they will stay on the job if the weather gives promise of clearing, provided they have such protection.

The strength and capacity of a ramp used for trucks that deliver equipment, supplies, or the steel itself should be checked. This may help prevent a failure of the ramp. Similarly, if a level trestle is being used instead of a ramp for trucks or cranes to operate, this should also be carefully checked. For both built-up ramps and trestles, the supports, footings, foundations, main material, and bracing must be adequate for the use. Deterioration or movement should be watched for constantly. An adequate, safe design must be used, with material in good condition—timbers not cracked, split, rotten, or otherwise defective; steel members not corroded, rusted, or in need of protective paint; and everything properly secured.

Cross-lot bracing, while not generally part of the steel erector's work, should be inspected to be sure it is sound, of adequate material, safely installed and maintained, and constructed to prevent collapse of the sides of the excavation being held by it. Collapse or failure can endanger the erector's men and the structure being erected. Care must be used in threading the steel members into place in order not to damage or dislodge any members of the bracing system.

Anchors for guys or tie-downs installed by others should, when feasible, be placed while a representative of the erector is present to ensure proper placement. Hairpin anchors should be of sufficient size for the stresses to be imposed, installed deep enough, but with the loop clear enough from the footings to permit fastening the erector's sling, shackle, or turnbuckle; and inclined in the same direction as the pull of the guy or tie-down to be attached thereto. The weight of the concrete in which it has been embedded should have been checked for uplift and sliding.

Split-end eye anchors should be used only in good, solid rock. The hole for the shaft should be just deep enough so that with the wedge in place in the split end, resting against the bottom of the hole, the eye will be just above the face of the rock when the anchor is driven. The wedge must spread the split end solidly against the sides of the hole. The diameter of the hole should permit the wedged end to grip the sides of the hole and yet permit grout of cement, sand, and water to be poured into the hole around the shaft, to give added frictional resistance. The angle of the hole should be in the direction of the guy or tie-down when stressed. If lashing has been embedded in the foundation, it should be buried deep enough and all parts arranged to take the applied load equally.

The first load lifted by a guy derrick off the delivering carrier should be picked slightly and the anchors opposite immediately inspected. The load should then be moved to where it is to be unloaded, still keeping it close to the ground, and all the guy anchors checked successively as the load is swung opposite each one. The load should be landed promptly if any anchor shows signs of insufficiency, defect, pulling out, or being inadequate.

Such anchors should then be replaced immediately with substitute anchors. When the first heavy piece is delivered the entire procedure should be followed again to make sure that all the anchors and tie-downs are adequate and safe. With a stiffleg derrick or a traveler, a similar procedure should be followed with tie-down anchors.

Shipments in cars, on barges, on trucks, or on other carriers should be inspected, and if unsafely loaded, the shipper should be instructed to remedy his unsafe loading procedures. Skids should be in place under drafts or loads; steel should be in place so that the loads will not shift, slide, or overturn in transit. The steel should be loaded so that no pieces are liable to move and trap a man as part of the load is removed. There should be room to place unloading slings safely. The slings used should be adequate for the weight and kind of steel being lifted. The unloading equipment should have the reach and capacity to handle the loads being lifted off the carrier and moved to the point where the material is to be landed.

Men should be kept clear of the steel as a load is lifted; should keep out from under a load that has been picked; and should be clear of any area where the load could swing unexpectedly, slip, or fall. Proper loading helps to avoid the need for men to try to move pieces in order to hook on to the steel. Such movement can cause unexpected and dangerous shifting, sliding, or rolling of the material being handled, or of other pieces underneath or alongside, and can trap the men.

When a number of trusses or deep girders are loaded upright in one car or on one truck, barge, etc., all but the one being lifted should be tied back. Most such loads are either tied together or otherwise secured to the carrier so that they will ride safely in transit. But they may be unable to stand by themselves as one after another is unloaded, unless they have been tied or braced temporarily to prevent them from falling over and endangering the men who are unloading. The piece to be handled is hooked on to and the rest must be secure before the piece is lifted, to prevent the balance of the car, or truck, etc. from falling over.

Floor planks and scaffold planks should be in good condition, without serious splits, checks, cracks, excessive knots, or other injurious defects. They should be unpainted because paint can hide injurious conditions. Linseed oil or a similar protective coating can be used as this is transparent.

Specifications for planks should cover the minimum strength requirements for their use, weighing the value of purchasing better planks with possible longer life against the extra cost and possibility of the planks being damaged and discarded before the extra value can be utilized. The strength required should be based on the span over the supports and the loads to be imposed of men, tools, equipment, steel, etc.; but they should be of such size, if possible, as to be light enough for two men to pick up, handle, lay, or move them. This will help the men avoid back strains or to drop the planks because of their weight. Where this cannot be accomplished, as in the case of heavier planks that are needed because of longer supporting spans, the weight should be reasonable for three or four men to handle them, with two men at an end using a short piece of wood between them under the end of the plank to carry it safely.

On a tier building, if the follow-up gangs keep close behind the raising gang, two complete floors of planks are usually sufficient, one of which is the working floor. On other types of structures, an area directly under men working above should be planked sufficiently to catch any man who falls, or any tools inadvertently dropped.

According to OSHA, where planking cannot be done safely or adequately, safety nets and/or safety belts may be required. In addition, the regulations for planking vary between derrick and crane erection and must be checked because the government sometimes changes them.

A sidewalk bridge or other protection for passing pedestrians is usually the obligation of the general contractor, since this will serve later as protection from falling material other than steel or tools, as the other trades complete their work. When the building is directly over an active railroad, the obligation to provide a covering to protect railroad cars passing under the steel erector is often included in the erection contract. This covering is usually installed before beginning to raise the permanent steel. Extreme care must be used in its installation, keeping in close touch with the railroad's representative, so that work can be stopped whenever an approaching train could endanger the steel erector's operation of installing the covering.

The raising gang should have an adequate supply of the correct size drift pins and erection or fitting-up bolts and washers. This will enable the connectors to connect members safely, leaving drift pins in place when advisable. Enough bolts should be used in connecting each piece, in a pattern to ensure that it will not roll if walked on, with a minimum of two bolts at each end. Enough bolts and pins should be placed in a column splice connection to withstand a wind load or the force of a piece inadvertently swung against it, or the column being jarred as a beam is swung into a connection on the column as part of the raising operation. All such connection bolts should be hand wrench-tightened before the piece is "cut loose" (the lifting sling or hitch removed).

Taglines should always be used to guide a piece within reach of a connector and to keep it clear of previously erected steelwork. Slings for raising steel should be provided; they should be light and flexible enough to grip the pieces but strong enough for the loads. Slings of several diameters (and strengths) should be provided to take care of lifting light and heavy pieces. The sling must not slide when erecting diagonal members and pieces that must be tipped by the tagline to thread them through other framing in sending them to connectors aloft. If it did, it could permit the piece to slip through and fall. Occasionally, with lightweight pieces, a round turn is needed around the piece before choking the sling and putting the other eye on the lifting hook. The extra turn provides additional friction and gripping of the piece.

Girders, extra heavy pieces, and awkward pieces may need special hitches that can be shop-assembled and bolted, or welded to the piece, to be left permanently in place. With a number of such members involved, one hitch is often provided to be field-bolted in turn as each piece is erected. When the hitch must be removed, if welded, it will probably be burned off. If bolted, it can be easily removed.

When no hitch is provided, and no sling of sufficient length or capacity is available, lashing is often satisfactory. In this case enough turns should be placed, using wire rope of adequate diameter and strength, with the ends fastened together by wire rope clips. Extreme care must then be taken to have all turns of the lashing take the load equally. When a sling or lashing is used and the bottom flange of the piece can cut or cause damage, a sling protector should be used. Provision must be made to prevent the protector from dropping when the sling or lashing is removed.

When girder dogs are used in place of slings, lashing, or hitches, some time can be saved in hooking on and cutting loose. The dogs must be placed so that they will not slide if the piece tips slightly. In addition, it is advisable to install screw clamps ("C" clamps) on each side of the dogs, secured to the member being lifted, to help prevent the piece from sliding through the dogs (see Fig. 3.7.25). Girder dogs are inadvisable if the piece is to be intentionally tipped out of level or is to be pulled laterally into place.

If the piece is laterally unstable when picked at its center, a balance beam is advisable, unless a pair of bridled slings can be placed far enough apart for them to be at safe lifting points. This can be done only if the pull due to bridling is not too great for the stability of the piece and the strength of the slings. The top flange of a truss, girder, or long beam may be temporarily reinforced with a structural member laid flat on top of the member and secured temporarily. Also, horizontal hogrodding can be installed on the compression flange as erected.

On deep girders, and even on some trusses, a safety "bar" running the full length will aid the connectors, fitters, bolters, and others working on the bottom flange or bottom chord to work in greater safety. A single $\frac{5}{8}$-in.-diameter wire rope, threaded through $1\frac{3}{16}$-in.- or $1\frac{5}{16}$-in.-diameter holes in the vertical stiffeners or members, about 3 ft. above the bottom flange and clamped at the ends with wire-rope clips, will usually be adequate. If the holes cannot be provided while the steel is being fabricated, or if the design requirements would prohibit them, a substitute should be used. Short eye bolts can be welded to the web of the piece at intervals, to be burned off and the surface chipped or ground to leave it smooth after all work on the piece has been completed. To do this cleanup safely the men should be provided with safety belts and lines, or should work from ladders hooked over the top flange, or a float can be hung from which to work.

Safety belts should always be on hand to be used whenever advisable. In areas in which the superintendent or foreman believes their use is necessary, the men should be required to use them. Whenever the men feel the belts would make a particular operation safer, their use should be encouraged. The lines should be chemically treated to resist mildew and rot. Certain dyes are available that form an opaque film resistant to the sun's harmful ultraviolet rays. In tying the lines care must be taken not to tie them where they can be cut by sharp edges of metal. Generally they should be tied not more than 6 to 8 ft. from the belt. A paint mark at the 8-ft. point is a good way to warn the men not to have more than that much line free. When the belts are issued, the men should be informed why the mark is there, how to tie off the line safely, and how to adjust, fasten, and unfasten the belt itself.

On a guy-derrick or climbing-crane job, the toolboxes used by the raising gangs should be moved to the new working floor each time the rig is jumped. On a mobile-crane job, the boxes should be moved as soon as the crane starts operating in a new area too far away for the men to reach the boxes conveniently. On a traveler-erected job, they are usually kept on the traveler platform. They should be used for storing slings, wrenches, manila lines, and other small material, as well as safety equipment, to aid in good housekeeping, since safety and a shipshape job go hand in hand.

The proper size, number, and spacing of wire-rope clips should be used, depending on the diameter of the wire rope. They should be checked as soon as the wire rope has been stressed, since the rope, especially if new, tends to stretch under applied load, which in turn may cause it to shrink slightly in diameter. The clips must then be promptly retightened to take care of this new condition. In addition, the clips should be inspected frequently to make sure that they have not slipped and are tight enough. A good rule is to retighten wire-rope clips on guys or beckets within an hour after they have been installed and the guys or beckets have gone into action.

The men must be taught the correct way to install clips. With Crosby types, which consist of a U-bolt and a saddle, the "U" portion must be on the dead end of the rope with all clips in the same direction, not staggered with one in one direction and the next in the opposite direction. Both nuts should be tightened equally. With Laughlin or Fist-grip-type clips using two identical parts, each combining a saddle and a bolt in one piece, it does not matter which way the two parts are applied. Here again, both nuts should be tightened equally, being careful that the nut is not tightened against the clip without the clip actually gripping the wire rope due to the wrong size clip being used.

The clip must actually grip the rope when the nuts have been tightened fully. If the wrong size clip has been used, there is danger, if it is too large, that the nuts are merely being tightened against the metal of the saddle portion and that the saddle is not being tightened against the wire rope itself. If a clip is too small the saddle may damage the rope.

When bolters, welders, etc. can work from a planked floor or a planked area, the work can be done safely and usually more expeditiously. When this is not feasible or is inexpedient, scaffolds or floats should be used. When several men will work in the same area without planking, needle-beam scaffolds are best, especially if one hanging of the two needle beams will serve several points on which work is to be done. When only one or two men will work at a point, floats, or ship scaffolds, or even a boatswain's chair can probably serve. Care must be used to tie them so that they cannot be pushed out and drop a man. Manila lines for needle beams and for floats should have sufficient extra strength and size so that if a strand is accidentally cut or burned or otherwise damaged, the remaining strength will be sufficient for the load on the float or scaffold.

When the men can work safely from the steel itself, this is preferable to hanging floats or scaffolds because it eliminates additional operations which, in turn, reduce the hazard of an accident. Lumber in floats and needle beams should be inspected for cracks, splits, burns, or other damage. Lines should be inspected frequently for cuts, wear, burns, or other unsafe conditions. The men should be cautioned to check their knots before getting on to a float or scaffold to be sure all the lines are in place and secured. The men should be warned not to jump on a float or scaffold, and to check any type of scaffolding before getting on it each time it is set up and each time it is moved.

If the lines cannot be tied directly to the steel, shackles or beam clamps or similar devices are advisable to hold the lines rather than using hooks which can come loose too easily. If necessary, a bracket for tying the lines can be welded to the steel and then removed when no longer needed. Screw clamps ("C" clamps) should not be used to hold the scaffold or float lines as they are liable to loosen from side pulls and let the scaffolding fall.

On a long, low structure, if ground conditions are adequate and suitable, it is often safer and more economical to use a rolling, built-up scaffold than hung scaffolds. The rolling scaffold can be fabricated from timbers or steel members bolted together, with rollers permitting the scaffold to be moved on the ground or on planks laid there. The height should permit men to work easily on the steel connections. The men should get off the scaffold each time it is moved. The scaffold should be wedged up off the rollers when it has been moved to a new point to relieve the rollers of the working load. A form of ladder should be secured to the side of the scaffold for safe and easy access to the top.

To aid men working on floats or scaffolds, as well as men in raising gangs or other gangs that use small material such as bolts or drift pins, adequate bolt baskets or similar containers with handles of sufficient strength and attachment to carry the loaded containers should be provided. They will encourage good housekeeping, a prime requirement for safety as well as for production, and will assist in keeping small objects gathered up (and prevent their loss). This reduces the possibility of small items being rolled or kicked and falling on persons below, either workmen or passing pedestrians or vehicles. It also prevents a man from inadvertently stepping on such an object, losing his balance, with a possible accident resulting.

The men should be trained to use such containers and to keep small tools gathered up and put away in toolboxes when not in use. Material must not be dumped overboard when a scaffold is to be moved.

Ladders should be used for safe access between the various levels of the work. The ladder should be set on a good, level, solid base. There should be a platform of some sort at the foot and at the top landing for men to step on safely when descending or ascending. The top of the ladder should be tied to prevent the ladder from slipping out of place. When feasible, ladders should not be vertical, but should be leaned one quarter of the length for safest use. Cracked, loose, or broken rungs or side rails should be watched for and promptly repaired or replaced. The ladders should be placed in readily accessible locations.

The men should be taught to clean their shoes of mud, grease, snow, ice, or other slippery material before climbing a ladder or going on the steel or a planked floor so that they will not slip or track slippery material for others to slip on. They should avoid carrying tools or other material when using a ladder; they should use a handline to raise and lower such material. They should look where they step when reaching the bottom of the ladder and step solidly on the platform provided there. Hooked ladders can be made of detachable hooks bolted to short wooden ladders. The hooks are usually provided with adjustments so that their parts can be bolted together to fit snugly over the top flange of the member on which they are to be used. Fixed hooks to fit various width flanges can be furnished to be bolted to the straight wooden ladders.

4.6 Protection to Be Provided

Every effort should be made to inculcate the men with the idea that safety has been built into the job, the erection scheme, the tools, the equipment; that protective devices have been provided to be used; that safe workmanship is required. Unless the men use them, hard hats, safety goggles, safety belts, and all the devices furnished to help make the job a safe one will be useless.

Safety hats should be worn as protection from falling objects, from a man bumping his head, and as protection to his head from more serious injury if he should fall. The hats must be worn properly to give maximum protection. The hammock straps should leave about 1 in. to 1½ in. of space from the inside of the top of the hat. The sweatband should fit the man's head snugly. Some sweatbands can be adjusted, while others must be selected to fit and must be snapped or otherwise fastened into place. The hat should be worn on the man's head so that the sweatband and straps fit directly and not over a cap or hat.

The men should be instructed about the need for wearing safety hats at all times and not only when they think something might fall. Furthermore, OSHA regulations make the wearing of a safety hat obligatory not only by the steel erectors, but also by all other personnel working on a structural steel erection job. In this way, it will become a habit and the hats will then protect them at times when they least expect danger. Provision should be made to clean a hat when a man leaves and before it is given to a new man if it cannot be returned to the toolhouse for refurbishing.

Winter liners, with or without earlaps, are advisable in cold climates since they give warmth while still letting the sweatband and hammock straps fit snugly on the head. Otherwise, the men may use wool or other warm hats or caps and perch the hard hat unsafely on top, which will tend to let the hat fly off if struck. It will neither stay in place nor protect the man unless it is worn properly. Chin straps can be used in windy locations, and while some men actually use the chin strap under their chins, others have found a way to secure the straps more comfortably against the back of their heads, near their necks.

Welder's helmets or shields can be secured to the hard hat (without a front brim) by a bolt and spring combination. This permits the welder to wear a safety hat and still be able to raise and lower his protective shield. The spring will keep the shield in place, whether it has been raised or lowered.

Eye protection should be provided whenever there is danger of foreign particles entering a man's eyes. If he has no protection, the exhaust from pneumatic tools can blow dirt, chips, or other material into a man's eyes. Some equipment can be fitted with exhaust deflectors which can help materially in preventing this type of eye injury. When drilling, reaming, grinding, burning, welding, or doing any work where particles of steel, sparks, or slag can fly, a man must protect his eyes. Protective eye equipment should also be used in very windy areas and on work where there is excessive dirt such as in dismantling old, dirty, dusty steelwork.

Safety goggles should be on hand but the men should be shown the correct way to wear them. Goggles should be adjusted to fit comfortably with the side pieces fitting snugly over a man's ears. If an elastic strap holds the goggles, it should be adjusted to fit over his ears and against the lower part of the back of his head to hold the goggles firmly in place. Safety goggles should fit as comfortably as ordinary prescription spectacles, which are normally adjusted and fitted when purchased.

In the case of burning, the correct glasses of the proper shade should be used to protect not only from sparks and slag but also from the harmful rays of the flame. In the case of welding, men working close by should have goggles with the proper shade of safety glass to protect them from flash injury to their eyes from the welding arc. The men actually cutting, burning, or welding should have the proper shade of safety glass in place in their goggles or shields and also a clear cover glass to protect the shaded glass from the spatter of the welding arc or the burning torch's action. The clear, less expensive cover glass can be replaced easily when pitted or coated with slag, thus avoiding replacement of the more costly shaded glass. Manufacturers of safety goggles provide charts giving the proper shade for burning, welding, and flash. These recommendations should be followed. Clear safety glass is usually adequate except when burning, welding, or using protection from flash.

Instead of safety goggles, a face shield can sometimes be used for operations such as drilling, reaming, or chipping, or where there is any chance of flying, injurious material. But here again, it should be adjusted so that it is comfortable and properly in place for protection. The men should also be cautioned to use eye protection when pouring hot metal or sulphur or the like into an anchor bolt hole or other hole. This is due to the danger of moisture in the hole or splash of the molten material, which might create a splatter that might burn a man's eyes unless protected.

Fire-resistant protective covering should be provided over operators stationed at equipment such as hoists, and also over equipment such as compressors and welding machines. This covering should be sturdy enough to prevent small falling objects from getting through to injure the operator or damage the machine. It will also serve as protection from falling sparks, slag, or anything that could start a fire, especially if the equipment is fueled by gasoline. If this protection also acts as a guard against adverse weather, the equipment will be maintained in better condition and thus will probably run in a safer manner. Similarly, protection of tool boxes and shanties from the effects of rain, snow, and the weather in general will tend to help keep the contents in a more usable and thus safer condition.

In addition to protection from fire, Protectoseal fill-and-vent fittings or a similar fire protective device should be installed on all fuel tanks. The men filling the tanks should be warned to leave the protective devices in place. This even applies to the small gasoline-driven starting engine used on some diesel-fueled machines. Safety fuel cans are advisable unless the tank on the machine is filled directly from a drum by means of a safety pump.

A fire-protection procedure should be set up if there is to be any flame cutting, burning, welding, or any operation that could start a fire from hot or molten material. Such hot metal must be checked immediately because such hot material can start a blaze after smoldering unnoticed in flammable material. The fire-protection program should provide a plan for quenching fires and for placing protective coverings over flammable material that could be endangered by any hot material used in the erection process.

Enough fire extinguishers, preferably not of the carbon tetrachloride type, should be in place at strategic points, suitable for the type of fire that might start. Extinguishers should always be in place in cranes, at hoists, at compressors, and at welding machines. Storage of flammable materials should be kept to a minimum. Men should be instructed in the proper use of the extinguishers on the job, as many types are operated quite differently—some work by pressure through a valve, some by turning upside down, some by puncturing a gas-producing cartridge, and some by means of an attached hand pump. An extinguisher should be made operative again immediately after it has been used.

Available fire-fighting facilities, alarm boxes, hydrants, fire-department stations, and fire boats and tugs with pumps and hoses (if the project is adjacent to water facilities) should all be determined before the start of the job, and the superintendent and foremen informed of their location. Where incidental flame cutting, burning, or welding is to be done, a man can often be assigned temporarily as a fire watcher. Part of the fire-protection procedure should include assigning someone to check at quitting time to see that all fires are quenched; that all slag from burning or cutting is safely under control; that electrode stubs are in containers that should be provided for each welder to protect not only against a fire but against a man slipping on a loose stub. A fire watch should be maintained where advisable for at least an hour after all of this work has ceased. The fire watcher should be equipped with a portable extinguisher and with means of communication in case he cannot control an incipient fire.

It is sometimes advisable to have watchman or guard service during the time erection crews are off the job. This can be a service with hourly or frequent rounds during the night, or on Saturdays, Sundays, and holidays. Adequate communication facilities must be available for reporting fires or emergencies needing aid. Vandals or other malicious persons can cause serious damage by tampering with equipment so that an accident is caused as soon as the equipment is used. In addition to fire watching, a protective service can often prevent such damage or alert the erector to it so that repairs can be made in time.

When possible and feasible, noncombustible material should be used for falsework, shoring, or temporary construction. If this is not practical, such material can be coated with a flame-resisting coating and recoated after exposure to weather for four or five months, unless a fire-resistant, waterproof paint can be used. When fire-protective coatings cannot be applied—as, for example, with creosoted piling fenders—the material should be kept wetted down frequently or covered with noncombustible protective material, if there is a danger of fire.

Extreme care must be exercised in storing any flammable material, especially gasoline, kerosene, diesel oil, oxygen and acetylene cylinders or tanks. This type of material should be kept segregated and protected. When there is danger to wooden piers, barges, work boats, etc., the supply of such flammable material should be limited to that needed for one or two days. Stored oxygen cylinders should be kept separate from acetylene tanks with a fire-resistant divider between them. Cylinders and tanks should be kept vertical and must be secured, if necessary, to prevent their falling or being knocked over. Their caps should always be in place when they are not actually in use.

For maximum safety, men handling oxyacetylene torches or welding equipment should be taught safe methods for handling such equipment and how to leave them at lunch time and at quitting time. Any canvas tarpaulins used over equipment or supplies should be treated for flame and weather resistance. Care should be taken when installing any temporary electrical wiring, heating oil device, or salamanders. They should be installed as safely as possible to prevent a fire or a man being burned or electrocuted by inadvertent contact with an unprotected live electrical or heating system.

Ovens provided for keeping low-hydrogen electrodes dry should be protected and the men cautioned against the hazard of being burned when working near them. If liquid petroleum stoves are used for heating, only Underwriters Laboratories or similar approved models should be used. They should be vented to avoid incomplete combustion and resultant carbon monoxide poisoning endangering men in the vicinity. Electric heaters that are safely insulated to prevent contact by persons or starting a fire in the surroundings are more advisable.

If radiography is being used to inspect welds, proper protective procedures should be set up and vigorously policed. The area where the radioactive material is used should be roped off with warning signs clearly posted. The men actually doing the work should always work as a team of two men. They should be equipped with film badges which should be checked frequently to determine if the men have been exposed to excessive radiation. With a two-man team, one man can do the actual radiography and the other can return the radioactive material to its container or turn off the machine in case of mishap to the radiographer.

If "safety solvents" are used, this should preferably be done in open spaces with good ventilation. If they must be used in confined spaces, a form of mechanical ventilation or respiratory protection must be provided. Certain filter-type respirators with chemical cartridges can be used, but an airline respirator is better. Safety solvents are nonflammable in themselves but create a flammable mixture if allowed to evaporate. They should, therefore, be kept in sealed or tightly closed containers when not in use. Some are toxic and may poison by inhalation, by swallowing, and even by absorption through the skin. When such solvents are used, the symptoms of poisoning should be watched for. These include headache, nausea, vomiting, fatigue, appetite loss, mental confusion, lack of coordination, depression, disturbance of vision, loss of balance.

In cold weather starting aids are sometimes used when internal combustion engines are difficult to start. Such starting aids should not be poured into the manifold or engine intake because of the explosion hazard this would create. Almost all are extremely flammable and explosive. A burning torch should never be used to warm the intake manifold while starting aids are being used. Ether should not be poured into the engine for cold weather starting. Heaters are available that use propane gas and are connected to the cooling water system so that they circulate hot water through the engine block. Electrical devices are also available to keep the engine block warm. Some approved type of starting capsule or a vaporizing spray could be used, but only in the manner specified by the manufacturer.

When the moisture in the air in pneumatic hose lines might condense and freeze, alcohol should not be added directly into the line. A special, injector-type lubricator should be used. It should be installed beyond the receiver so the air will have a chance to cool below the alcohol flash point. If an air line becomes clogged with ice or even dirt, goggles should be worn when cleaning it out to prevent the possibility of anything flying out and injuring a man's eyes. No man should ever be allowed to look into an air line when the air is turned on.

When ground conditions are poor (as when covered with mud, water, snow, or ice) or the area is soft, greasy, or oily, the space where the men will work and where equipment will be used should be protected. It should be planked over with 2- or 3-in.-thick planks or corduroyed with heavy timbers. Timber mats can be used or adequate fill dumped to cover the dangerous areas.

When wire or manila rope blocks with hooks are used, the hooks should be moused to prevent slings on them from coming loose inadvertently. (Mousing is the name given to a single strand of multistrand manila or similar type of rope used on the hook for the purpose. To mouse a hook, this single strand of rope is wound around the point of the hook and the back of the hook several times, and then the ends are tied together. This, of course, is done only after the hook has been fastened into a sling, shackle, or other form of loop to which it is to be secured.)

If the work is to be done in excessively hot areas, heat tablets of sodium chloride and dextrose should be provided in dispensers. The men should be cautioned not to take them in excess but to use them only when needed. They should be warned not to drink too much cold water, especially if overheated and working in hot places, or during exceptionally hot weather. It is better to drink a little water frequently rather than too much at one time. Some men like to have a little raw oatmeal added to the water.

The men should be instructed about the danger of, and precautions to be taken against, heat exhaustion and heat prostration or sunstroke. Heat exhaustion may cause a man to collapse and fall. If a man feels weak from the heat, he should stop work before he is overcome and get into a shady area before going back to continue working.

In hot weather the men should be cautioned to eat moderately and to eat easily digested foods, avoiding fats and reducing their intake of such proteins as meats, eggs, and cheese. The symptoms of heat exhaustion should be explained; namely, a cold, clammy, pale face, excessive perspiration, a feeble and rapid pulse.

If a man is overcome by heat exhaustion, he should be moved promptly to a comfortable, shaded place and covered with blankets to keep him warm. This is just the opposite of the treatment for heat prostration or sunstroke. In the latter case the man will be hot, red, burning up, and should be moved promptly to a cool place and cold water applied to his body. In both cases the chances are that the man may need hospitalization.

The men should be checked frequently if there is any chance of their being overcome by heat. They should be persuaded to wear shirts in hot weather as protection from sunburn, heat exhaustion, or overexposure to the sun's ultraviolet rays, as well as protection against minor scratches and irritants to the skin.

In areas where there may be danger to the men from gas, a frequent check should be made with gas meters suitable for testing the gas involved or suspected. Such a check should always be made in a possibly gaseous area before starting work in the morning and again after lunch time. When frequent checks cannot be conclusive, face respirators should be used, using an airline type if a canister type does not give complete, positive protection. If airline respirators are used, the intake air must be from an area that is absolutely sure to be free of gas. Nothing should be assumed—everything should be checked to prevent the possibility of any man being subjected to toxic fumes, nontoxic fumes, or explosive mixtures.

If there is a hazard of red-lead vapors, as in heating, burning, or cutting old steel in an alteration or demolition job, respirators should be provided and used. On new steel, the fabricator should have been cautioned to use nontoxic paint on connections to be welded. On such connections, lacquer is best on the faying surfaces, with red oxide or similar paint on the other surfaces.

When the men must work in dark areas, as, for example, construction inside existing buildings, or where the work is to be done at night, there should be adequate lighting. This lighting should be arranged to avoid the casting of shadows; this can be done by having the lights shining from all sides. Acetylene or electric lights are usually adequate if of sufficient intensity. They must be located to eliminate the hazard of men working near them being burned or electrically shocked. Wires to electric lights should be strung so that there will be no danger of their being shorted or breaking, and so that they are not apt to be fouled by the operations. Acetylene lamps should be firmly secured so they cannot be knocked over.

If the structure is such that the steelwork .is at a considerable height above the ground, or above the next lower level of other steel or construction, nets strung below the high steel may be the logical way to protect the men in case anyone should lose his balance or be caused to fall. Otherwise, the men must wear safety belts and use the lines whenever feasible. The disadvantage is that a man must untie himself at a point in order to move to another point at which he can again "tie off." Elevator shafts in a building can usually be covered over a reasonable distance below where men must work. Many erectors and general contractors have used small nets in such openings, rather than using planks or other covering because of the safer installation and removal of such nets.

If cranes or derricks are mounted on barges for erection from water, the underdeck construction must be checked thoroughly to be sure the supports under points of stress are adequate and in satisfactory condition. Additional supports may be found necessary, and all supports should be checked frequently. If the barge could tip during the erection procedure, the freeboard must be watched and the allowable, which should be stated on the erection scheme drawings, must not be exceeded. Steel loaded on the deck of a barge should be limited to the safe amounts and locations on the barge.

5

The Erection Scheme

5.1 Preliminary

After a job has been estimated, the proposal or bid submitted, a contract awarded, and the contract documents, specifications, and drawings checked against the proposal—against what was bid—work must be started promptly to prepare a safe, efficient, economical erection scheme. The scheme should aim at expediting fieldwork as much as safely possible but weighing any additional costs to do so against the saving in time. This is quite necessary if the allowed contract time is "tight," and especially if it is a penalty contract.

If little time has elapsed since the site was visited when preparing the bid, it may not be necessary to revisit at this time. Occasionally a considerable time has elapsed since the visit, and conditions may have changed. Foundation work may have been started, involving digging, so that the planned access route is no longer available. The general contractor may have erected offices or shanties or may have set up his equipment in places that will interfere with the delivery of equipment, tools, or steel. Telephone or power wires may have been strung that may interfere with the use of equipment planned when bidding. All such possible contingencies must be considered, and if there is any doubt as to changed conditions, the site should again be inspected before going too far with erection planning. After checking the site and surrounding conditions, a plan must be decided on that may or may not be the one anticipated when the estimate was made. A check visit should always be made after the scheme of erection has been decided and drawings prepared, schedules given to the fabricator, and other preliminaries completed.

Usually a particular type of equipment will be best suited for a project being studied. But sometimes any one of several different types may be as

110

safe, economical, and efficient. Then the decision rests on the availability and cost of providing each one. With derricks or travelers as possibilities for erecting a job, this must be compared with the use of mobile cranes and climbing or fixed tower cranes. Short-boom heavier-capacity equipment must be compared with long-boom lighter-capacity rigs.

Attention must be paid to the possible use of a gin-pole, basket-pole, dutchman, gallows frame, or jigger stick. Some work may be best suited to simple handline operations. When water is available, floating equipment must be considered: derrick boats and derricks or cranes mounted on barges. The possibility of combining two different types of erecting equipment must not be overlooked, such as derricks with cranes, poles with derricks or cranes, or other combinations.

The motivating power of equipment must be studied to decide whether to use diesel fuel, gasoline, electricity, or coal- or oil-fired steam boilers. On handline work it must be decided whether to use pneumatically or electrically powered small hoists, manually operated crabs or winches, or manila rope powered by hand or by a spool on an available powered hoist. The decision may affect the type of auxiliary equipment selected. If an electrical installation will be used for a hoist, then electrically powered compressors, welding generators, transformers, or rectifiers may be chosen. With no electricity needed, the welding machines, compressors, hoists, etc. may all be run on diesel or gasoline fuels. These decisions can affect the weight of the equipment chosen which, in turn, may involve design considerations of the areas where the equipment is operated.

Many features of the work must be studied when deciding on what equipment to use. Foundations and ground conditions may be the determining factors. Many sites are so cluttered with pits or raised footings that a crane or traveler could not move safely or economically without fouling the footings or without costly bridging over the pits. Local police restrictions may prohibit one type or dictate another type of erecting equipment. Overhead exposed power lines that cannot be moved or deenergized may restrict the erection method. The capacity required to handle the heaviest piece must be considered.

Finally, the time needed, the cost, the efficiency, and the safety of the erection method using one type or a combination of types of erection equipment must be weighed against one another, and the one that gives the desired result in the time allowed, by the safest means, and usually the lowest cost must be selected. Often there is no hard-and-fast answer concerning what equipment to choose.

5.2   Selecting Erection Method

A study of the contract drawings and a review of the site conditions will usually lead to a decision on the equipment and method to be used. The method selected depends on the speed required and on equipment available—whether owned, to be purchased, or rented. The relative costs of each and many other factors must be taken into account. The method depends on site conditions, on areas available for operating the equipment, and on the hazard of one scheme in comparison with another.

Regardless of whether the ground conditions will permit the use of either cranes, derricks, travelers, or other types of equipment, alternative methods should be studied to determine which is best, all things considered. For example, a low, heavy-membered building might be erected by a guy derrick, a stiffleg derrick, a high-capacity crawler-mounted crane, a heavy-duty truck-mounted crane, a tower crane, or even a traveler.

The time for setting up a derrick must be compared with the time for delivering a completely rigged crane. The cost of delivering a truck crane on its own wheels is usually far less than that of delivering a crawler crane on a carrier. Similarly, the cost of shipping, unloading, assembling, setting up, and later dismantling and loading out a derrick may outweigh the advantages of the derrick over the crane.

Mobile cranes require travel space on the site, thus reducing the areas available for unloading, sorting, and distributing the steel, whereas a derrick permits the entire area around it to be used for doing the work. When an erector owns a suitable derrick but has to rent or purchase a crane of sufficient capacity, or vice versa, there may be no question about what to use since not only time but cost must enter into the decision if the erector expects to make a profit and stay in business.

The type, size, and height of the structure, possible interference from other trades, surrounding highway and pedestrian traffic that might delay delivery of steel or restrict the area to which material can be delivered at the site must all be taken into consideration. Frequently local police regulations limit the time of delivery by truck. Then it is important to have heavy-capacity equipment to unload large drafts of steel quickly.

For a high tier building it is generally better to use a guy derrick, jumping it upward tier by tier, than to use a lower capacity climbing crane or an excessively long-boom mobile crane. In some cities a crane cannot be operated while standing in the street. The crane must then work inside the building lines. This means that part of the structure must be left down from ground to roof as the crane erects portions and backs away, then erects another portion from bottom to top, and continues to back away and erect. This interferes with the completion of the building since no floor is complete until the crane has finished erection. With a guy derrick or a climbing crane, complete floors can be worked on by other trades as soon as the derrick or crane has jumped and the fitters and bolters or welders complete their work on each floor.

The weather, the possibility of flooding, high winds, and the like must be considered. In a deep excavation a sudden cloudburst or steady, continuous rainfall can so flood the "hole" that a mobile crane cannot operate on the ground, whereas a derrick or climbing crane can be jumped to an upper tier and be ready to work the instant the rainfall ceases.

Surrounding structures can affect the decision as to how to erect and what equipment to use. If the new building is narrow and hemmed in by old buildings or other structures, a crane may be the logical erecting rig to use. A derrick may have such close guying that not only is it unsafe, but it may be difficult to turn the boom under the steeply sloping guys. When a guy derrick must be used, a mast that is 20 ft taller than the boom can be used, instead of the usual mast that is 10 ft longer than the boom.

If the location is such that available ironworkers are experienced only in crane erection, this factor may affect the decision to use a derrick. On the other hand, this factor should be balanced against the cost of transporting experienced derrick men to the site to offset the slower or costlier crane erection.

Always endeavor to use the method that entails the least risk to the men and to the equipment. Accident prevention is of prime importance inasmuch as a minimum accident rate usually goes hand in hand with maximum production and minimum cost. The speed anticipated must be tied in with the speed with which the fabricator will be able to fabricate and load, with the speed with which the carrier can deliver the fabricated material, and with the speed of unloading and erecting by the erector's equipment.

5.3   Crane Erection

When the site will probably provide good ground conditions suitable for operating mobile cranes, either with or without the use of timber mats, planking, or corduroying a road through the area, cranes may be selected. If there are pits or openings, can they be spanned to support a crane? Will footings, foundations, or walls interfere with the safe movement of mobile cranes? Will overhead obstructions interfere? These questions must be answered. If the structure is not too high for the reach of the booms in available crawler or truck cranes, crawler-mounted, or truck-mounted tower or guy-derrick cranes, and if the weights to be lifted to the various heights are within the capacity of the crane, this may be the equipment to use.

A crawler crane must usually be delivered on a railroad car or on a truck carrier, since it is liable to damage a highway if moved by its own treads and often requires dismantling for clearance in transit. Therefore, the cost of dismantling and loading and of the subsequent unloading and assembling must be reviewed. If delivered by carrier, must the boom be completely or partially dismantled, or can it be delivered completely assembled and in place on the crane?

A truck crane can ordinarily travel over the road under its own power, requiring the removal of heavy counterweights if the load is over the legal weight limit. It may also require the dismantling of the boom. With the advent of high-capacity crawler-mounted as well as truck-mounted cranes, with single boom, with combination boom and mast, some even with guys on the mast, with a tower and boom, it has become necessary to dismantle the crane even more. The crawlers may need a separate truck or car for shipment; the truck carrier may need to be separated from the body of the crane; the boom, mast, and tower may need separate carriers; and the counterweights usually must be delivered as a separate shipment. As a result, such cranes of excessive total weight will require additional equipment to unload and assemble, ready to operate, such as a smaller capacity mobile crane that can be delivered completely assembled and can operate on arrival at the job site.

When a delivering railroad has tracks leading into the site, and especially when permanent tracks will be installed over the site itself, a locomotive crane may be the logical type of erecting rig to select if the tracks can clear obstructions. Temporary beams can span pits to support rails and ties, and if the walls or piers are not too high or too close for the locomotive crane body to clear in swinging, this type of crane may be best. Most locomotive cranes have fairly high capacities, even with a long boom.

The general shape of the structure, the weights of the members, the ground conditions, overhead obstructions, and vehicular and pedestrian traffic interference must all be studied before a decision is made to use cranes in preference to derricks or travelers, and which type of crane to use.

At this time the tower crane is finding favor as the erecting rig for certain types of structures. Currently they are available as crawler mounted with capacities of almost 100 tons on about a 200-ft tower and 100-ft boom at the minimum radius and a high boom. With towers about 250 ft high and using a 200-ft boom that can lift 10 tons at 200-ft radius, the crane can stand outside the building line and erect well over 25 stories high and 200 ft in depth.

With a truck-mounted tower crane, models are now available with capacities of about 20 tons at the minimum radius on a 100-ft boom with a tower almost 200 ft high, and can handle about 5 tons with a 150-ft boom on about a 200-ft tower at maximum radius. A tower crane of the static or fixed type requires exceptional footing since it must be heavily counterweighted or tied down to compensate for the excessively high overturning moment, but it generally has higher capacities than the mobile tower cranes, provided the foundations are sufficient for the uplifting forces as well as for supporting the vertical load.

Truck cranes are now available with maximum capacities on a basic 70-ft boom of 300 tons at minimum radius, and able to handle considerably lighter loads on as much as a 330-ft boom and 100-ft jib. The weight of some of these high-capacity truck cranes can be almost 200 tons plus the weight of the load being lifted; this must be taken into consideration since ground conditions may not be adequate for the truck crane to operate safely.

Crawler cranes are available to lift as much as 500 tons at 21 ft on an 80-ft boom, but they are limited to about 10 tons on a 300-ft boom at almost 200-ft radius. They can use booms with jibs that can reach almost 500 ft with light loads. Such a crane is also quite heavy, weighing as much as 380 tons plus the load.

Ground conditions must be compared in selecting crane type, since the concentration of the loads from the truck crane wheels is usually greater than from crawler treads.

The structure must be adaptable to fixed- or climbing-tower crane erection for this type to be selected instead of a truck crane, crawler crane or even a truck- or crawler-mounted crane. On a long structure, space must be available alongside to use a tower crane on a platform mounting sliding or rolling on rails laid on the ground.

Some crawler and truck cranes (called "Ringer®"* cranes) can be mounted to rotate on a special circular, blocked, temporarily fixed foundation by means of auxiliary devices attached to the crane supporting

* Manitowoc Engineering Co.

frame; they can then handle as much as 600 tons on a 140-ft boom at minimum radius and 100 tons on a 300-ft boom at 250-ft radius, but then they are limited to working from this temporarily fixed foundation.

In deciding on mobile cranes to erect a tall structure, one must consider the extra hazard to the connectors and others working aloft. If the steel is raised directly from the ground to its position aloft, through previously erected areas, there is no protective covering under the connectors. With a guy derrick, a fixed crane, or a climbing crane, there will be planked floors under the men, usually a maximum of two or three floors below.

The alternative is to pick the material clear of previously erected steel and swing it into the connectors piece by piece. This permits the use of planks close under the areas where they are connecting and where the fitters and then the bolters or welders will follow, but it is a slow method. Instead, small areas of floor can be planked and steel members lifted in bundles to be landed on the planks, sorted, and erected piece by piece from the floor, with these areas remaining planked over until all work directly over them has been completed.

Some structures such as those in a deep excavation lend themselves to erection by a mobile crane standing in the street, erecting a panel or a bay across the front, up to street level. (A bay is a series of panels across a building.) Then the crane is moved onto the erected steel, on suitable mats, timbers, or temporary supports in order to erect the next panel or bay in the building. The crane then repeatedly erects ahead up to street level and moves onto the erected steel. As soon as it reaches the rear of the building, it can be moved back out to the street on the erected steel, and guy derricks can be set up to erect the upper portions of the structure as it backs out. The crane may also be used to erect steel above as it moves back toward the street.

For this use of a crane, the permanent steel structure must be checked to make sure that the members are adequate to support the load of the crane, traveling or erecting. If any members or connections must be made stronger than the original design shows, the fabricator's drafting room must be notified in ample time to change the detail drawings and before the fabricator has placed his mill order. At the same time, the extra cost for this strengthening of material should be determined and an agreement reached to decide whether the erector, fabricator, customer, or owner will pay for it. The use of falsework or shoring under the permanent steel members involved, may make it unnecessary to strengthen the pieces to support the erecting equipment.

When the structure starts at a level so that the crane can move directly off the street onto the site and ground conditions are favorable, a mobile crane can move to the rear of the site to erect bays across the structure from ground to roof. Then, backing and erecting repeatedly, it can erect bay after bay, finally coming out to the street to erect the last bay at the front of the building. This, of course, means that other trades cannot complete entire floors until the crane has finished erecting all the steelwork.

As an alternative in the case of a truck-mounted or crawler-mounted tower crane, if the building is not too deep, if the tower is high enough to clear the top of the completed structure, and if the boom is long enough and has sufficient capacity at the maximum reach required, the mobile tower crane can be located just clear of the front of the building line. By swinging around to the street to unload the steel from carriers there, and then swinging the loads into the area where they belong, it can proceed to erect an entire tier. It continues to erect the upper tiers, one by one, swinging the steel from the unloading point out onto the completed steelwork level, erecting in a manner similar to a guy derrick or climbing crane. Instead of jumping as a guy derrick must do, the tower crane continues to erect with the tower and boom as initially installed and positioned outside the building area.

The cost of shipping and assembling each type of crane must be studied, together with the time differences in the erection speeds. The need to block the outriggers of a truck crane, heavy-duty crawler crane, truck- or crawler-mounted tower crane, or guy-derrick crane, and locomotive crane, when handling capacity loads, must be compared with the saving in time in using other types of cranes that require no outrigger type of blocking.

A study will determine whether the time saved, considering the extra cost, warrants one or more cranes to unload and distribute material over the site while one or more cranes follow to erect all this material.

The unloading cranes and the erecting cranes can unload and erect, respectively, and can later combine to raise any pieces that are too heavy for either one alone. A lighter capacity crane can maneuver more easily and erect much faster than a heavier one. This scheme also permits one crane to be assembling trusses, girders, or preassemblies before they are needed in the erection schedule. It also permits the permanent fastening of such assemblies to be made close to the ground instead of aloft, thus reducing the cost and the hazard.

When there are a few exceptionally heavy pieces and the balance consists of relatively light members, it may be better to have two cranes of equal capacity to unload and set the heavy members jointly. Then each crane separately erects the lighter material in adjacent areas. Although this will expedite erection and will require less sturdy ground support, it means double the costs of shipping, unloading, assembling, and later dismantling, loading, and shipping two lighter capacity cranes instead of one of greater capacity. A better scheme might be to use a truck crane working with a crawler crane to handle the heavy pieces, the truck crane being used to unload all the light material ahead of the erecting crawler crane.

If trusses are to be erected a truck crane can hold the first of a series of trusses immediately after it has been erected. Cranes of greater capacity then erect an adjoining truss and fill in enough bracing to make the two trusses stable and self-supporting. The truck crane is then released. This eliminates the cost and need to guy the first truss to make it safe for the erecting crane to "cut loose." Anchorages for such guys are often difficult to obtain readily, and this replacement of guys by the extra crane makes the entire operation safer, quicker, and often more economical.

5.4   Guy Derrick, Climbing and Fixed Tower Crane Erection

If the site conditions are unfavorable for erection by crawler, truck, or mobile tower cranes, the equipment selected may be a guy derrick, stiffleg derrick, traveler, fixed or climbing tower crane. If the building is too high for the available crawler, truck, or mobile tower cranes to reach, or if the loads are beyond their capacity, a guy derrick may be the logical selection.

On a very tall building, many erectors erect the first tier of steel with a truck crane or a crawler crane, if it can be rented locally and delivered economically, ready to work. On top of this steel, the crane sets up guy derricks with which to erect the balance of the structure. It is an efficient procedure to use a crane to unload and assemble equipment, to set base plates or grillages, and then to erect the first tier. This eliminates the installation of derrick guy anchors and avoids the extra expense and time of changing the lengths of the guys after jumping out of the hole. When a foundation contract has been completed before the erection contract is let, the elimination of substitute anchorages for the guys is a big saving.

To decide on the size and a logical location for a derrick, climbing crane, or fixed tower crane, it is necessary to divide the structure into areas and tiers and to note the height of the tiers. Then, if the building is not too wide for the equipment to work from one side to the other, the steel framing is studied and an approximate spot is selected for the derrick or crane. Preferably it should be about midway between the point at which a truck or other form of carrier will deliver the steel alongside the structure and the back of the building opposite the delivery point. Then the boom can reach equally to unload from the street and to erect the rear steel. This, however, may give undesirable, unequal guying for a guy derrick, and in that case a compromise will have to be reached.

The derrick, fixed crane, or climbing tower crane should not be located in an elevator shaft or a stairwell since this will interfere with the installation of elevators and stairs, which should follow the steel erector as closely as possible in order to reduce the number of stories the men must climb on temporary ladders to reach the working floor. The location should be such that the lead lines for the boom and load will not foul lower level steel as the rig is jumped higher and higher.

The rig should be located so that it will clear the permanent steel during jumps; otherwise, too many members will have to be omitted until after the jump. When feasible, the location should be such that the boom can reach over surrounding headers to erect members framing into them in the panels around the area where the rig will be closing itself in just before jumping. If these pieces are not erected before jumping, it may prevent the permanent supporting steel from having the necessary lateral support from beams that it was calculated to have when the strength of the structure to support the working rig was checked.

A reasonable area in the other direction (the length of the building) should be selected in deciding if one derrick, climbing crane, or fixed crane with a long boom can handle the entire floor, or if two or more rigs with shorter booms should be used. The areas covered by several rigs should be balanced so that each will have approximately the same amount of work to do. Otherwise, one will get too far ahead of the others with resulting complications.

Supporting steel for jumping beams supporting a guy derrick (Fig. 5.4.1) or climbing crane must be checked, and if the strength of the permanent structure is insufficient for the load, temporary supports must be designed, usually framing special members directly to the columns around the derrick position.

If guy-derrick erection is selected, after the number, location, and particular derricks to be used have been determined (based on capacity and length of boom and mast), the location of the guys, preferably eight in number, should be decided on and the columns to which to fasten them should be selected. These should be spaced angularly as nearly equally as is feasible, and as far from the foot of the derrick as necessary for adequate guy distances, determined from capacity tables. An effort should be made to keep the guys for two adjacent derricks from interfering with their operation. The guys should be located so that at least two will be acting in back of the derrick when unloading steel and at least two when set-

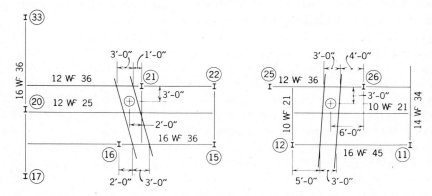

FIG. 5.4.1. Typical sketch showing location of jumping beams. + indicates location of derrick.

ting the maximum weight piece. Short guys on one side and long guys on the opposite side, in any position of the boom, should be avoided because this tends to make the derrick difficult to turn.

Outriggers can be used if some guys will not be far enough from the foot of the mast to safely handle the required loads due to the limits of the steelwork. Outriggers consist of steel beams or heavy timbers securely lashed to the steelwork and cantilevering out beyond the edge of the building. The guy is fastened to the outer end of the outrigger, which is then tied in diagonally to the structure below, to take the vertical stress from the guy into the structure below the working floor.

The framework of the working floor must be analyzed to make sure that it will transmit the horizontal stresses from the lower ends of the guys back to the foot of the derrick. Temporary struts are sometimes needed for this purpose since some buildings have long, open panels with no structural steel bracing between columns in the direction needed to bring the guy stresses back to the foot of the derrick.

The footings of the columns to be used for the guys are checked for size and weight to withstand the horizontal and vertical reactions from the guys with the derrick in its initial setup on the ground (or on the false-work tower, if required because of cross-lot bracing interference). If the footings are satisfactory, a sketch (Fig. 5.4.2), photostat, or copy of the foundation design drawing, or merely a diagram of the column locations is made showing the locations of the anchors to be embedded for securing the guys, together with the location of the derrick. This is sent to the foundation contractor or to the general contractor for the foundation

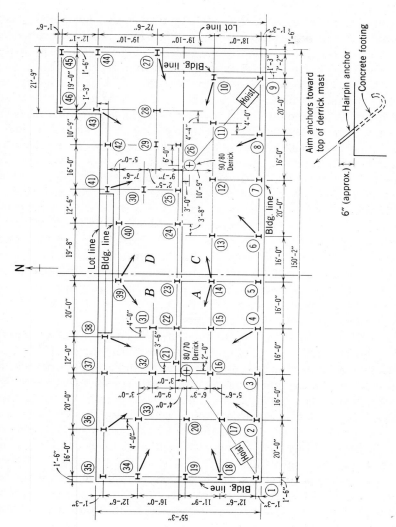

FIG. 5.4.2. Typical sketch for placing derrick guy anchors.

121

contractor to install the anchors. A copy must be given to the field superintendent so that he can check to see if they have been properly installed. This sketch should show how the hairpin anchors, if they are to be used, should be placed.

If lashing is to be used instead of hairpin anchors, the sketch must show how it should be installed. A reinforcing bar is usually indicated to be left in the bottom of the loop of the lashing, in the footing. The toolhouse should be alerted as soon as it has been determined when the anchors or lashings are to be delivered. In the case of lashing it is advisable to prepare the various turns in their final shape with light wire wound at several points to help hold the various turns in exact alignment with each other, since all turns should share the load equally. This should be indicated in a separate sketch to the toolhouse superintendent, giving the size of wire, number of turns, and dimensions of the lashing loop.

If the footings are found to be inadequate, and if good rock is available on the sides of the excavation or in the ground, split-end eye anchors can be used instead. Here, again, a sketch should be sent to the general contractor and given to the field superintendent, showing the location, slope or direction, and size of the holes to be drilled, their depth, and the exact method of installing the anchors with their wedges in place at the lower ends. It is safer to have the foundation contractor merely drill the holes and seal the tops temporarily, using the erection gang to drive the split-end eye anchors in place, and then test them. In this way the erection superintendent will know they have been installed correctly.

If no rock is available and the footings are inadequate, counterweights may have to be added to the footings to increase their weight. If a crane will be used to set the grillages and/or slabs, or to set up the derrick, it can be used to place counterweights, which may consist of some of the heavier steel members, placing them so that they will not foul the setting of the grillages or slabs in final position. The fabricator must be informed which pieces will be required at the site ahead of time for use as counterweights because he may have to fabricate them out of order to have them ready to ship at the proper time.

As a last resort, heavy stakes can be driven into the ground for use as anchors. This is somewhat precarious since there is no certain way of knowing what force they will resist; it depends on the consistency of the ground and on the depth and size of the stakes.

The load under the derrick or climbing crane is calculated to make sure the permanent steel on each working floor is adequate. This reaction must take into account not only the weight of the derrick or crane, but also the lead-line pull of the boom and load falls, the jumping beams, and any planks, steel, etc. in the immediate area. In the case of a guy derrick the vertical mast stresses from the guys due to the critical loads at their maximum reaches must also be included.

The permanent steelwork should be analysed for strength to support any drafts of steel that may be landed on it before being distributed about the floor. If this is critical, detailed information should be included in the instructions to the field. The erection-scheme drawings should show where to unload safely. Drafts of steel are unloaded in concentrated bundles, and because it is not known where the raising-gang foreman will distribute them, a uniform floor load should be used in calculations, based on a total of one and a half times the weight of the tier of steel being studied plus the weight of the floor planks and skids, toolboxes, and men.

The structure supporting the jumping beams under the derrick or climbing crane should be checked for bending moment, shear, web buckling, and strength of end connections in bearing and shear. At this time it may be necessary to relocate the jumping beams (Fig. 5.4.3) (and the rig) closer to the columns at either end of the supporting beams since the moment may be too great where originally planned. If the two opposite supporting beams in that panel are still too weak with the jumping beams moved, it may be possible to place the beams diagonally (Fig. 5.4.4.) so

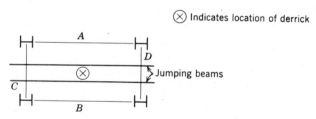

FIG. 5.4.3. Arrangement of jumping beams. Normal arrangement; beams **C** and **D** are adequate for loads at their centers.

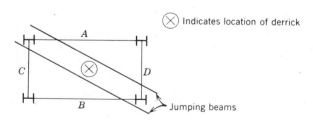

FIG. 5.4.4. Arrangement of jumping beams. Beams **A, B, C,** and **D** are inadequate in bending but adequate in shear near columns.

that one end of each jumping beam is on each of the four beams framing into the two diagonally opposite columns in the panel. Or, after one has checked the supporting beams together with those one floor below, it may be better to shore between the permanent beams on the two floors instead of moving the jumping beams; or a pair of jumping beams can be used close to the columns (Fig. 5.4.5) at opposite ends of the supporting members on either side of the panel, placing a second pair of jumping beams at right angles on top of them, close enough to each other to support the footblock or climbing crane. It may be advisable to furnish a pair of framed beams that are actually connected to the columns, fabricated from stock beams usually found in the toolhouse storage of falsework material.

In checking the supporting steel it is necessary to consider the unsupported lengths of the members and reduce the allowable stresses accordingly. This is particularly important in checking the supporting beams at right angles to the direction of the boom when unloading steel and when setting the heaviest pieces. This is necessary because it is always possible that the beams may roll, even though the footblock kickers of a guy derrick are in place and assumed to be holding the foot of the derrick firmly; the beam rolling can affect the value of the kickers in holding the footblock. From this analysis, the supporting steel and connections are checked for safety. At this time it is important to notify the fabricator's drafting room what the end connections must carry so that they can be detailed and fabricated accordingly.

When temporary support steel must be framed to the permanent columns because the permanent beams are of insufficient size, the drafting room must be informed so that the connections for the temporary members can be detailed on the columns before the drawings are sent to the shop for fabrication. If the fabricator is to supply this temporary steel, he should receive the details in time to ship these erection members with the tier of steel ahead of the one for which they are to be used.

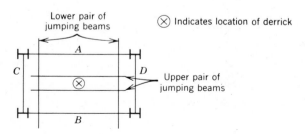

FIG. 5.4.5. Arrangement of jumping beams. Beams **A** and **B** are adequate if jumping beams are placed near columns, and **C** and **D** are inadequate for loads.

The topmost tier of steel must be checked after it has been decided how the erecting rigs will be dismantled and removed. If a Chicago boom is to be used, the column to which it will be connected should be studied; similarly, any "heavying-up" of the steelwork for a jigger stick or jinniwink should be noted. If the permanent steel is to be used to dismantle the rigs, the steel members involved are analyzed and a selection made of those members that can serve the purpose of dismantling from overhead.

5.5 Setbacks

If guy derricks are used, setbacks must be checked to make sure that the derricks can continue to operate in the same locations on the upper as on the lower floors. If setbacks require elimination of the steel framing to which the guys were fastened, a decision must be made whether to move the derricks laterally into the building to a position for safe guying or to shorten the boom and mast, leaving the derricks in the same relative positions. In either case, if the boom is now too short to reach the street for unloading, an auxiliary rig must be used to land the steel on the setback. The setback must be checked for strength to support the steel that will have to be raised from the street and temporarily landed there, plus any rig placed on it for unloading. There will usually be a much larger concentration of unloaded steel than for the lower part of the structure because of the smaller areas available. The hoist powering this auxiliary derrick, which is usually located on the same setback level, must also be taken into account.

A relay derrick can be a guy derrick or a stiffleg derrick. In the case of a guy derrick, since there will be little room between the edge of the building and the steelwork to be erected above the setback, a triangular, horizontal frame of structural members should be designed to connect the top of the mast to the face of the steel to be erected above that level, with a provision for holding the gudgeon pin in place of the spider and guys. In addition, the column to which the triangular framing holding the gudgeon pin will be attached must be checked for the horizontal forces that will be imposed.

A light stiffleg derrick, if available, may be selected instead, assembling it with the erecting derrick on the setback steelwork before the derrick has filled in the steel between. This stiffleg derrick must be adequately tied down to the structure in order to resist the uplift forces that will be developed in raising loads of steel from the street and landing the material on the setback framing.

If the guy derrick is to be moved laterally, or a relay derrick is to be set up on the setback instead of using a crane on the ground, or if the setback is to be reinforced for use as an unloading platform, or if the hoist is to be moved to an upper floor or to the setback, any or all of these must be shown on drawings for the superintendent and his foremen to use.

Columns above and adjacent to the area must be analyzed if they are to be used to secure a Chicago boom to dismantle and load out the auxiliary derrick and its hoist. That would probably be the safest and most economical method to remove them.

5.6   Hoisting Engines

If the building is very high the hoisting engine should be moved to an upper floor to reduce the weight of lead-line ropes. Otherwise, there may be too much rope on the drum, especially while erecting the lower floors. By reducing the weight of rope between the foot of the derrick and the hoist below, the weight of the overhauling ball can be reduced. The ball is needed to overhaul the load falls in order to permit it to go down to the street empty. In any case the hoist should be moved out of the excavation, where it is usually located at the start of work, up to street level, either inside the structure or out in the street if permitted by local ordinances or police regulations, as soon as street-level steel has been erected. This will make it much easier to remove the hoist at the end of the job; it can be moved out on rollers instead of trying to get it out of the "hole."

The hoists to be used are selected, having decided on the lead-line pull and speed of the drums required, together with the number of parts in the topping lift (or boom falls), main load falls, and runner (if used). Care must be taken that when all the wire rope is on the drums with the derrick on the highest level above the hoist, the hoist will still have sufficient power on the maximum layer of rope to operate the boom and the load falls lines when picking the maximum load at that elevation. Remember that loads will be picked with three or more parts in the load falls when picking from the street, or from the setback if the steel is being relayed from that level. With three parts there will be three times the height of the tip of the boom above the street (or setback) to be taken in on the drum when picking a draft of steel up to the working floor. This is another reason for moving a hoist to an upper floor when erecting a tall building. The condition of maximum rope on the drum with the derrick set up in its initial position should also be checked for sufficient lead-line pull.

5.7   Traveler and Stiffleg Derrick Erection

A structure that is starting in a deep excavation can be erected by a mobile crane as described. However, a traveler may be better than a crane in this situation. The traveler is either assembled at street level clear of the excavation and adequately counterweighted to permit it to erect a bay ahead from the excavation to street level, or a crane in the street is used to erect the first bay to street level and to assemble the traveler on this steelwork. The traveler is then tied down to the structure and proceeds to erect ahead of itself up to street level, repeatedly erecting bay by bay and moving forward. At the far end the traveler then erects steel above, backing out as each bay is completed, until finally it is out on the street again where it can be dismantled and loaded out. Any requirements for reinforcing the structure must be given to the fabricator before the steel involved is fabricated.

Hangars, train sheds, convention halls, and similar structures all lend themselves to erection by traveler rather than by cranes. Certain types of traveler can take the place of falsework, which would be required if cranes were used, for temporary support of the roof trusses, girders, or arches. But the ease of moving cranes about the site, even though falsework may be needed, should be weighed against the cost of assembling a traveler and the possible difficulties of moving it.

The stiffleg derrick is the most suitable type of derrick to be mounted on a traveler platform. When it can be set up in one position from which to erect the entire structure, instead of on a traveler platform, such use will be found to be very satisfactory. Adequate tiedowns and anchorages are best, but if the latter are not available or feasible, sufficient counterweights must be provided to resist any uplift from loads to be handled. Most stiffleg derricks have a greater capacity at longer reaches than the average guy derrick, but they lack the mobility of a high-capacity, long-boom crane. The cost of shipping, handling, setting up, and dismantling, and the time required for getting it ready to work must be compared with corresponding costs for a guy derrick or a crane.

Frequently a stiffleg derrick can be supported on a tower high enough so that the boom will clear the completed structure. With a boom long enough to encompass the entire area involved, with the stiffleg derrick securely connected to the top of the tower, and with the tower tied down or counterweighted, such a rig can erect a structure without imposing any erection loads on the permanent steel. (This condition would be the same if a fixed tower crane of comparable capacity were used.) In the case of a building with welded connections this could be ideal. Were such a welded structure to be erected by a guy derrick or climbing crane, the erected tier would have to be sufficiently welded or fitted up before the rig could jump safely and start to erect the next higher tier. The stiffleg derrick on a tower would permit erecting a complete tier, and start erecting the next tier as soon as enough connections had been welded, or erection bolts placed to support only the loads of steel for the next tier.

If the boom can reach the full depth of the steelwork but not the length of the building, the portion it can reach can be erected completely. Then the tower and derrick on top of it is skidded to a new position from which to erect another section completely, from front to rear, repeatedly skidding the tower after each section is erected until the entire structure has been erected. A traveling tower crane can be used in a similar manner if the length and capacity of the boom are adequate.

5.8   Erection by Miscellaneous Equipment

If it is planned to use only one crane and there are only one or two pieces of excessive weight to be erected, a crane with the capacity to handle all the rest of the steel may be most economical, with a gin-pole being used in conjunction with the crane to set the few heavy pieces.

When the structure is in an extremely out-of-the-way location, poor highways leading to the site or bridges en route inadequate to transport heavy equipment may be deciding factors. Or the amount of steel involved may not warrant the shipment of cranes, derricks, or similar equipment. A gin-pole alone may be the logical rig to use in such a situation.

If a gin-pole is to be used either alone or in conjunction with other equipment, the erection-scheme drawings should show or state the location for assembling or fitting the pole; where to install the anchorages for tripping the pole the first time; and the anchorages, if needed, for lowering the pole on completion of erection. All moves of the pole should be indicated or a note given calling for the main erecting rig to set up the pole, and later pick, move, and again set up the pole for the next joint operation.

All details for the use of a pole should be clearly given in detail, since many erection superintendents are not too familiar with the use of a pole for erecting and must be guided correctly in the operation.

The use of a dutchman should be limited to low structures with very lightweight material. The design of the rig should be such that all its members can be handled by one or two men, and a small gang should be able to move the assembled rig.

When there are a few heavy pieces a truck shipment may exceed the legal limits over the highway. The delivery must then be made by rail to a point that may be inconvenient to the rigs that will erect such pieces. The method of unloading must be decided in preparing the erection scheme. Frequently the shipper will load several heavy girders, trusses, or beams, in one load. The erector may not need all those pieces at one time. He may need them only at intervals and have insufficient room to store them near where they will be needed in the structure. The equipment for unloading the bulk of the steel may not have the capacity to handle these heavy pieces. In such cases a gallows frame may be the best piece of equipment to unload the heavy pieces and store them until each is needed (Sec. 9.14).

Helicopters are presently available with moderate lifting power. When an alteration or an addition is to be made on the roof of a tall, existing building, they can be used advantageously to deliver and set steel members if the weight is within their lifting capacity. This eliminates setting up erecting equipment, which can be quite difficult and costly, in the case of an addition to the top of a structure already in use. Protection of an existing roof for supporting erecting equipment can be a serious problem.

5.9   Study of Contract Drawings and Specifications

A study of the specifications should be made for any restrictions that must be taken into account in preparing the erection scheme and in selecting tools and equipment. Are there any provisions involving sequence of erection; coordination with other trades; tie-ups during the course of erection to permit other necessary work to be done before proceeding further; location of hoists, compressors, and welding machines? The contract should be reviewed for any items affecting the actual setting in place of the steel members and work incident thereto, such as a requirement that some work of other trades must be completed at a certain point in the erection sequence.

When reviewing the contract and specifications, everything that should be brought to the field superintendent's attention should be noted, even though it may not affect the preparation of the erection scheme directly. This would include citizenship requirements for the workmen; residence; union or nonunion employees; minimum wage rates to be paid; wages to be paid in cash or by check; limitations on hours to be worked per day;

days per week permitted; overtime payment requirements; inspection; use of electricity, gasoline, or diesel fuel; use of burning or cutting torch; facilities provided by others, such as light, heat for shanties, watchman service. After this detailed information has been noted, the contract drawings must be carefully studied. The sizes of awkward pieces must be checked; the weights of heavy pieces must be noted; the area must be studied for dividing into shipments and a decision made on the capacity of the equipment to be used.

The heaviest piece to be set and the heaviest drafts to be unloaded need to be determined. A check is made to see if a derrick or crane with the capacity to pick such loads at the reach required, on a boom of the length as calculated is available. If not, a shorter boom to handle the loads and additional equipment may be needed, reducing the area intended for one rig to cover. If two adjacent rigs can jointly unload and set the heavy pieces, this will be more economical since the lighter the derrick or crane, the quicker it can usually operate and the more steel it can erect in a given time. But two lightweight (and capacity) derricks or cranes can cost more to ship, handle, assemble, set up and dismantle than one heavier, greater-capacity piece of equipment with a longer boom. The cost of setting up two plus the faster erection is then compared with the lower cost of setting up one heavier rig plus the slower erection. A study of the steel, if it is intended to support a derrick or climbing crane, may disclose that it is better able to support two rigs at separate locations than one heavy rig.

The drawings are then reviewed to see what pieces should be preassembled in the shop. Assemblies may be limited by the fabricator's facilities, his loading limitations, railroad, truck, or barge clearances and capacities, delivering carrier's equipment, and the capacity of the unloading and erecting equipment selected to handle such assemblies. Trusses should be shop-assembled as completely as possible, even if it necessitates using a "deep-well" railroad car or a special truck movement. The slow movement of a deep-well car as usually required by the railroad, or the expense of special truck movements (often involving a police or other escort) must be weighed against the advantage of being able to set one heavy piece in place instead of assembling, fitting, and bolting or welding many smaller pieces at the site. Often such assembling at the site may require falsework with additional expense, time, and possible danger.

Very deep girders must be studied. Some may be too deep or too long to transport safely and thus will require a horizontal or vertical splice, as the case may be. In other words, the opposite of assemblies must also be determined. Additional pieces will be handled, fitted, and permanently connected, all of which can increase erection costs and delay completion.

In addition to the design drawings for shop or field assemblies, attention should be given to heavy, deep, and awkward pieces, and a decision should be made whether to have the fabricator install hitch angles or weld plates for lifting such pieces, or whether to omit them and have the erector use slings or other types of lifting devices. Hitch plates or angles may be desired on heavy slabs unless the connection angles for the columns are on the slabs, in which case holes in the upstanding legs may suffice.

Small slabs should generally be shop-assembled to columns, provided that anchor bolts will be in place. Otherwise, it may be dangerous to erect a free-standing, slab-connected column. Grillages should be shop-assembled as completely as feasible, leaving the grillage slab loose if the weight with it in place is too great for the equipment to handle. Double beams or girders should be checked for shop or field assembly. It should be decided whether to ship hung lintels on their spandrel beams or whether they could be so damaged in transit that it would be better to ship them loose and preassemble them at the site just before erecting the beams from which they hang.

If column-setting hickeys will be used, the fabricator must be given the size of holes for the pin through the splice plates, which must then be on the upper ends of the columns. These holes will usually vary from a 2-in. diameter for lighter weight columns to as much as 2½-in. diameter for heavier columns (Fig. 3.7.42). Some designs call for the splice plates to be on the lower ends of the columns. This prohibits the use of this type of lifting device and makes it awkward to upend the column for erecting. In such cases a timber of some sort must be used so that, as the column is brought to an upright position on the timber, the splice plates will not foul the floor on which the column is being rotated. The use of this timber should be noted for the field, and the timber should be included with the tools shipped to the job.

On some tight column splices for which no fillers are provided, it is advisable to have the shop omit the upper one or two rows of fasteners or part of the welding, so that the splice plates can be sprung slightly to enter the upper column more easily. Similarly, on a girder or truss splice, if the first row of bolts, or part of the shop weld nearest the joint is left for the field to complete, it will usually aid the erector materially, even though he must then do some work originally intended for the shop. If the erector performs some of the work the shop should have done, this will usually be balanced by the fabricator against additional work requested by the erector, such as furnishing lifting hitches or erection seats.

Any complicated splices should be reviewed to be sure that the splice material will be fastened to one portion of a truss or girder so that the pieces can be assembled without too much difficulty in the field. When feasible, the splice can be made so that part of the splice material will be on one side of one piece and the rest on the other side of the adjoining piece. The connection can then be made by swinging either piece into place sideways. Occasionally it will be advisable, for ease in connecting, to ship all the splice material temporarily bolted to one side of the splice point, even though this means additional work in the field to permanently secure the material originally intended to be shop-assembled.

5.10   *Lateral Stability*

Trusses and girders must be checked for lateral stability when picked at the center, at the ends, or at two intermediate points, as the case may be. (Excellent technical information on checking lateral stability can be found in an article entitled "Strength of Beams as Determined by Lateral Buckling" by Karl de Vries, F. ASCE, printed in Paper No. 2326 of the *American Society of Civil Engineers Transactions,* Vol. 112, 1947, pp. 1245–1320.) If the trusses and girders are laterally unstable, it must be decided whether to heavy-up the members, to add temporary horizontal bracing of structural angles or channels, to use strap, tie-rod, or wire-rope hogrodding, or to secure a temporary beam with the web flat on top of the top flange of the piece in trouble. Hogrodding (Fig. 5.10.1) is tying the two ends of a piece together by means of tie-rods, straps, wire rope, or similar material over intermediate struts extending outward on the com-

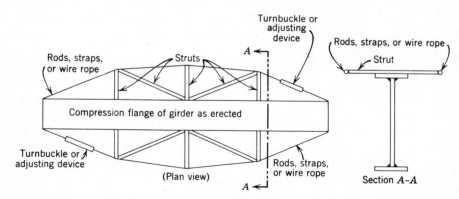

FIG. 5.10.1.   Hogrodding. Schematic arrangement showing hogrodding of laterally unstable girder.

pression flange of a piece as erected. The tie-rods with threading to permit adjustment, or the straps with some provision for change in length, or the wire ropes and turnbuckles are stretched over these supports. Actually, with the piece itself, this hogrodding forms a type of lateral, horizontal truss. Wire rope is not recommended for hogrodding because there is danger of its stretching unevenly on one side or the other, especially if the rope is new. This can cause a beam or girder or truss to fold over in the center or at one or both ends, or to collapse. Girders and trusses should be checked to see if they must be shipped or hauled upright or if they can be laid down without excessive deflection or deformation in transit or when rolled upright for erection.

5.11 Details of Connections

Occasionally design drawings indicate connections that are so difficult to make that a minor change will result in easier erection and frequently more economical fabrication. Details should be studied for easier construction without increasing the fabricator's costs. The fabricator's drafting room should be asked to change complicated details as shown on the design drawings if it will simplify and expedite erection.

For example, a one-sided, double-angle connection (Fig. 5.11.1) with the angles back to back on a header beam will permit swinging a filling-in beam into place easily. This is a better connection if properly designed than a pair of end connection angles on the web of the filling-in beam (Fig. 5.11.2) and is easier to erect than a knife connection. The same amount of shopwork may be required in either case. Generally these filling-in beams are fabricated as "bull-eye" pieces, meaning that they are only punched for a connection and need no fittings. Bull-eye beams need not be moved to the fitting skids during fabrication, whereas the header

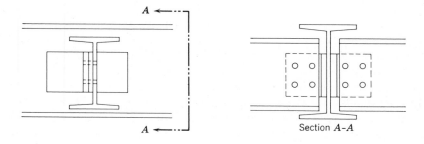

FIG. 5.11.1. Double-angle connections.

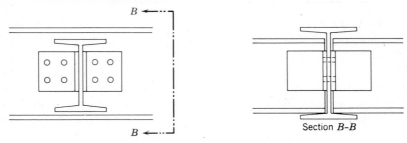

FIG. 5.11.2.   Connection angles on filling-in beams.

beam is usually a framed beam (or girder), which therefore must be sent to the fitting skids.

With the connection on the filling-in beam the connector must align the holes in the angles and in the web of the header. When the beam framing in on the opposite side of the web is to be connected, the erection bolts must normally be pulled back far enough for the connection angles on the second beam to clear the ends of the bolts, so that the first beam is hanging on the bolts without nuts, with the bolts held merely by the header web.

If this type of detail cannot be changed, for safety, an erection seat should be requested on which to land the beams while the bolts are withdrawn to connect the pieces to the header web. This seat can be just a small angle but must have sufficient shop connectors or weldment to hold the weights of the member and the man who is connecting the piece. When no bottom connections are provided on a column under beams framing into the column, erection seats should also be provided for the safety of the connectors. In some cities this is a legal requirement.

One side of a back-to-back angle connection should be paint-striped or marked on the erection diagram in order to ensure connecting on the correct side of the two angles.

Occasionally, one of a pair of connection angles is permanently fastened to the header in the shop (Fig. 5.11.3), and the second angle is left loose, being temporarily bolted for shipment. Then the connector has a loose angle to hold while making the connection. This introduces the risk of the loose angle being dropped and injuring someone below, but it eliminates the danger of having to pull the erection bolts for making the connection on the other side of the header and does not require an erection seat. In this case the erection bolts connecting the first beam to the fixed angle remain in place and no temporary bolts are needed through the header web.

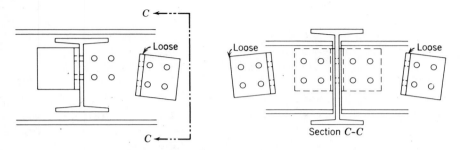

FIG. 5.11.3.   Connections with one loose angle.

Most drafting rooms prepare "general design" drawings of complicated connections and conditions. These should be obtained by the erector before they are used to make the details for fabrication. Then, if there are any objections because of increased costs or difficult connections that cannot be made easily, the detail can be changed before it is too late.

5.12   Installments

The fabricator should be given the division of the steel into shipping installment areas and tiers required as soon as a decision has been made on desired column splices and number, capacity, and type of equipment to be used. This information must be incorporated on the detail sheets from which the steel is fabricated. The lengths of the columns depend on the location of splices, and until this is known material cannot be ordered from the mill by the fabricator.

Shipping installments should be selected so that carloads or truckloads will cover such areas that a preliminary sorting of the steel will have been done automatically by the shop. The weight of these areas should be kept within reasonable amounts so that no installment will be less than a minimum required carload or truckload.

With a guy derrick, the areas should approximate, whenever possible, the space between two or three guys to eliminate unnecessary booming up and down with loads to be moved from an area between one pair of guys to another area between any other pair of guys. Occasionally it is only reasonable to divide a derrick area into four parts as formed by lines at right angles through the position of the foot of the derrick.

For mobile crane erection, both the weight of an installment and the area covered are important. It would be ideal to have an installment cover only the area the crane boom can reach from one position. When the structure is long and not too high, and a crane is to be used, the entire area should be divided for shipping and erecting so that an installment can be unloaded and erected with as little maneuvering of the crane as possible.

As a start, an area about 100 × 100 ft should be investigated, and if it will result in better than minimum carloads or logical truckloads, this will help the erector and will probably not be objectionable to the shipper. Fasteners such as bolts and rods are divided into larger installments covering greater areas so that they can be shipped as truckloads or carloads, unless the fabricator is willing to ship the fasteners for a particular erection installment with the steel for that installment.

The exception to dividing the area as suggested would be a long, mill-type building, which is normally erected by a mobile crane. The columns and bracing along each side should be separate installments since this steel can be erected before the balance of the material between the side framings arrives. For the convenience of the fabricator in shipping, the material between the sides should be divided into installments which will enable him to ship transverse girders or trusses, etc., separately from the filling-in members and bracing.

5.13   Tiers

Unless the design drawings specifically call for something other than two floors to a tier, a two-floor tier is advisable for most buildings erected by guy derrick, climbing crane, or fixed tower crane. Some designers use one section (size and weight per foot) of column for three floors, in which case it is logical to splice the columns in three-floor-tier lengths. This increases the number of pieces that must be unloaded, sorted, and distributed on the working floor in the case of a guy derrick, climbing, or fixed tower crane; or on the ground in the case of crawler, truck, mobile tower crane, or guy-derrick crane erection. A three-floor tier frequently results in accidents to the men since there is much less room for them to move about. With a two-floor-tier there would be only two-thirds as much steel on the working floor, or in the working area, for each tier. In the case of a guy derrick a three-floor jump with average floor heights of 10 ft is often more difficult and costly than a two-floor jump. If the floor heights are much greater than average, the jump may have to be made in two steps, which is costly and more hazardous than a normal one-step jump.

If the erector insists on two floors to a tier when the design drawings show three, the fabricator is forced to add additional field splices, which mean extra milling and splice material. A shop splice must also be made at the alternate third-floor levels if the sections change at those points. This undoubtedly would not have been anticipated by the fabricator when he bid his portion of the work. As an alternative, he may decide to eliminate these extra shop splices and provide the heavier section of column for the full tier lengths. This also costs him more for the extra weight, not anticipated in his bid, because many engineer-designers will refuse to approve the extra material for payment by the customer or owner.

When a three-floor tier must be maintained, it may be necessary to have an additional field splice in the columns next to an interior erecting rig one or two floors above the working floor. This is needed for the boom to clear, in erecting steel beyond these columns, especially if the rig has been located very close to those columns or to the header beams between them.

The exception to objections to a three-floor length of column in a tier is when long-length columns are desired for crane erection from the ground, if the steel will be laid out on the ground and the crane can handle the longer and thus heavier columns.

Sometimes the fabricator may be asked to shop-splice some levels designed as field splices. For a column-core building the longer the columns can be fabricated in one piece, the more economical the erection will be, provided the crane or derrick has the capacity and reach, and provided the column can be upended without buckling. (A column-core building is one in which the only structural steel members are the columns, with occasionally some light struts between the columns. The rest of the structure is of reinforced concrete. By using steel columns a smaller floor area is required in the final permanent structure than if concrete columns were used for the same loads, usually quite heavy, that are to be carried.)

Elimination of field splices in this case reduces fieldwork by eliminating the cost and time of making the permanent splice connections at points designed for field splicing. Loading, shipping, unloading, and delivery limitations may affect the length of shop-spliced columns that can be used.

5.14   Erection Diagram

There should be liaison with the drafting room to make sure that the erection diagrams will be usable by the erector. An erection diagram (Fig. 5.14.1) consists of a line diagram plan of each floor of a structure, and an elevation or side view when needed for locating girts, struts, and similar pieces. The side view is always needed when there is side framing on the structure.

The erection diagram shows the dimensions between columns, between columns and intermediate members; locations of spandrels, hung lintels, etc.; the elevations of floor members, generally as a dimension below the finished floor level unless most of the members are at the same elevation, in which case a note will state that fact and only exceptions will be noted on the diagram.

The size of pieces, such as 14WF228, which is like those shown in the example, should be given. (This description is based on the formerly accepted designation of a beam known as a 14-in.-wide-flange beam, weighing 228 lb to the foot. This designation was subsequently changed to "W 14 × 228" but still retains the same size and properties. For the actual size of such a member, one should consult the handbooks published by most steel manufacturers or the *Manual of Steel Construction* published by the American Institute of Steel Construction, Inc. For example, a "W 14 × 228" beam is actually 16 in. deep with flanges measuring 15⅞ in. wide. However, some sections will no longer be rolled since the American Society for Testing and Materials (ASTM) has revised some sections and eliminated others to help make the steel mill rolling operations more efficient.) This information is given to help the erector find particular pieces.

The most important item on the erection diagram is the number assigned to each individual piece. The numbering system should follow a logical sequence on the diagram. If several pieces are identical and interchangeable in the structure, the same number is usually duplicated for all those pieces. This number appears not only on the erection diagram but also on the detail sheet from which that piece is to be fabricated, and it will be painted on that piece by the fabricator. When the piece must be put into place in a definite direction, the piece mark is usually painted near one end and shown on the same relative end on the erection diagram.

When a piece can be installed either way, the piece mark will still be painted near one end, but can be shown on the erection diagram at the center of the line on the diagram representing that piece. Some detailers will not bother where they show the mark—at the end or at the center of the line on the diagram regardless of how the numbered end should be erected. In this case they will add a small "x" or an "x" in a circle at the end on the diagram where the painted mark should be erected. This mark on the diagram is usually considerably smaller than anything else on the sheet, and with the rough usage an erection diagram is given in the field, the small "x" mark often becomes so blurred or obliterated from dirt or creases in the sheet (usually a linen drawing) that an erector often has extreme difficulty finding it. Work is frequently done when there is poor light, early in the morning before the sun is well up in the sky or late in the afternoon after the sun has gone down. The sheet is often left out where rain, snow, and wind can wreak havoc. It is far better to omit this extra mark and show the piece-mark designations in heavier numbers at the end corresponding to the end on which the mark will be painted in the shop.

Tie-rods, especially when there are many similar ones, are often marked only with an "x" followed by a number equal to the length in inches, with a separate note giving the diameter of the rods. When bracing rods are unusual, they will be given individual marks and the rods themselves tagged with a metal tag on which these piece marks will be punched, unless the rods are large enough to have the marks clearly painted on them.

Some draftsmen use the detail drawing number as part of the piece mark on the erection diagram. This eliminates the need to make a drawing index sheet showing piece marks against drawing numbers. This practice should be resisted vigorously by the erector because it will not only clutter the erection diagram but make it doubly difficult to find pieces when sorting, distributing, and raising. The number would include a drawing number and the number of the piece on that drawing. In some cases this may mean six or seven digits and letters that the man searching for a particular piece must remember.

In addition to all of this information on the erection diagram, complicated connections should be shown in detail as separate sketches. For example, if three or four pieces must be erected in a particular sequence, this should be indicated. When an angle or a channel can be put in place incorrectly, as with the outstanding leg of an angle either up or down, or a channel facing one way or the other, some indication should be given either in a note or a small cross-sectional sketch next to the piece on the drawing. Cross sections are usually shown where they are important to erect the steel correctly or where the connection is too complicated to be shown on the line diagram itself. If camber must be maintained, this should be noted, usually by a separate sketch, to caution the erector to hold the camber built into the piece.

To save making an extra drawing, a detailer will occasionally include information on the steel erection diagram for placing reinforcing rods or reinforcing trusses, or decking, for use by the concrete floor contractor or other contractors. This is not objectionable and is acceptable provided such information is shown relatively lightly, with the steel erection piece marks heavied-up enough to stand out clearly.

The cleaner the diagram the easier it will be for the erector in the field to find the erection mark for locating a piece on the floor or on the ground, and vice versa; if he has a piece it is that much simpler for him to find where it goes by locating the mark on his diagram quickly.

To save money and effort, a fabricator often will not make a new erection diagram but will reproduce design drawings on tracing cloth or paper (sepias) and add erection marks to this reproduction. This clutters the diagram so much or the reproduction is so poor and faint that the field force is unable to use it satisfactorily. If the foreman does try to use it, however, he may spend too much time trying to decipher the marks amid these details that are only necessary for the detailer. Thus, what the fabricator saves in his drafting room is lost many times over by the erector. The erector is entitled to a diagram that is easily usable, not cluttered with unnecessary details or information, is distinct, and that will aid him in erecting the structure correctly and quickly. A cluttered diagram impedes erection; a clean, distinct, well-drawn diagram expedites it. A satisfactory erection diagram is just as necessary as a good erection-scheme drawing. The piece marks on the diagram should stand out in comparison with all the other information such as dimensions and beam sizes.

The erection piece marks are used for ordering steel to be shipped. An entire group of pieces may be ordered by areas, the areas being shown in some fashion on the diagram, for instance, divided by long dashed lines or dot and dash lines. A large letter should be added to the diagram to identify each area (Fig. 5.14.1).

When the steel is sorted or distributed at the site, the piece mark is frequently transferred to the top flange of each piece with keel because several beams standing close together with just enough room to install an erecting sling make it too difficult to read the piece mark painted on the web of the beam. If the fabricator can be persuaded to add the mark on the top flange of the member it will save extra work in the field. This is an added expense to which he may object, even though it would also help him if he stores the beams close together before loading and shipping.

In addition to the piece number painted on the web, the fabricator usually adds his contract number, the number of the drawing on which the piece was detailed and from which it was fabricated, and an inspector's mark. If the steel has been divided into areas for the erector's convenience, such a designating mark will be shown on the erection diagram, on the detail sheets, and on the piece itself.

When there are a number of similar but different areas, such as elevator shafts, cupolas, or dormers, some detailers try to make one sketch and add a number of different erection marks on the same piece as shown, noting each mark for the area to which it belongs. This can be so confusing that the erector is justified in demanding a separate sketch for each area. This does not apply when a floor plan is similar for a number of floors and the piece mark is noted for the floor to which it applies.

The mark painted on the piece must give not only its number but also the floor on which it is to be erected. A single line on the diagram may be marked 42 (the piece number) 3rd fl. and immediately below it 42 4th to 10th fl. In this case the mark on one piece would be shown as "42" with a small "3," generally in a circle, and the other pieces would all be marked "42" with a small "4–10," again generally circled.

In this case any of the beams marked "4–10" can be erected on any of those floors. Given one drawing and the same sequence of numbers for several floors, a skillful raising gang will know where to erect particular pieces without referring to the diagram after they have erected a few floors.

The size of erection diagrams desired, the quantity of each, and the type (paper or linen) should be specified. The number of sets of detail drawings and sketch sheets required, usually one set for office use and one set for the field office, should be stated.

5.15   Grillages and Slabs

The method for setting grillages and/or slabs needs to be determined. Shims can be used for reasonable weight slabs, using three or four piles of shims of steel plate about 4 × 4 in. or 3 × 3 in., one of $\frac{1}{2}$ in., one of $\frac{1}{4}$ in., and two of $\frac{1}{8}$-in. thickness for each pile for the first inch of grout; additional $\frac{1}{2}$-in.-thick shims will be needed for more than 1 inch of grout.

Using $\frac{1}{8}$ in. minimum thickness will permit the slabs to be set within plus or minus $\frac{1}{16}$ in. of the required elevation. If the pile of shims is between $\frac{1}{16}$ in. and $\frac{1}{8}$ in. low, the addition of a $\frac{1}{8}$-in. shim will bring it to less than $\frac{1}{16}$ in. high. Similarly, if between $\frac{1}{16}$ in. and $\frac{1}{8}$ in. high, the removal of a $\frac{1}{8}$-in. shim will bring it to less than $\frac{1}{16}$ in. low.

Three piles of shims (rather than four) will make it easier to level the top of the slab and bring it to final elevation since it is more difficult to have all four piles at exactly the same elevation. This is because of the usual unevenness of the top of the concrete on which the piles of shims are placed. The shims can be placed ahead of time, and grout of sand, cement, and water can be placed around each pile to hold it securely after the grout sets. The slab can be placed on loose shims, and the foundation contractor or the general contractor usually builds grout boxes around the slabs after they are finally placed in exact location. Then grout is poured under the slabs inside the grout boxes.

Some erectors prefer small mounds of grout of relatively thick consistency, placing one ½-in. shim on top and tapping it down into the grout to the correct elevation of the bottom of the slab. The grout is permitted to set and then it is firm enough to land the slab on the shims later, merely sliding it to the correct position. This can be done only if the slab has been milled top and bottom, which is especially advisable if the slab has been shop-assembled to the column.

Overrun or underrun in thickness of slabs on previously set shim packs can cause extra work and expense. Underrun requires lifting the slab or grillage assembly, adding shims, and replacing it. Overrun requires not only removal of the piece, but tearing down the previously set shims or screed angles, and the cost of resetting the piece back in place.

For heavy slabs it is better to have the fabricator weld two nuts on each of two opposite sides of the slab, with a long bolt with a cone point threaded through each nut, to rest on two heavy shims (½ in. or 1-in. thick). Two shims are advisable since the lower one will stay in place on the concrete footing while the upper one can slide on it, the bolts generally digging into the upper shim.

Screed angles are advisable for grillages, set like an inverted "V." Two angles, one for each end of the lower tier of grillage beams, 1½ × 1½ × ³⁄₁₆ in.—6 in. longer than the extreme outside dimension—are satisfactory for lightweight assemblies; or 2½ × 2½ × ⅜-in. angles for heavier assemblies. Shims about 3 × 5 in. are needed to support the ends of the screed angles and to bring them to correct elevation. The shims can be one of ½ in., one of ¼ in., and two of ⅛-in. thickness for each end of the two angles, for the first inch required. Additional shims of ½-in. thickness should be provided, the need depending on the amount of grout specified between the top of the footing and the bottom of the grillage. The height of the angles with their two legs resting on the shims must be taken into account.

The slabs or thin leveling plates can be shipped in advance for others to set to line and grade. If the footings have been brought to final elevation ahead of time, the screed angles and/or shims will not be needed. This condition is ideal for the erector, but the fabricator must make the total height of the grillage and/or slab exact and the erector must take the precaution of checking the setting of such plates before he starts work.

If anchor bolts are required, an agreement must be made with the fabricator as to whether he or the steel erector will arrange with the customer when they are to be delivered. It is advisable to have washers for use under anchor-bolt nuts shipped with the first tier steel to be sure that they are not lost or mislaid if shipped with the anchor bolts. If anchor bolts are to be used, the drafting room should be informed how much larger the diameter of the holes in the slabs should be to allow for discrepancies in setting the anchor bolts in the field. An additional $\frac{5}{16}$ in. is usually satisfactory for large-diameter bolts, but the approval of the designing engineer must be secured if this differs from the dimension shown on the design or contract drawings. Similarly, if boots are required on the columns to connect to the anchor bolts, the fabricator needs to be informed how much larger the diameter of the holes should be. An additional $\frac{9}{16}$ in. is usually advisable for the erection as well as providing for inaccuracies in the shop or field. A tapered pilot nut will help thread the bolts through the holes in the boots. An effort should be made to have the anchor-bolt holes located as far from the center of the slab as feasible to give maximum resisting force against the possibility of the column overturning.

When connection angles or plates for connecting columns to slabs are shop welded, the fabricator should be cautioned to develop the welds for the full weight of the slab and grillage if connected to it, and to resist a lateral wind force on the first length of column. Otherwise, the field erector must be instructed to guy each column as erected, leaving the guys in place until a completed panel of framing has been erected and fitted, to make it safe to leave the columns without guys.

Failure of the weld when lifting a slab or slab and grillage, or when the column is left free-standing, can cause a serious accident. The fabricator should be requested to have axis lines shown by center-punch marks or scribe lines near the four edges of each slab to aid in the final, correct setting in the field to theoretical column lines.

5.16 Bolts—Welds

At this time it is also expedient, if no standard has already been established between the erector and the fabricator, to notify the latter's drafting room what percentage of excess bolts or electrodes should be added to the list of connection fastenings to be furnished. This is to take care of losses, unusable electrode stubs, damaged bolt threads, and a small amount that can be used for erection purposes since most contracts for fabricated structural steel include an excess for which the customer is expected to pay. This percentage can vary from 2 to 5 percent depending on sizes, diameters, and type, but must be kept below that for which the fabricator will be paid by his customer. The heads of bolts and the nuts for the same diameter bolts should be the same size so that the same wrench will fit both.

When connections are to be made with bolts, this will be shown on small sheets usually termed bolt lists. The lists should give the location of every connection, such as beam number to column number, or beam number to another beam number in the case of a filling-in beam framing into a header. They will also give the size of the bolts for that connection, with their length, diameter, and type, such as H.S. (for high-strength) A-325, H.S. A-345 or H.S. A-490, ribbed bolts, turned bolts, machine bolts, etc.

In the case of welding, welding lists are furnished that also give locations of connections, together with welding details, electrode size and type (such as E6010, E7010, E6011), and type of weld (such as fillet or butt). If the connections are complicated, or if it is important to run one pass or bead on one side and then one pass on the opposite side or on another member to frame in to the connection before completing a second or third pass at the first point welded, sketch sheets of each of these complicated procedures and connections should be furnished to the field erector.

5.17   Overrun or Underrun

Provision must be made to take care of overrun or underrun caused by fabricating inaccuracies or mill tolerances of a long line of beams, girders, or trusses. Otherwise, the building will end up too long, too wide, too short, or too narrow, as the case may be. If too short, shims or filler plates can be used if the connections are framed connections. If the connections are not framed connections, the holes can be reamed whether there is an overrun or an underrun, and larger diameter connectors can be used. If welded, undersize-diameter fitting bolts can be used in the holes provided for erection purposes, and the welds will then hold the members in their adjusted positions. With framed connections, if there is a possibility of overrun, an agreement should be made between the erector and the fabricator to deliberately shorten some of the pieces that may cause trouble and provide shims or filler plates equal to the intentional underrun to be used if needed.

5.18   Ramps and Trestles

If a ramp is required to bring equipment into the excavation (the "hole"), this must be designed by the erector unless the general contractor will provide one. He may do this since the foundation contractor and other contractors will probably also need it. Unless a solid dirt ramp is used, the design of a ramp provided by others should be checked for safe use by the steel erector. A notation should be made on the erection-scheme drawing for the superintendent to make sure that it was constructed as designed.

When the steel erector is to build the ramp, separate construction drawings should give the superintendent complete details. In that case the required material should be furnished from the toolhouse or purchased to be delivered when needed. If a derrick is to be used to erect, a truck crane or crawler crane can unload steel from the street into the hole, eliminating the need for a ramp. If a crane is to be used to erect, a ramp is generally needed unless the crane can erect from the ground around the excavation.

With a building that has many floors below street level, the use of auxiliary unloading equipment may be necessary if a ramp cannot be built without blocking many of the column footings. If there is a very deep excavation the general contractor or the foundation contractor may build a level trestle from the street out over the excavation from which to work. It should be strong enough to support a crane for erecting the steel or for setting up a derrick on the bottom of the excavation and feeding steel to it until the derrick has erected enough steel to jump up to street level. When loading facilities in the adjoining street are restricted by traffic conditions, such a trestle may be the only means of delivering not only the steel but all other materials for the structure as well.

5.19   Power Houses

In preparing the scheme for erecting a power house and adjoining boiler room, it is expedient and economical to start a guy-derrick-erected job with a crawler or a truck-type crane if it is of sufficient capacity. The crane will set all the grillages and/or slabs and the first or first and second tier steel of the boiler room. The derricks are then set on top of the crane-erected steelwork. By proper scheduling the crane can be used profitably on the low power-house steel until the derricks are ready to be loaded and shipped away. It is better to use two lighter capacity derricks to erect most of the boiler-room steel separately, and use the two together to unload and erect the heavy girders. If the two derricks cannot be located to unload together, the crane, if it is of sufficient capacity, and the boom, if it is long enough, can help one derrick to unload and raise the heavy girders to the working floor. Then the girders can be moved in until the two derricks can take hold and erect them. The crane returns to continue erecting its portion of the power house. All of this must be shown on the erection-scheme drawings to be sure that the superintendent follows the planned, safe procedure. Otherwise, he might overload the derricks or the crane or the permanent structure, which might result in an accident.

5.20   Column-Core Buildings

In selecting a method for erecting a column-core building, the steel columns need to be analyzed so that the drafting room can be informed promptly as to the lengths in which they are to be fabricated. The designed splice level must be considered, the weight the equipment can handle, the length that can be shipped and delivered through city streets, and finally the stability of the column section as it is rotated from its horizontal delivered position to its upright position.

If permanent struts are not included in the contract, temporary timber or steel struts or bracing must be designed between the columns. The struts should be framed into the columns a reasonable distance above the finished floors so that the men can remove them safely later, while standing on the concrete floors after they have been poured. Otherwise, saffolding of some sort will be needed, which adds unnecessary hazards and expense to the erector's work. The connections needed for these struts or braces must be transmitted to the fabricator's drafting room so that they can be fabricated in time.

If the struts are to be furnished by the fabricator, the cost to the erector should be determined in advance. It may turn out that the erector could save money by purchasing the material and fabricating the struts in his toolhouse.

With columns standing free for extreme heights, and only light struts between, both horizontal and vertical (diagonal) guys are used. The horizontal guys, like the struts, should be above the floors as poured, in the plane of the struts. The vertical guys will be concreted in and ruined unless the concrete contractor boxes them where they pass through the floors. This should be arranged in advance.

If the total height of the columns can be handled by an available crane, this will probably be better than using a guy derrick. But if the building is high, or if, as sometimes happens, the core columns extend only part way up with normal construction of a beam-and-column structure above, a guy derrick is best. Arrangements must be made to have an opening left in the floors as concreted that is large enough for jumping the derrick.

If derricks are to be used, the erection scheme should be planned so that the derricks erect the first tier as selected by the erector, which may be four or more floors. The derricks are then tied up until those floors are poured and the concrete sufficiently set and strong enough to permit working on it. Then the erector must return, jump the derricks, and erect another length of columns; or he must start the beam-and-column portion above. Hitches should be added to the core columns to make sure the erecting foreman picks them at safe lifting points to prevent buckling or collapsing.

As with the ordinary building, the location of the guys at particular columns should be selected and arrangements made for securing column anchor slings to those columns, possibly using lashing embedded in the concrete floors as poured, and burning the lashing off later above and below the floors.

5.21   Open-Panel Construction

For open-panel construction the study must include the material of the working floors. Open-panel construction is the name given to a building frame with large distances between the columns, which are connected only by members framing directly into the columns, and with no intermediate filling-in beams framing into these column-connected members. A concrete slab is poured on the entire panel, which may be as large as 20 or 25 ft by 25 or 30 ft.

The usual 2 × 12-in. floor planks will be unsafe if any steel is unloaded on them or for the men walking across the planks on such a span since there is no intermediate support steel, as in the case of normal beam-and-column construction.

There are several ways to solve this difficulty. The best is to use plain beams, such as lightweight 8- or 10-in. wide-flange beams, as temporary intermediate supports. These can be sold later at almost their purchase price because they have a high salvage value with no connections or connection holes in them. These beams, spanning the openings, should be selected to carry the floor planking and any permanent steel unloaded on the floor.

The columns to be erected should be lifted from the ground separately and set directly in place. If this is not feasible and they must be temporarily landed on the floor, they should be placed diagonally on skids placed to carry the load directly to the two beams framing at right angles into the same previously erected column.

The erection-scheme drawing should show the distribution of unloaded steel on the floors, the spacing of the temporary steel support beams under the floor planks, and all information necessary for the erecting foreman to follow the plan devised by the erector's engineers. When jumping the floor planks, they should be piled off their supports. The temporary support steel is gathered to make a bundle of beams for any one panel, lifted by the crane or derrick to the new working floor, and spread out again. The floor planks are shifted to the new level and laid on the support steel.

As a substitute, some erectors will use heavy wire rope and turnbuckles hooked over the beams at opposite ends of the bay, with two or three lines in each bay. The hooks are secured over the top flanges of the spandrel beams at each edge of the building if the flanges are strong enough. Preferably, the individual ropes and turnbuckles are connected to separate, bridled slings forming a "V," with the ends of the slings secured to the two columns alongside the bay at each end of the building. The turnbuckles are taken up (tightened) until there is just a little slack in the ropes, and then the floor planks are laid on them. It is quite difficult to calculate the stress in the ropes, which is one reason for advising against this method. In addition, the floor will be excessively springy and the men will move slowly.

Skids should be placed to try to bring as much as possible of any loads landed on the floor directly into the permanent structure in each panel. In tripping columns landed on the floor to a vertical position (if they have not been lifted from the ground and set directly in place as recommended previously), care must be used in erecting them not to overload the floor planks or ropes below them. A long timber or steel beam that spans the entire panel should be used on which to rest the lower end of the column as it is upended. Such timber or beam should be clearly indicated on the erection-scheme drawing.

With either method of supporting the floor planks, it may be advisable, depending on the span, to increase the size of the planks from the usual 2 × 12 in. to 3 × 12 in., and possibly their length from the usual 22 or 24 ft. This will make the planks too heavy for two men to carry safely, and so additional men will be required to handle them. All these details together with any necessary precautions should be placed on the drawing itself and in the written field instructions.

5.22   Bridges Between Buildings

Sometimes bridges between buildings must be erected over streets with heavy traffic. Although a crane may be ideal for erecting such bridges, provided the height and weights of the members are within the reach and capacity of the crane, traffic restrictions may prohibit its use. Then the guy derrick may be the logical piece of equipment to use.

The simplest scheme utilizes a pair of heavy steel beams or timbers to act as outriggers, cantilevered over the edge of the structure at the level of the bridge, and securely lashed down at their inner ends to the steelwork already erected. A guy derrick is set up on these beams or timbers, either guying it to the two structures being connected by the bridge, or using a support for the gudgeon pin similar to the one described for use on setbacks.

5.23   Pier Sheds

When the structure, such as a pier shed, is directly adjacent to a body of water, and steel can be delivered by water, a derrick boat may be preferable for erection. Also, the boat may deliver the material to the area where the erecting rigs can use it efficiently. In some cases the fabricator cannot fabricate and deliver material as rapidly as an efficient erection schedule would require. Then a derrick boat can deliver the material as shipped and unload it in approximate positions over the site. After enough steel has been delivered, an erecting rig can start and not be delayed by nondelivery of the balance; such a scheme may work out to the best advantage of the erector, the fabricator, and the owner.

In preparing the erection scheme, it must be decided whether to rent a derrick boat with a boom long enough and of sufficient capacity to unload the steel and land it far enough from the boat to place it where it should be landed; or it might be more economical and satisfactory for the erector to rent a barge on which to set up a derrick and hoist, or even to move a crane onto the deck of the barge. If a derrick is to be mounted, it may be more efficient for best utilization of its capacity to lash two barges together to spread the footing under the derrick. Or a tower can be built on one or two barges secured together and a stiffleg derrick installed on top of the tower. The tipping of the barge or barges when the rig is handling the heaviest piece at the reach required, or a lighter piece at maximum reach, must be investigated. The allowable freeboard will probably be the limiting factor.

The erection-scheme drawings must be explicit on how far out the boom can operate with various loads, since the further out it goes, the more the boat or barges will tip. They may tip so far that the overturning of the rig can be critical.

If the pier is very wide the steel may have to be unloaded and passed to an inner aisle by an auxiliary crane or other equipment. Most pier sheds are three aisles wide and lend themselves to three rigs, one in each aisle, with additional rigs sorting, passing, and distributing ahead of them, as well as assembling knocked-down trusses so that the latter can be erected as single pieces. With such an arrangement the area over which the steel is distributed must be watched in order to keep the aisles clear for the erecting rigs to travel as each erects. These areas should be outlined on the erection-scheme drawings.

Steel can be delivered on railroad cars on a carfloat, shipped on a barge directly from the fabricating shop, or unloaded onto a barge by the railroad at a railroad yard elsewhere. The derrick or crane then unloads the delivery barge alongside and swings the material over to the erection site. The sequence and cautions limiting the weights to be picked, the distance to be boomed out, and the spotting of the derrick boat for each installment should all be shown on a drawing. The shipping schedule should be carefully coordinated with the erection-schedule requirements.

Facilities for moving the boat from point to point should be taken care of and shown, using either tugs to move it, or mooring lines to an auxiliary hoist or to spools on the main hoist that actuate the boom and load falls and the slewing lines or bullwheel.

With a tower-mounted derrick the boom may be high enough to clear the completed structure and thus may be able to erect the entire structure without the need for other erecting rigs on the foundation—cranes, derricks, or travelers.

5.24   *Wall-Bearing Steel*

When there is much wall-bearing steel, it must be decided whether to shore it with steel or timber supports, or to leave the wall-bearing material down until the walls have been built. The latter requires return trips after the main steel has been erected. On these trips, if the material is light enough it can be placed by handline, or if a crane is available that can be used for short periods at a reasonable cost, it can be used if the boom is long enough to reach the area involved (see also Sec. 6.16).

5.25   Erection-Scheme Drawings

After the job has been analyzed and the type of erecting equipment selected, the erection method or scheme is developed. As soon as the scheme is definite, erection-scheme drawings are made for the field to follow.

In working up the erection scheme difficulties may be encountered prohibiting the use of assemblies already requested, or requiring heavier connections of steel to support the erection equipment finally selected, or necessitating other changes from the details arranged with the fabricator. It is vital that such changes be communicated promptly to avoid backcharges if the fabricator has to remake drawings or undo fabrication already undertaken.

It is advisable to inform the drafting room of the loads to be carried by the steel if it is to support erecting equipment operated on permanent steel. This must be done to make sure that the connections will be heavy enough when fully fitted or fastened to carry the derrick, crane, or traveler. It is not always possible to make these connections with permanent connectors such as high-strength bolts or welds before the equipment is jumped or moved. A minimum based on using fitting bolts, preferably heat-treated bolts, and some drift pins should be counted on in checking the strength of the connections. This should be noted on the drawings prepared for the field to show the location of a derrick or climbing crane on the various floors, or on the sheets showing the procedure for a crane or traveler operating on the structure.

Prints of a basic sketch (Fig. 5.25.1) showing only the locations of columns can be very useful. They can be used to show details of the erection method not shown on erection-scheme drawings. These sketches will be useful in the office and in the field to keep track of erection progress. They can serve to inform the drafting room about installment areas and the details of location of special connections required. The field engineers can use the sketches to keep records of variations in column locations from their correct positions. When a specified area has been assigned by the general contractor or customer for shanties, offices, material storage, etc., this can be shown on such a sketch sheet or indicated in instructions instead of being given on the erection-scheme drawing.

The erection-scheme drawings should show in detail not only the scheme

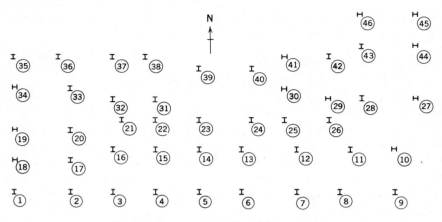

FIG. 5.25.1.   Typical sketch for general use.

decided on, but also any unusual conditions to which the field must be alerted. When the erection is normal and nothing unusual will be involved, erection-scheme drawings may be replaced with a simple set of written instructions describing the procedure to be followed, giving all details necessary to alert the field superintendent to the scheme planned by the erector's engineers.

Capacity tables should be given for the equipment to be used. The maximum permissible reach for setting heavy pieces should be shown, as well as the maximum load that can be lifted at maximum reaches (with a flat boom). In preparing the erection scheme an analysis should be made of the weights of critical picks, together with the weights of steel in each tier, the number of pieces, weights by areas, number and type of connection fasteners, and any details affecting the work.

The length of boom needed for an erecting rig is determined by calculating the distance from the foot or heel of the boom to the top of the furthest column, allowing for drift above that point. (Drift is the name given to the height required by the upper and lower load falls blocks when in their closest position, plus the overhauling ball, hook, and sling or hitch between the hook and the top of the piece.)

If trusses or girders are to be assembled at the site before raising them as completed members, the assembly areas should be marked so that no other steel will be unloaded there, which may foul the area. The amount of blocking on which to assemble the individual pieces, if assembled flat, should be given; it should be enough to permit men to work safely for installing the permanent bolts or welding the members together while the pieces are still on the blocking. If a truss or girder is to be assembled aloft, the necessary falsework or shoring must be designed to support the members being assembled, and their stability in place must be checked. The procedure should be given in detail.

If any members are to be lifted by two adjacent rigs, or if there is a limitation on picking any particular piece at a particular location, this must be added to the erection-scheme drawing, or a separate drawing should be made to cover this situation. With mobile-crane erection, the movement and location of cranes picking critical loads, sequence of erection, and similar items should be decided and shown in detail on erection-scheme drawings (Figs. 5.25.2 and 5.25.3).

Schools, churches, and theaters generally have auditorium areas. These must be carefully checked to see if the roof trusses, girders, or other members directly over the auditorium area can be erected as single pieces, or if falsework will be needed. Cantilevered balconies or canopies generally require some shoring or falsework with provisions for jacking so that the outer ends can be raised or lowered to the correct elevations before the permanent fastening is done. When tie-rods, sag-rods, or rods with turnbuckles hold the steel permanently, these can sometimes be erected first and the steel members fastened to them, which eliminates falsework or shoring. When necessary, adequate falsework can be designed and detailed for fabrication in the toolhouse or in the field, or possibly by the fabricator of the permanent steel. This must be indicated on the erection-scheme drawings.

Hangars with cantilevered trusses overhanging interior supports usually require temporary supports under each truss until all are erected, adjusted, and secured. Door tracks suspended from the ends of these trusses require more than normal accuracy in setting them to line and grade. The trusses supporting the tracks must be accurately positioned. Due to fabrication tolerances, some trusses may need to be jacked up, some lowered, or shims used to achieve a final straight and level line at their outer ends. The instructions should include a caution to check the elevations of the ends of such cantilevered trusses before fastening them permanently in place. Temporary vertical hogrodding on the top chord of each truss may be needed to hold the outer end if falsework or shoring is inexpedient. All this must be shown.

If cross-lot bracing is used to shore the sides of the excavation, it must be decided if this bracing will interfere with the operation of cranes or the logical placement of derricks. If the bracing will interfere, cranes may not be able to work properly. If a derrick is selected, it may be necessary to design a falsework tower on which to stand it the first time. If at all pos-

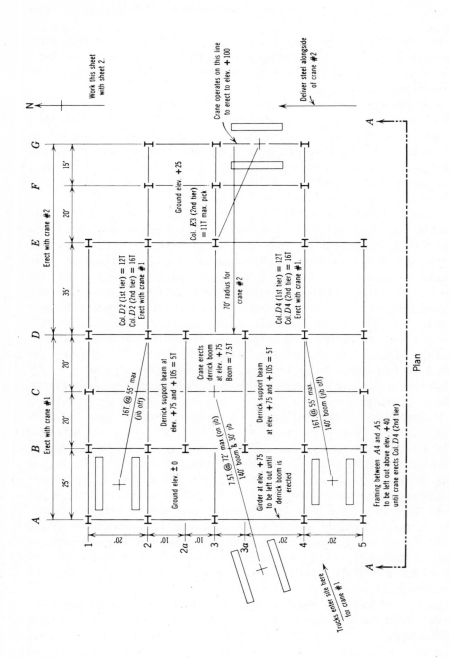

Work this sheet with sheet 2.

Crane operates on this line to erect to elev. +100

Deliver steel alongside of crane #2

N

Erect with crane #2

15'

20'

35'

20'

20'

25'

A   B   C   D   E   F   G

Erect with crane #1

Ground elev. +25

Col. E3 (2nd tier) = 11T max. pick

Col. D2 (1st tier) = 12T
Col. D2 (2nd tier) = 16T
Erect with crane #1

16T @ 55' max. (jib off)

70' radius for crane #2

Col. D4 (1st tier) = 12T
Col. D4 (2nd tier) = 16T
Erect with crane #1.

Derrick support beam at elev. +75 and +105 = 5T

Crane erects derrick boom at elev. +75
Boom = 7.5T

Derrick support beam at elev. +75 and +105 = 5T

Ground elev. ±0

7.5T @ 72' max (on jib)
140' boom & 30' jib

16T @ 55' max.
140' boom (jib off)

Girder at elev. +75 to be left out until derrick boom is erected

Framing between A4 and A5 to be left out above elev. +40 until crane erects Col. D4 (2nd tier)

1   2   2a   3   3a   4   5

20'   10'   10'   20'   20'

Trucks enter site here for crane #1

A

A

Plan

154

Erection Notes

1. Rig Crane #1 with 140 ft. boom, 30 ft. jib
2. Rig Crane #2 with 110 ft. boom, no jib
3. Place Crane #2 as shown on G line. Use timbers and mats as road between F and G line footings to provide firm, level roadway.
4. Trucks to deliver steel to cranes as shown.
5. Assemble derrick boom on ground. Set on temporary support steel at Elev. +75.
6. Assemble derrick mast on ground with crane. Set in place by derrick boom at Elev. +75.
7. Jump derrick to temporary support steel at +105 in normal manner.
8. Use Columns B1, D1, E2, E4, D5, B5, A4, A2 for derrick guys.
9. Erect steel to Elev. +100 with cranes. Erect steel above +100 by derrick.

General Notes

Cranes must be level, derrick mast must be plumb, load falls must be vertical. Check ground and use mats to keep bearing pressures within safe limits
No change is to be made in erection scheme as shown without checking with Main Office.

FIG. 5.25.2. Erection scheme drawing—Sheet 1. Work this sheet with Sheet 2.

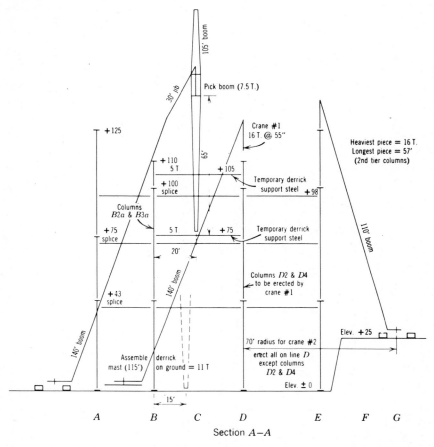

FIG. 5.25.3. Erection scheme drawing—Sheet 2. Work this sheet with Sheet 1.

sible, this tower should be built with some of the permanent members from an upper tier. This will eliminate the cost of shipping temporary falsework and handling it at the toolhouse. In designing the tower provision must be made for the bullstick man to operate safely on top of the falsework tower unless a bullwheel swinger is used. It will be necessary to determine if permanent members in the structure will have to be left out temporarily because of interference until the bracing is removed. All this information should be included in the instructions to the field or clearly shown on the erection-scheme drawings.

If the fabricator does not have enough space to store steel before it

Capacities in Tons

Crane #1

Radius (ft)	20	30	40	50	60	70	80	90	100	110	120	130	140
100-ft boom	50	35	25	20	15	10	8	6	5				
30-ft jib		12	12	12	11	10	8	6	5	4	3		
120-ft boom		35	25	20	15	10	8	6	5	4	3		
30-ft jib		12	12	11	9	7	6	5	4	3	2	1	
140-ft boom		35	25	20	15	10	8	6	5	4	3	2	1
30-ft jib		12	11	10	8	7	5	4	3	2	1	1	

Reduce boom capacity by 2T when 30-ft jib is in place

Crane #2

Radius (ft)	20	30	40	50	60	70	80	90	100	110
100-ft boom		30	15	10	7	5	4	3	2	1	
30-ft jib			8	8	6	4	3	2	1	0.5	
110-ft boom		25	15	10	7	5	4	3	2	1	
30-ft jib			8	8	6	4	3	2	1	0.5	
120-ft boom		15	10	7	5	4	3	2	1	0.5	
30-ft jib			8	6	4	3	2	1	0.5	0	

Reduce boom capacity by 2T when 30-ft jib is in place

should be shipped in proper sequence and at the proper time for an efficient erection schedule, he may ask the erector to unload it ahead of time. The area and method of storing, blocking between pieces, heights of stacking, etc. should all be shown on a drawing. (The extra cost for this service would normally be borne by the fabricator.)

Any obstructions to be guarded against, or any other hazards, should be indicated clearly, and the precautions to be taken should be shown on an erection-scheme drawing and included in the written instructions for the superintendent. The sequence and direction of erection should be shown clearly as well as the installment areas.

When a piece will be too heavy for one rig, two rigs may be needed. When the two rigs are of equal capacity, they can hook onto the member

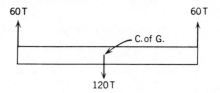

FIG. 5.25.4. Two cranes of equal capacity; the girder laterally stable when picked at its ends.

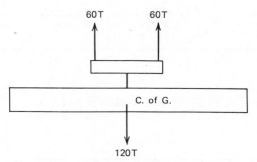

FIG. 5.25.5. Two cranes of equal capacity; the girder laterally stable when picked at its center of gravity, using a balance beam.

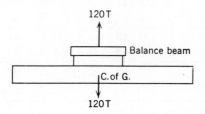

FIG. 5.25.6. One crane picking girder with balance beam, lifting at two points equidistant from girder's center of gravity, with girder laterally stable when picked at those points.

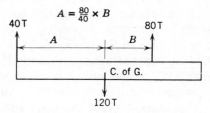

FIG. 5.25.7. Two cranes of unequal capacities dividing load in direct proportion to their capacities; girder laterally stable when picked as shown.

158

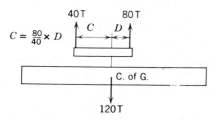

FIG. 5.25.8. Two cranes of unequal capacities dividing load in direct proportion to their capacities by means of balance beam; girder laterally stable when picked at its center of gravity.

at points equidistant from its center of gravity if the piece can be lifted safely at those points (Fig. 5.25.4) and if its lateral stability has been checked and found to be satisfactory. If the piece can be picked at its center, a balance beam could be used (Fig. 5.25.5), each rig hooking on to one end of the balance beam, and the piece then lifted by a hitch at the center of the balance beam. This is safer than picking at separate points, since the piece will remain level even if one rig lifts its end of the balance beam ahead of the other. If only one rig is used, a balance beam will permit spreading the lifting points if the girder would be unstable if picked at one point (Fig. 5.25.6). If the two rigs are not of equal capacity, the points where each should hook onto a piece must be calculated and then shown (Fig. 5.25.7) on a drawing. These points are determined by dividing the total weight into the weights each will lift in direct proportion to their capacities. If the member can be lifted at its center of gravity, but the two rigs lifting are of unequal capacity, a balance beam can be used. The lifting hitch is then at a point on the balance beam to divide the load between the two in accordance with their relative capacities (Fig. 5.25.8).

6

Job Under Way

6.1 Yard

After a check has been made of the status of foundations, footings, or piers to make sure that the scheduled date for starting work in the field will find the site ready, steel should be ordered shipped to arrive at the correct time. If a yard is to be used for unloading and storing shipments before they are needed at the site; or for the assembly of trusses, girders, or other components of a group of members to be assembled prior to delivery to the site, such a yard is set up in time to be ready for the first steel delivery.

Steel is normally shipped by rail, water, or highway. Railroad cars are sometimes delivered to a dock on carfloats, or the steel is unloaded elsewhere by the railroad on the deck of a derrick boat with a flat deck area for the steel from the cars. A yard away from the site of the structure is often rented for unloading the steel regardless of the method of delivery. A storage yard may be a necessity if the fabricator must ship steel before it is needed at the site. In some areas the railroad has facilities for unloading steel from cars into ground storage for a fee, or monthly charge, and then reloading carload lots as needed for delivery to the erector. (Part-carload lots can also be obtained out of ground storage at an extra charge by the railroad.)

The erector may need a yard to sort the steel into areas or installments smaller than the fabricator was willing to do, if minimum economical shipping installments are larger than the installments the erector may have desired. Some cars require a minimum weight to be paid for transporting regardless of the weight actually loaded in the car.

Tools and equipment to set up the unloading equipment should be delivered far enough in advance of the first shipment of steel so that every-

160

thing will be ready to unload on arrival. The schedule for unloading and assembly must mesh with the schedule for arrival of the steel.

If the yard is on railroad property with railroad tracks, a locomotive crane can usually be rented from the railroad company or the erector can ship his own. The cost of shipping his own to and from the yard and the cost of assembling and dismantling must be compared to the charge by the railroad for delivery and removal plus the daily rental charge for the railroad crane. The railroad's crane normally arrives reeved and ready to go to work. The erector's own crane must be assembled and reeved and prepared to work.

Shanties should be unloaded and set in place for the use of the men to change clothes or to eat, and a separate shanty set aside for use as an office if considerable tonnage will be handled. Otherwise, a combination change shanty and office can be used. A supervisory force must be lined up and men must be hired to unload and set up the shanties, and then to set up unloading equipment.

When it is inexpedient to use a locomotive crane because of no spare tracks in the area selected or because of cost or nonavailability, crawler or truck cranes can be shipped or rented locally for delivery to the yard. If shipped by carrier or by rail they are unloaded, assembled, reeved, and made ready to work (Secs. 9.5 and 9.6). If the ground is satisfactory and the material light enough to be handled by a truck crane, this may save transportation charges. It is usually more economical to drive a truck crane over the highway than to load, ship, unload and assemble, etc. a crawler or locomotive crane.

Occasionally a guy derrick or a stiffleg derrick can be set up for unloading in the yard when a considerable tonnage is to be delivered there, but the area over which they can operate is limited. When a few heavy pieces are involved and the ground conditions are favorable, a gallows frame is sometimes erected directly over a track on which the heavy pieces will be delivered, a piece is picked off its delivering car, the car pulled out, and a truck or truck and trailer placed under the piece suspended by the gallows-frame mechanism. Then the piece is lowered onto this carrier for delivery to the job site. Lighter capacity equipment is then used to handle the balance of the steel shipments. The gallows frame can similarly be set up over a road if the heavy pieces arrive by truck.

If steel arrives at a yard, the material should be unloaded on timber blocking on the ground. Steel loaded in cars or on trucks should be on blocking to permit installing lifting slings. The material can then be unloaded safely and this blocking used on the ground. Old railroad ties can be used to good advantage as blocking. Blocking will make it easier to land material to leave it in safe condition against falling over, as well as provide a space under it for removing the slings used in unloading and for placing the slings later for reloading. When pieces are piled one on top of another, or girders or trusses are stood upright, care must be used to brace or block such material or land it so that it cannot fall over on the men working. It should be landed so that slings, girder dogs, or other lifting devices can be used later without additional handling of the pieces.

Grillages, slabs, base plates, and similar material should be stored so that they can be accessible for first delivery to the site. Columns should be landed so that if piled two or three layers high, with timber blocking such as 4 × 4s between the layers, the top ones will be those wanted first and the bottom ones those wanted last. Similarly, beams, etc. should be piled with blocking between drafts in such a way that a good-sized draft can be picked at one time to be landed on a truck for delivery to the job site.

If any assembly of trusses or girders is to be done, a fairly level, unobstructed area should be used, with timber blocking laid so that the truss or girder will be level and the pieces can readily be assembled. The blocking should be high enough for the men to work safely on the connections, installing permanent bolts, or performing the necessary welding on the top, bottom, and sides as the assembly has been laid out.

There must be a close liaison between the job and the yard so that material needed will be loaded on delivering carriers far enough in advance to arrive at the job site when needed. One of the purposes of the yard is to ensure the arrival of material at the job on time so that the raising gang is not delayed waiting for steel. A telephone in the yard office shanty is of great value in enabling the superintendent or foreman in charge of the yard to keep in close contact with the job. It also gives him the necessary facility to check on incoming shipments and to make prompt changes in scheduling if this becomes necessary.

Occasionally bad weather prevents work in the yard or on the job while steel continues to be delivered from the shop. This can build up demurrage costs appreciably, and the fabricator should be informed in time to permit him to slow down shipments. A telephone can be worth its cost in case of an equipment breakdown by permitting prompt action to make repairs or replace parts, or by securing prompt medical aid if an accident should occur, an obligation a steel erector should automatically accept and foresee.

Once all steel has been delivered to the yard and subsequently reloaded and sent to the job site, the unloading equipment can be loaded out and shipped away, unless it will be needed to load tools and equipment from the job itself at the completion of the project. The shanties are dismantled and loaded out together with the rest of the tools and equipment at the yard. If required, the area is restored to the condition in which it was received.

6.2   Job Site—Preliminaries

Assuming a yard has been set up and is in operation, or no yard is to be used, work begins at the site. The superintendent in charge should have been in close touch with site conditions so that he is sure everything will be ready on the scheduled starting date.

He should have checked any cross-lot bracing to be sure it would not interfere with erection; to see that the setting up of a derrick, or the movement of a crane or traveler (as the case may be) will not be prevented by the bracing. If the structure is to be erected by a derrick and the bracing will interfere with setting up the equipment at ground level, a falsework tower may be needed, so constructed that its members will clear the cross-lot bracing and the derrick set on top of the tower. Such a condition should have been foreseen and the erection-scheme drawings prepared to show the location and construction of the tower.

Occasionally, only a few permanent steel members will have to be omitted if not actually needed to proceed with steel erection, and when the cross-lot bracing has been removed later, these pieces are erected, usually by a handline gang, unless openings above the locations of the pieces have been left in the floors to permit the derrick to lower erecting pendants through the openings. In this case the missing members together with their fastenings must be stored so that they will not be damaged or lost, until needed.

If a ramp is required for access to the site area where steel erection is to start, or to deliver tools and equipment into the hole, and possibly for steel delivery later, this ramp should be in place if it is to be constructed by others. It should be checked to make certain it will be adequate for the erector's requirements. If the ramp is to be constructed by the steel erector, tools and equipment will be needed in ample time, and the material should be scheduled for building the ramp so that it can be completed before the first steel shipment is due.

Depending upon the size of the job and the type of tools and equipment needed for actual erection operations, the superintendent should arrange to have the tools and equipment delivered just far enough ahead so that when everything is set up and ready to work, the footings are completed and the site is ready.

The superintendent should check to see that the site is satisfactory for the movement of cranes or setting up derricks, and for trucks to deliver tools and equipment as well as steel later. He may find that conditions have changed from previous site visits; the ground may be muddy or wet and in need of timber mats or corduroying the area, i.e., laying individual timbers next to each other, each long enough and heavy enough to permit safe travel over them, and the length helping to distribute the loads over the ground satisfactorily.

Mats of four or five 10 × 10-in., 10 × 12-in., or 12 × 12-in. timbers bolted together, long and wide enough to hold the crawler treads of a crawler crane, or all the wheels of a truck crane, or a crawler- or truck-mounted tower crane, or a guy-derrick crane, are advisable if there is any question of the loads from the treads or wheels respectively being too great for the ground to support without the crane sinking in. By using such mats the load of the crane can be reduced to a safe load per square foot on the ground.

These mats should have pockets cut in their top surface and shackles or wire-rope lashing (in the form of loops) secured in these pockets so that sling hooks can be inserted in the shackles or lashings but so that the latter can drop back into the pockets as the cranes or trucks ride over them, thus avoiding damage to them. As the cranes move into new areas, the mats left behind can be picked and swung ahead for continued travel. The shackles or loops make it unnecessary to pry up the mats to install slings around them for handling when they are to be moved.

The areas in which work is to be done should be inspected and planned for space to unload steel, for assembling trusses or girders if required, and to be sure the steel can be sorted and distributed close to the location in the structure where the various pieces will be raised. Steel should be delivered as close to the cranes or derricks as feasible; the delivery areas should be checked against the capacity of the equipment at the reach involved between the point where the erecting rigs will be standing and the point where the carriers can deliver. With crane erection, the closer to the crane the material is delivered, the less lost motion will be involved in moving back and forth.

If locomotive cranes will be used to erect, the superintendent should be sure that tracks have been laid at least between the delivering railroad and the area where the crane will start work. He should check to see that arrangements have been made for the railroad or others to lay the tracks before they will be needed in the various areas over the site. If the erector plans to lay his own tracks, rails, ties, and splice material should be delivered in ample time to be installed and ready for use when needed. If the tracks will cross highways or pedestrian paths, he should make the necessary arrangements for protection when the cranes will cross.

If the erecting tracks will be close to active railroad tracks the superintendent should develop the precautions necessary for safe operations. The railroad may require operations to stop when a passing passenger or freight train is due. In some cases the railroad may require the use of a flagman or a work-train. The superintendent should check the arrangements needed to provide these services at the proper time.

6.3 Personnel

The superintendent should determine the availability of skilled men to do the work: structural ironworkers, apprentices, hoist and crane operators, and oilers or firemen (if coal-fired steam equipment is to be used). A superintendent will usually have key foremen or pushers who follow him from job to job. If not, or if he needs more, these men must also be found. If not available locally, they may have to be secured elsewhere and asked to come to the site. At the same time, it is well to observe any outstanding structural ironworkers who may be potential foremen. By putting them in charge of small gangs, under close supervision and observation, the superintendent can develop good foremen or pushers.

Foremen or pushers usually start as apprentices or ironworkers. With experience and skill, demonstration of their personality and ability to handle groups of men, and ability to read drawings and follow erection schemes, they are placed in charge of gangs. Those with the necessary ability then graduate to general foremen or superintendents. A raising-gang foreman should be a good rigger, have a logical mind for sorting and distributing steel, and have a good memory for recalling where particular pieces have been landed temporarily.

A welding foreman should be able to weld in all positions so that he will know if the work is being done correctly. He should have sufficient knowledge of generators, transformers, or rectifiers—whichever are used—to instruct and check that his men are using them properly, as well as being able to recognize any malfunctions of the machines. He must be able to read welding symbols on the drawings in order to know whether the proper size and type of weld are being made. He must be able to recognize defective workmanship, and above all, be able to handle men and get them to produce satisfactorily.

On a large job it is advisable to have foremen in charge of the raising gangs, plumbing and fitting gangs, handline or detail gangs, bolting crews, and the welders. On a small job a foreman can usually take care of the plumbing and fitting gangs as well as the bolters. One foreman can take care of the handline or detail gang that erects small material left down by the cranes or derricks, or corrects errors in the steelwork, together with some of the other work. On a job with many cranes, it is advisable to have a general foreman to assist the superintendent. With a number of derricks working simultaneously, a "floor boss" or assistant superintendent can also be of great value.

To start, a minimum or skeleton crew is advisable, building up to the force needed, as required. For example, a normal guy-derrick gang includes a foreman, hoist operator, oiler (if several hoists are operating simultaneously), signalman, bullstick man or bullwheel operator, two connectors, and two floor men (one on the tagline, the other to hook on to the piece being raised). More than two connectors may be needed if heavy connections are to be made, if the floor heights are excessive, or if there is an excessive number of floors to a tier. The use of only two connectors to climb back and forth over excessive heights may delay erection so much that it is more costly than the wages of additional connectors.

A crawler, truck, tower, locomotive, or similar crane gang would normally consist of a foreman, crane operator, oiler, signalman (unless the foreman will give signals, but this is not advisable because it takes him away from his most important duty—the supervision of his entire gang), two ground men, and two connectors, with extra connectors if needed (as described above for the guy-derrick gang). In the case of the truck crane and the truck-mounted cranes, the oiler usually drives the truck in addition to his normal duties of lubricating the crane and filling the fuel tanks.

A stiffleg derrick gang or a gang operating a traveler would be similar except that when the bullwheel swinger is controlled by the hoist operator no bullstick man is needed. For moving a traveler it is usual to borrow extra men from gangs normally doing work other than raising steel when it is time to move the traveler. Thus these extra men, needed only for the move, are gainfully employed the rest of the time.

When erecting with a gin-pole, the gang would probably need only a foreman, hoist operator, two connectors, and two ground men. Here again, as in the case of the traveler, other gangs should be drawn upon for the extra men needed when the pole is moved.

Normally, a bolting gang consists of two structural ironworkers, with an apprentice ironworker or helper feeding bolts to several gangs, and a foreman supervising the gangs with an extra man (compressor engineer) often taking care of the compressor if pneumatic tightening wrenches are used.

If there are only a few gangs, a foreman can supervise the bolting as well as other operations.

Welding usually requires several men working individually doing the actual welding, with a foreman supervising the fitting-up and welding, and a gang of two or more men setting up and moving scaffolds for the welders, plus a welding engineer (operating engineer) or other man taking care of the welding generators, transformers, or rectifiers.

If the work is to be done by union workmen the local rules in some localities require additional men in the gangs, the minimum then being clearly stated in the union agreement. Under some agreements, a "compressor engineer" is required to start and stop the compressor and see that it runs properly. Similarly, with an electrically operated hoist, compressor, or welding machine, some union agreements require an electrician to throw the switch to start the motor, disconnect it at lunchtime and at quitting time, and take care of minor electrical troubles.

When working under union conditions, the superintendent should locate the hiring hall unless hiring is to be done at the site. He should check with the union business representative (B.A.) and arrange in advance for the men he will need, when they are to report, and where they can be interviewed to check their experience and probable ability. If the job is to be open shop, he should determine the availability of men, or notify those who have worked with him before that he wants them to come to work. He should check on the location of governmental employment agencies.

Men should be hired on the basis of their skills and experience, taking into account the actual needs of the job. If the erection is simple, less experienced or less skillful men can be used, but the final cost will probably be greater than if more experienced men were used. Connectors and riggers should be watched, when the work is close to the ground, to determine if they show the skill needed later aloft.

In hiring welders, if local ordinances require certification, the men should be checked to see if their certificates are current. In addition, it is always best to test welders in advance of the need in order to avoid delay later. The tests depend on the type of welding to be done, i.e., horizontal, flat, vertical, overhead, butt, fillet, high-strength electrodes.

If there are minimum-age laws in the area, men suspected of being minors should be carefully screened and perhaps be required to show birth certificates, baptismal certificates, or other documents for proof of age. But care must be taken in asking for proof of age so that the request does not conflict with discrimination laws pertaining to age.

As already mentioned, most superintendents have key foremen who in turn usually have a following of organized gangs. These should be utilized when feasible since organized gangs will work together much more harmoniously and efficiently than a gang that is assembled indiscriminately. Care must be taken in ordering men to come to work at the job site, especially if this means crossing state lines. The instruction to report should state clearly that hiring is to be done at the job site if the man is found to be satisfactory. Otherwise, the erector may be liable for workmen's compensation claim if a man happens to be injured in transit. If the contract or specifications stipulate restrictions on the men to be hired, this must be carefully checked. It is also advisable to determine any apprenticeship requirements as well as any applicable state or federal nondiscrimination laws in the area.

If cranes are to be used, unless the superintendent will use an operator he knows is capable, he should make sure that the operator hired is experienced in the operation of that particular type of crane. The operator selected should also be familiar with the type of erection involved, such as heavy lifts. His past record plus a check with previous employers, if he is unknown to the superintendent, will usually develop assurance of the operator's ability or his inability to perform the job in question safely.

In addition, as soon as the crane is in operating condition, the operator should be obliged to demonstrate his ability to operate safely, with no load on the load hook, and then with a light load. Later, before the operator handles a heavy load, the superintendent should see that he is competent with such a load—that he has checked his brakes, his frictions, his air pressure (if an air-controlled machine), and that he is demonstrating his qualifications to operate the equipment safely. Similarly, with a hoist for a derrick or traveler, the operator should be checked for his ability to handle the machine safely. All operators and signalmen should be familiar with the signaling system to be used—whether hand, bell, light, or sound signal system.

Regardless of what form of carrier delivers the tools and equipment, it is often desirable to secure a truck crane of sufficient capacity, or even a crawler crane if it can be obtained easily and ready to work, to unload and assemble the derricks, cranes, travelers, or other erecting rigs, and to unload shanties, small tools, and equipment such as compressors, hoists, or welding machines.

6.4   Shanties—Offices—Trailers

It is best to unload and assemble the shanties first, as these can then be used for storing small tools or material such as manila rope and electrodes that should be protected, and they provide the men with a place to change or eat or secure protection from bad weather. For a normal-size job there should be an office shanty for the superintendent, engineers, and time-keeper to work. On a bigger job, involving more engineering and a larger labor force, a separate shanty for the engineers may be advantageous. There should be a good-sized one for the men and as many as needed for storing material that must be protected from the elements or from petty thievery.

There should be facilities for a tool man to service small tools and sharpen drills or reamers, to prepare float and scaffold lines, or to do other work on tools as necessary, such as servicing pneumatic equipment or cleaning safety hats turned in by men leaving the job. When the general contractor does not furnish toilet facilities, small portable units can be rented that utilize the septic-tank principle. These are then serviced by the vendor.

In place of shanties, many erectors use office trailers that have one axle (Fig. 3.7.77) with two or four wheels and a support (landing gear) that is lifted out of the way when the trailer is hauled over the road while secured to the "fifth wheel" on a truck-tractor ("horse"). The landing gear is lowered to support the front end of the trailer after it is unhooked from the tractor. Some trailers used for job shanties are of the house-trailer type with two axles (Fig. 3.7.78); they are completely self-supporting and merely need to be moved from job to job or pulled around the site by hand or by truck.

The trailer can be loaded with desks, chairs, drawing racks, files, and can be wired for electric lights before being shipped from the toolhouse. It is thus ready to serve as an office immediately on delivery to the job site.

Similar office trailers are often used for the men and are equipped with benches and hooks for their clothes. Tool trailers can be equipped before shipment and have powered grinding wheels, vises, supply racks, and equipment in place when delivered. On some larger projects trailers of this type have been fitted up in advance as first-aid stations with medicines, connections for water and lights, and supplies in place.

Some erectors use covered automobile trucks with worn-out engines that are no longer able to function or are not worth repairing. These can be purchased very economically and can be rolled from place to place on the job where convenient to the work. They are pulled from job to job by means of a truck, which is tied up only for the time of the move. In many states such vehicles must have auto trailer license plates.

Using trailers eliminates the need to assemble knock-down shanties (Fig. 3.7.75) or the need for powered equipment to handle completely assembled shanties (Fig. 3.7.76), which are usually small, heavy, and of a fixed size.

The floors of the knock-down shanties are in sections about 4 ft wide and of a length equal to the width of the finished shanty, about 10 to 12 ft. The floor is unloaded and laid on level blocking, planks, or timbers, and the sections are bolted together to form the entire area to be used. The sides and ends are then unloaded and laid on the ground, with the bottom edges close to the floor on the proper sides of the floor. The ends are usually in two pieces that are then bolted together. The sides are also in about 4-ft widths, and as many as desired can be bolted together to make a space 10 or 12 by 8, 12, 16 ft, etc., long. After the side sections have been assembled they are tripped upright by "brute strength and awkwardness," held in place by the men, the ends similarly tripped up and the sides and ends bolted together. The roof sections, also in 4-ft widths (except the two ends that are wider so as to overlap the end supports), are next laid on top of the end and sides of the shanty starting at one end. As each section of roof is added it is bolted to the one before and down to the sides of the shanty proper. It is advisable to lay roofing paper over the roof sections to prevent rain, snow, or ice from leaking through between the sections.

Although the knock-down type of shanty gives area flexibility, it is costly to unload, handle, assemble, and later dismantle and ship out when compared to the trailer on wheels. Although the latter represents a higher initial cost, it will produce an overall saving and will be far more convenient in case a shanty must be moved as the job progresses. Trailer shanties are also available on a rental basis.

On a tier building, a shanty in one piece can be picked up bodily and placed on each new working floor. This gives the men a place to change clothes and to eat as well as to get out of the rain in the event of a short shower; otherwise, they are liable to "knock off" work and go home since they normally prefer not to continue work in wet clothes. On a tall building, unless there is elevator service, it is quite a hardship for the men to climb down many ladders and stairs to the ground at lunch time. The shanty on the working floor helps to eliminate this.

With the shanties in place, small tools sorted and stored ready for use, equipment such as hoists, compressors, and the like set up ready to work, and erecting rigs assembled and working, the steel should be in sight. Erection diagrams should be given to the raising-gang foremen to study and plan on how the steel will be distributed for the easiest, most economical, and safest erection sequence. Bolt lists or welding lists should be given to the respective foremen so that they can sort bolts or electrodes in advance and plan on how the correct sizes, lengths, types, etc. will be marked at the connections to guide the men doing the actual work.

As soon as the erecting rigs are completely assembled and in working condition, they should be thoroughly inspected and checked to make sure all parts are assembled correctly, all rigging properly made, all wire ropes in safe condition and of correct size and length for the use on that job, all splices properly made and tightened, sufficient lubrication where needed, fuel for engines lined up to be safely stored and arrangements made for safely fueling as needed, counterweights or tie-downs for stiffleg derricks or travelers in place, foundations for static, climbing, or traveling tower cranes adequate, and all equipment in good, safe working order. Slings and lifting devices should be checked and lined up to be ready when needed, the sizes and types depending on the material to be erected.

6.5   Grillages—Slabs—Base Plates

Normally while all this has been going on the engineers have been completing surveys to establish center lines on the piers, checking the location of anchor bolts, elevations of setting shims, screed angles, or leveling plates, ready for grillages, base plates, and slabs. With an auxiliary crane to unload tools and equipment and assemble derricks, cranes, etc., grillage material should be scheduled to be on hand at the same time. Then the crane can be unloading and setting grillages or slabs in approximate positions in between the times it may be needed to assemble the equipment. A crew of one or two structural ironworkers and an apprentice, with a resident engineer, field engineer, junior engineer, or surveyor, operating a level and/or transit, can be moving the slabs, grillages, or other base material into exact alignment and level.

When shims or screed angles are used these are brought to the correct elevations and grouted far enough ahead so that the grout of sand, cement, and water will have set sufficiently to hold them in place. Where the slabs are quite heavy, leveling devices are advisable instead of shim packs or screed angles.

While the grillage material was being put in place, as soon as the rigs are ready, the first steel to be erected should be ordered in and the unloading started. If a crane is being used to erect, the space in which the crane will have to travel in unloading, sorting, distributing, and then erecting must be noted, and no steel should be unloaded in those areas, unless the crane can back away to unload material to be erected at an earlier time. Then, as the area on which the steel has been unloaded becomes free (as the crane erects that material), the crane can reach the balance of the unloaded steel, erect it, and continue moving into the cleared area. With a derrick, this is no problem since the only area that cannot be used for unloading is the area around the foot of the derrick, which must be kept for the bullstick man or for the bullwheel to turn.

6.6   Starting Erection

For unloading from the carrier at the site, two eye-and-eye slings ("street slings") should be used, heavy (strong) enough and long enough to handle the drafts within the capacity of the rig at the reach involved. The slings are choked around the load at two points such that when the load is lifted it will be balanced and safe. To choke the load with a sling around it, one eye is passed through the other; the free eye is then hooked or shackled into the hoisting spreaders on the main load hook or block, preferably a safety-type hook. A tagline is then used on one end, or one on each end, to steady the load as it is lifted off the carrier and swung into the erecting area for sorting, distributing, and then erecting the various members in the load. The tagline controls the swinging of the load and, when the steelwork rises higher and higher, it also helps to prevent the loads from striking the structure. The line can be pulled in by the erectors at the upper levels to control the load on arrival at their working-floor level.

Loads should be landed on timber skids: 2 × 12 planks piled two or three high, 4 × 4 timbers, or blocking from loads as shipped in cars or on trucks. These skids permit easy removal of the unloading slings and easy placing of the slings later if the load must be moved. They also make it safer in case a man has inadvertently put his foot under a load as it is slacked down. Skids also prevent the slings from being crushed if the load were landed directly on the ground or working floor. Judicious placement of the skids on the working floor later helps make sure that the stresses from the loads are transmitted properly into the permanent supporting steel.

Steel beams, channels, and angles are normally unloaded in bundles.

Columns, if light enough, are similarly unloaded in groups; otherwise they are handled singly. Trusses and girders are best handled singly so that they can be stood up on blocking and immediately braced or guyed to prevent them from falling over. They can also be laid flat, preferably on level blocking so that they will not be distorted. Some trusses and girders may be so limber and unstable laterally that they should not be turned from upright (as usually shipped) to horizontal positions. In such cases the erection-scheme drawing will clearly indicate this.

As soon as the bundles of beams, etc., have been landed and the delivering carrier emptied, if time permits before the next load is delivered, an attempt should be made to sort out the load, placing material for each general, small area into separate piles. These are to be picked later to be moved into the particular area where they belong in anticipation of actual erection into the structure.

Some erectors like to have columns delivered ahead of filling-in material; others prefer the reverse. If beams, etc., are delivered first, sorting and distributing can be done more easily since no space is taken up by columns landed on the ground. The columns are then not in the way of swinging a boom to move material from an area where it was unloaded to an area where it belongs, which would be the case if the columns had been delivered and erected first. Then, when the columns are delivered later, they can be erected as needed, and erection of the filling-in material that has been distributed follows.

If a sorting yard has been set up, or if there is good control of material delivery, as in the case of delivery by truck directly from the fabricating shop (which is almost ideal since the material for a bay or a few panels can be delivered as needed), the four columns framing the four corners of a panel are erected and the steel framing between is erected at the various floors connecting to these columns. This permits the plumbing-up of these columns and fitting-up of the connections to proceed promptly, providing an area that can be planked and a place to which to set up ladders.

When a panel is erected, it is faster and safer to start erecting the beams in the adjacent panel at the level on which the connectors worked last; for example, if, in the first panel, beams framing between the four columns were erected on the first floor, then the connectors climb to the second floor to erect the beams framing between the same columns directly over the steel previously erected. Then the next steel erected should be at the second floor level in an adjacent panel. The connectors then climb down to erect the first-floor steel in that panel. They stay at the first floor to erect again at that level in an adjoining panel. This saves additional climbing for the connectors, saves time, and will result in safer and quicker progress. Several panels can be erected on one floor before requiring the connectors to climb to the next floor where they would erect steel directly over those panels before proceeding to erect elsewhere.

Planking of the upper floors erected proceeds as quickly as possible, using 2 × 12-in. or 3 × 12-in. planks to span two or three filling-in beams or temporary support beams, but of a length light enough for two men to handle safely yet strong enough to bear the loads to be imposed on them. Erecting continues while planking, plumbing up, fitting up, and bolting or welding are being done in the areas erected. When possible, the permanent connections of the upper floor should be completed before a guy derrick or climbing crane jumps (if such equipment is being used for erecting). The connections of the floor below can be completed later. In any event the connections of the members supporting the rig after it jumps and the connections transmitting the stresses from the bottom ends of the guys to the foot of the mast (in the case of a guy derrick) must either be permanently fastened or temporarily secured before the derrick is permitted to operate again after jumping to an upper floor.

Steel should be delivered by derrick or crane areas, split when feasible into smaller subdivisions, as between guys on a derrick or into sections for a crane-erected building. This can be achieved by the fabricating shop cooperating in their shipments or by using an unloading and sorting yard at the delivery point.

With truck delivery and several rigs erecting, it is advisable to schedule the work so that only one rig is unloading at a time while the others are erecting. This enables the trucker to concentrate on delivering an entire tier or area to one rig quickly. Then that rig can start an entire area or tier without being delayed by the trucker feeding steel elsewhere on the job. By the same token, the delivering carrier or the superintendent in the yard should be clearly informed what steel is wanted, and when, and how fast.

After loads of steel have been delivered, the material should be distributed so that each piece is approximately under its final position in the structure. At the same time, by checking the erection diagram, any pieces missing can be ascertained, and if they are vital, arrangements should be made to have the missing pieces rushed to the site in time for their erection as needed. If they are not vital the shop can forward these pieces in a later shipment, and they can be erected by the detail or handline gang by hand or with powered tackle. If any pieces are damaged, a prompt report should be made to the carrier for later claim for the cost of repair or replacement. If the damage can be repaired easily at the site, this can sometimes be done before the piece is needed in the erection sequence.

Sorting is usually done by using lightweight sorting spreaders consisting of a pair of eye-and-eye slings with a sorting hook in one end of each, either spliced into the eyes or shackled to them. These hooks are designed so that the point can be placed in a hole in the end connection, or the hook itself slipped around the web and under the top flange of a beam or channel (Fig. 3.7.47). The other ends of the eye-and-eye spreaders are shackled together to the load block hook, preferably a safety-type hook. With this arrangement, a beam can be picked out of a load and laid in a pile to be moved close to its correct location; there, the individual pieces are again picked out and stood right side up under their location in the structure. At that time, the erection mark should be added to the top flange so it is readily visible when erection actually begins. Yellow or black keel is usually used for this purpose. Space should be left between the pieces for the erecting slings to be placed around them later.

To erect, an eye-and-eye sling is ordinarily choked around a piece at its center with a tagline on one end. Then the piece is lifted to the connectors waiting for it on the columns or other steel, and the tagline is used to guide the piece to the connectors. The line is pulled down when the piece should be tipped to clear the steel on a lower floor already in place on its way to an upper floor.

Care must be taken to use a sling that is light enough so it will grip the piece as it is tipped, tight enough so that the piece does not slide through, and yet strong enough for the weight being handled. A single pendant of wire rope with an eye in one end to fasten to the shackle on the overhauling ball, and with a safety hook spliced into or shackled to the lower end, is a great help in safely erecting a two- or three-floor tier area. The pendant should be long enough so that when the lower end is at the working-floor level, the overhauling ball is still above the top-level steel being erected in that tier; in this way, the ball does not have to be lowered through the steel.

With a guy derrick or a climbing crane erecting a tier building, the first panel erected should be at the back corner. Then proceed to erect the adjoining panels across the back, followed by the panels on each side of the area. Next the panels between the rig and the back of the structure are erected, and finally the panels across the front and toward the derrick or crane. When all the steel in that tier that can be erected without interfering with jumping has been erected, the rig is ready to jump. The exception to this sequence occurs when two rigs are working next to each other. Then the first steel erected should be those panels adjacent to the dividing line between the two rigs.

On the large derrick job, if the shop is unable to keep up with several derricks working at the same time, one gang can be used on two or more derricks to complete the erection of a tier with one derrick up to the point of preparing to jump. Instead of jumping they will tie up this derrick and move to the second, erecting that steel until that derrick has been filled in ready to jump. Meanwhile, the bolters or welders have been permanently fastening the steel around the first derrick and planks have been placed on the upper, new working floor. The gang then returns to jump the first derrick and proceeds to erect the next tier with it. Meanwhile the steel around the second derrick is being permanently fastened.

This procedure will help to expedite the final completion of all work since it permits the permanent connections of the first area to be completed while the raising gang is working elsewhere. In this way the floor planks to be placed on top of the completed area can be kept off longer to permit installing the permanent fastenings in the connections on the upper floors without having to move floor planks to expose the connections, as might otherwise have to be done. This is not only more economical but safer. With such a procedure steel must be scheduled so that steel for the correct derrick arrives in proper sequence and at the right time.

Steel for tier buildings is normally divided into two-floor tiers, except for the last tier, which is often three floors to save jumping to erect just one floor, which is usually of a much smaller area than those below. In the case of three-floor tiers care must be observed in distributing this additional floor of steel on the working floor to be sure the permanent structure is not overloaded by the extra weight. This usually requires setting columns directly instead of landing them on the working floors, and landing bundles of steel well away from the midpoints of the supporting steel members. In comparison with two-floor tiers, three-floor tiers usually take longer for erection, are less safe, involve more danger of overloading floors, and require more steel to sort and distribute in the same area.

When erecting a tier building with a crawler crane, truck crane, or crawler- or truck-mounted crane, it is usually necessary to erect all the panels across the end bay of the building away from a delivery ramp or the delivery entrance to the site. This bay is usually erected from ground to roof, and the crane backs away to erect panels across the next bay, backing away each time toward the delivery end of the site. If the terrain is such that a crane can work on two sides of the site, two cranes are sometimes used to reach the centerline of the structure, erecting from the two sides simultaneously, or one crane will erect one-half of the structure from one side of the site and then proceed along the other side to erect the remainder of the structure.

A mobile tower crane standing clear of the building line with a boom long enough to reach clear across the structure can erect an entire tier before starting to erect the next tier above. The advantage of this type of equipment is the elimination of erection loads on the permanent structure. This type of crane also permits completion of an entire tier, including raising, plumbing up or aligning, fitting up, and permanent fastening. A static (fixed-foundation) tower crane has this same advantage, but it is not as flexible as a derrick or mobile equipment and its use is limited for most steel-frame erection.

When ground conditions alongside the structure permit, a traveling tower crane mounted on wheeled trucks running on rails laid on the ground close to the building can be used instead of crawler or truck cranes, mobile tower cranes, or static or climbing tower cranes. With a derrick mounted on a fixed tower, or the tower crane mounted to roll on rails alongside the structure, the sequence of erection can be varied somewhat from either guy derrick or mobile crane erection. Generally, an entire bay at the end of the structure is erected first, followed by adjoining bays right down the length of the structure, but erecting the steel by tiers. The panels across the back of the building can be erected for a length that can be covered by one position of the tower crane. This is followed by the line of panels in front of that which has been erected, and finally closing in across the front.

The tower is then moved to a new position to erect a similar area next to the completed one. Generally this is done by tiers so that entire floors or large portions of floors can be given over to other trades to follow with their work. Upper tiers are subsequently erected in similar fashion, tier by tier. The exception would be when one position of the tower crane or derrick on a tower permits erecting a substantial area sufficient to provide a more-or-less self-contained section of a structure.

Frequently a series of smaller areas may be designed with bridges between at several levels. In this case each area, excluding the bridges, is virtually a separate building. In such instances it is more economical and better for the general contractor to have the steel in each of the smaller units of the building erected complete from ground to roof. This saves moving the tower crane more than is absolutely necessary. A derrick can erect an entire section when a tower has been set up to support it. Then, by means of skids of heavy timbers or rolled steel sections laid flat, the derrick can be skidded or moved on rollers to a new position and the next area erected complete, repeating the moves as often as necessary. This method has many advantages for a welded structure or for a lightweight structure that cannot support a guy derrick or climbing crane.

With either the fixed or the traveling tower crane, each tier can be delivered to be unloaded and sorted on the ground, raising each piece singly for erection into place. This procedure is very slow, expensive, and inefficient. Instead, as the steel is unloaded, it can be landed in bundles on a working floor aloft on the steel already erected, there to be sorted, distributed, and erected far more efficiently, speedily, economically, and safely.

While a climbing crane can erect like a guy derrick, it ties up more of the floors needed to support, brace, and steady it than are required by a guy derrick. In addition, this type of crane requires a considerable amount of electric power, which is not always readily available. The cost of setting up and dismantling such a crane is probably more than that for a guy derrick, but the cost of jumping can be less, although auxiliary equipment is usually required, which can be burdensome. In addition, whereas a guy derrick can be set up initially on the ground and immediately erect, a climbing crane presents a real problem in starting erection and until it can build a structure around itself to use as an anchorage and against which to block itself. A special concrete foundation with anchors is occasionally necessary on which to set up the static or the climbing crane.

6.7   Erection by Locomotive Crane

If a locomotive crane is to be used, after it has been assembled with boom, jib, and reeving in place, the keeper plates holding the trucks to the bolster pins between the crane deck and the trucks should be checked to be sure they are in place before starting work. Similarly, they must be removed on completion of the job before shipping the crane away. In addition, the entire crane should be checked to make sure the boom sections have been correctly assembled, the splices tightened, the reeving correctly installed with safe wire rope of the proper diameter, the engine in working order, lubricated where needed, counterweight in place, brakes functioning correctly, kingpin and rollers in good condition, track clamps installed, outriggers rolling in and out easily, trucks and bearings in working order and lubricated, and traveling mechanism functioning correctly.

The locomotive crane is of value only where a siding is available or can be installed from a delivering railroad to tracks in a yard for unloading or into the building area itself. Laying ties and ballast and installing tracks can be done by the railroad for a fee, or by a subcontractor employing labor whose wages are less than those of ironworkers, or by the erector with his own forces. In any case the tracks should be laid into the area where the locomotive crane is to work and should be checked first by a visual inspection and then by running the crane over the tracks slowly, with no load in the falls. A second run should then be made carrying a light load. Finally, when a heavy load is to be handled, the track, ballast, ties, rails, splices, etc., should be inspected to be sure that they are in safe condition. Then the crane is moved slowly over the rails, and any unusual sagging or sinking of either or both rails is watched for. If any defect is observed the crane should be stopped immediately and remedial action taken. If there has been an excessively heavy rainfall during the time the job was temporarily tied up, such as over a weekend, this same procedure of inspection and trial run should be followed if there is any possibility whatsoever of the ground under the ties having been weakened by the rainfall. This is essential in order to avoid the possibility of the crane over-turning as it moves over a soft spot or the crane causing the track to sink excessively.

Heavy steel can be erected quite efficiently with a locomotive crane, even if the crane erects only the lower, excessive weight material. Lighter capacity derricks can then be set up on this erected steel.

For an open-hearth type of building, where permanent tracks will be laid through the building and connect with an operating railroad, loco-motive-crane erection is usually efficient and economical. Many of the girders and columns are too heavy for an ordinary crawler or truck crane operating on the ground. These pieces usually exceed 100 tons each. The locomotive crane can erect up to the charging floor and then tracks are laid on the floor. The columns and girders can be brought in—still on the de-livering cars—on this level to be unloaded and set in place directly by the crane. It is often necessary to leave the jib off and use a short boom to permit lifting these heavy pieces safely within the capacity of the crane and boom.

When the heavy girders and columns are all set, the boom can be laid down as for the original assembly, the necessary splices disconnected, extra boom sections installed, and the jib put in place. This gives the additional reach needed to erect the roof steel. The crane starts erecting the roof at the end of the building away from the siding, erecting and filling in and backing away as it erects, finally moving out of the completed building, to be dismantled and loaded for shipment back to the toolhouse or to the next job.

For the most efficient use of this method of erection, the steel should have been loaded so that a minimum amount of sorting and distributing is required at the site; whenever possible the material should be lifted out of the delivering cars and raised directly into place, especially the heavier girders and columns. Small material such as roof purlins, bracing, or charging floor material and the like will have to be unloaded and spread out to erect the individual pieces.

Regardless of the type of equipment selected to erect the structural steel frame—whether derrick, traveler, or crane—the basic method involves delivery of steel, sorting it into areas, distributing or moving the sorted steel into areas close to final position, and then raising each piece or assembly into place. Steel is delivered as described, either directly from the fabricating shop or from an unloading or storage yard.

6.8   Mill-Type Buildings

If the building is a mill-type or a long building to be erected by a mobile crane, columns can be delivered for erection along one side, either part way or the entire length, together with bracing and crane girders or other main, longitudinal framing. Then the same type of material is delivered for the other side. Between loads the erector should endeavor to erect the columns as fast as they are delivered, upending them and setting them in place. If the slabs are shop-assembled to the columns, shims or screed angles or setting plates should be set far enough in advance to receive the columns and slabs, unless the piers have been brought to finished grade, thus eliminating the need of shims, etc. If the slabs are loose they should be set far enough ahead to be grouted before the columns are placed on them, especially if there are no anchor bolts, so that the grout will help hold the slab.

When there are no anchor bolts, the columns should be temporarily guyed with manila ropes to prevent being knocked or falling over, unless they are so heavy and wide that they will be stable without such guying. Next, the bracing should be erected between columns, and crane girders and longitudinal connecting material erected. By following this sequence the ground has been kept clear for easy operation of the crane erecting the two sides of the structure. Next, the framing across the building should arrive to be sorted and laid approximately under its final place in the building, but space or aisles should be left for the crane or cranes to travel.

An ideal setup, if there is enough steel involved to warrant it, would be to have two cranes, each erecting one side and later one-half (longitudinally) of the building, with a third crane unloading ahead of the erecting rigs and laying out the steel for speedy raising. In this way the entire area can be used for unloading since the unloading crane will back away over the unused area as it unloads, and the erecting cranes will clear paths to operate as they pick and set the material ahead of them.

With a derrick mounted on a traveler platform, columns, beams, girders, trusses, etc., must be delivered to permit erecting one or two bays across the structure, depending on the area over which the derrick boom can operate. As the bays are completed, the traveler moves back to unload, sort, distribute, and erect subsequent panels and bays, always moving toward the point where it can be dismantled and loaded out. When the derrick is mounted on a tower on the traveler platform, it has a far greater reach with the same length boom than it would have if not mounted on the tower. In this way a large portion of a tier can be erected equal to the area the boom, high in the air, can cover before erecting the next tier above. While this may not give complete floors, it does give the general contractor or other trades far greater areas of the lower floors to work on, as the traveler erects above, before moving back.

With hangars, train sheds, and convention-hall types of structures, the roof trusses or arch ribs are usually unloaded ahead by auxiliary equipment, to be assembled into either entire or partial sections. These are laid on blocking arranged so that the individual parts will conform to the predetermined shape, and the permanent fastening of the parts is completed on the ground. These sections can be assembled in such places alongside the traveler so that the steel between can still be delivered to the traveler without fouling these assemblies, and so the traveler can pick them within its rated capacity. Then either falsework is erected or the tower of the traveler is used for supporting the sections temporarily. The engineers should have established the exact heights of blocking or shims or jacks well in advance so that the sections, when landed, will make a complete truss or arch at the correct elevation when all parts are connected. When entire trusses or arches have been preassembled before picking, it is important that they be assembled correctly to the camber and shape specified on the drawings.

6.9 Ladders

Ladders should be on hand to set up as soon as any steel is erected in a tier, in order to provide safe access between the ground or working floor and the upper level of steel erected. They should be stood on solid ground or on safe planks with enough planks for a man coming down to land on safely. Similarly, there should be enough planks on the upper level for a man to step on safely upon reaching the top. Connectors are trained to be able to climb columns for moving from one level to another, but for safety, economy, and efficiency, ladders should be set up for their use and the use of other men as promptly as possible after starting the erection of each new level of steel. The ladders should be secured at the top to the steel so that they cannot be dislodged or slip off their supports when a man is climbing, or be knocked over if inadvertently struck by a swinging piece of steel. They should extend above the working floor to permit a man coming down to grab hold of the sides before stepping onto the rungs.

In some cities and states the minimum distance that ladders must extend above the upper floor is definitely specified as a legal requirement, and this should be taken into consideration when ordering ladders for the job. On a tall building, if the permanent stairs are not installed as soon as the raising, plumbing up, fitting up, and fastening of a tier has been completed, ladders must be left behind until the stairs are installed or elevator serivce is provided, or until the general contractor provides other means for the structural ironworkers to reach their work.

6.10   Signals

With crane erection and with the start of derrick erection the signalman is usually within sight of the crane or hoist operator. Later, when the derrick or climbing crane has jumped to an upper floor, this is no longer the case and signals are then transmitted by the signalman to the hoist operator by a bell, light, or voice system. When the signals can be given directly, visually, a standard set of hand signals is customary.

The signals commonly recognized in steel erection are shown in Appendix B, Table 12.

6.11   Plumbing

As soon as the first panel of steel is erected, the plumbing-up gang must get busy installing plumbing-up guys, if needed. A properly run toolhouse will ship guys ready for use. Otherwise, the field force must take the coils of wire rope, $\frac{5}{8}$ in., $\frac{7}{8}$ in., or 1 in. in diameter, depending on the requirements, form a loop on one end around a special hook and install wire-rope clips to hold the loop. Similar hooks are inserted in one end of eye-and-eye turnbuckles of a size to match the wire rope guys, which can be new or old rope provided the latter is strong enough for the purpose. The wire rope is cut to the length needed.

The hook, often called a "crow's foot" because of its shape, consists of a rod bent into a loop with the ends brought close together and then bent over about 1 in. from the ends to slightly less than a right angle to the plane of the loop (Fig. 3.7.64). The bend is such that it can be slipped over the edge of a beam flange, or over the guy itself, and held in place against the pull of the turnbuckle on the guy.

The hook on the guy is either placed over a beam flange on an upper floor, or the guy is wrapped around a column at the upper floor and the bent end of the hook is laid over the guy. The turnbuckle hook is placed over a beam flange on a lower floor, generally one tier below or at the working floor, one or two panels away. With the turnbuckle extended almost its full length, the free end of the guy is slipped through the free eye of the turnbuckle, the guy is pulled tight by hand, and wire-rope clips are installed to hold the wire-rope guy in the turnbuckle eye. A second guy is then similarly installed in the opposite direction to form a vertical, diagonal cross (Fig. 6.11.1), the upper end at an adjacent column at the same upper level but directly over the lower end of the first guy. The

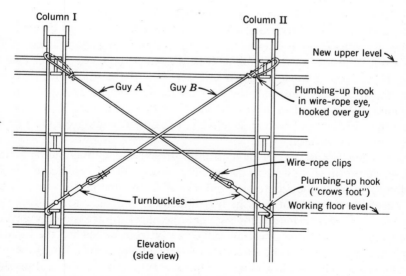

FIG. 6.11.1. Arrangement of plumbing-up guys. Taking in (tightening turnbuckle) on Guy **A** and slackening (loosening turnbuckle) on Guy **B** will move the tops of Columns I and II toward the right. Taking in on Guy **B** and slackening Guy **A** will move the tops of the columns toward the left.

second turnbuckle is hooked in place at the lower level directly below the upper end of the first guy.

Then, by tightening one turnbuckle and loosening the other, the tops of the columns can be moved one way or the other after the connections between them have been pinned (with drift pins) and the erection bolts installed.

A lightweight plumb bob on the end of a piece of chalk-line, or a light flexible wire with a heavy bob, is lowered from the upper level to the lower level. It is held in place away from the face or web of the column at the upper end by a rule that the plumber-up man has notched at a fixed distance (such as 6 in.) from the end. As he holds the end of the rule against the column with the string in the notch, the man on the lower floor reads the distance from the line to the same face of the column. If it is the same as the notch distance in both directions, i.e., east-west and north-south, the column is considered to be plumb. If it is more or less the turn-buckles are either tightened or loosened respectively to pull the tops of the columns until they are plumb. If the columns have been placed accurately on their base slabs originally, then both columns using this set of guys should theoretically be plumb and in their correct locations. Since this is not always the case, each column must be checked separately.

Under windy conditions causing the bob and chalk-line or wire to sway back and forth, a pail of water can be used around the bob to dampen the sway. In some cases it is easier to set up a transit in the street and sight on a rule held against the column. If held against the face, the distance to the column center must be taken into account. If the rule is held against the web of the column, half the web thickness must be added. If all the columns being plumbed have the same depth between column faces or the same thickness of web, respectively, the direct reading on the rule by the transit can be used.

The transit must be set at a definite distance from the building line. Then, based on this distance the reading on the rule is calculated for the columns to be in their correct positions. With this method it is necessary to check the corner columns in two directions and then use a steel tape between the intermediate columns since the check by transit only ensures their plumbness in one direction, either away from or toward the building line, at right angles to the direction of sight of the transit. Since the mill is allowed a tolerance in rolling column sections, and the fabricating shop is permitted additional tolerances by most specifications, it may be impossible to locate the columns exactly where they should be. Accordingly, there must be some tolerance for the erector in plumbing the columns.

Ordinarily the permissible tolerance in mill, fabrication, and erection will balance each other instead of being cumulative.

The contract or specifications will usually stipulate the tolerance permitted or will indicate compliance with a code such as the American Institute of Steel Construction (AISC) "Code of Standard Practice," which permits a deviation from plumbness for the columns, and out-of-level for the beams, girders, etc., depending on whether the columns are outside face, elevator shaft, or intermediate columns. The important columns to plumb accurately are those on the face of the building and those around the elevator shafts.

When a metallic skin will be installed on the face of a building, the plumbness of the outer columns becomes increasingly critical, and the superintendent must ascertain the tolerance he will be permitted in plumbing those columns.

Normally, as soon as the floor steel and the column splices have been permanently fastened, the plumbing-up guys can be removed. Because some structural frames are so light, or because heavy winds are liable to be encountered, removal of the guys before the permanent floors are installed may rack the steel structure out of the plumbness achieved by the guys. In such cases, and also where the stiffness of the structure will be secured only after the permanent floors are in place, the guys should be left in place. Then the general contractor should be instructed to have the guys boxed where the permanent floors would otherwise encase them or their fastenings to the steel.

On some buildings, contracts are awarded at different times so that the steel erector may complete his entire work before any floors or walls are even started. Then, it may be highly advisable, especially if the structure may be unstable without the guys, to leave all guys in place and notify the owner or the erector's customer that he must be responsible for protecting the guys, turnbuckles, and hooks, and cautioning him that tampering with or removing the guys too soon may cause the structure to rack and move out of its correct position. Usually, under these conditions, it is the responsibility of the owner or his general contractor or subcontractor to remove the guys and turnbuckles at his own expense and return them intact to the erector's toolhouse, reimbursing the erector for any damaged or lost material.

On completing the plumbing of each tier, it is always advisable to record the exact location of the tops of the columns on the outer faces of the structure and around the elevator shafts, submitting such a record to the customer to establish the locations of the columns as left by the erector at the time. Then, if subsequent contractors cause any dislocation through improper forming or pouring of floors, or the use of equipment that might cause movement of the steel through no fault of the erector, the record has been established to show that the erector fulfilled his obligation within the permitted tolerances.

6.12   *Fitting*

Sometimes the steel has been fabricated a little long or a little short. If long, the erection bolts may have to be removed and smaller-diameter bolts installed. The drift pins are removed and the columns pulled in by the guys. If the steel is short, again, smaller diameter bolts may be needed. Then, by wedging between the ends of the connecting beams and the columns, and slacking the guys, the columns can be pushed out to where they should be. In either case the bolts are then tightened, the connection holes reamed, and the connections permanently fastened. In the case of bolting, the holes reamed are the permanent holes, and oversize bolts will be needed to fill the reamed holes. If the permanent fastening is by welding, the fitting-up bolts will be in the erection holes. Normally the holes will not need to be reamed since the welding will hold the connection in its shortened or lengthened position.

To complete this permanent fastening, if the holes match without further treatment, fitting up proceeds after plumbing or coincidentally. If the holes are only slightly mismatched, the fitting-up gang will use a reamer to "spear" the holes, the reamer usually being powered pneumatically or electrically. With this reamer, the slightly mismatched holes will be faired enough for permanent bolts of the correct size and diameter. When the holes do not match properly because of fabrication discrepancies, the fitting-up gang will ream the holes to a larger diameter. As soon as the holes have been matched, fitting-up bolts and drift pins are used to pull the plies of steel in the connection together.

When an excessive amount of reaming becomes necessary due to fabricating or drawing-room errors, it is usually advisable to notify the fabricator, since the cost of this excessive reaming can be so great that a backcharge against him may be desirable. In such a case he is entitled to observe the condition before it is corrected; this will avoid any argument later on as to the actual need for the excessive reaming backcharged. It is also advisable to notify the fabricator in case of any errors in fabrication or on the drawings so that he can send a representative to inspect the conditions. He may elect to specify how the condition shall be remedied, since he will probably be responsible for the corrective work. If the connections are butt connections, shims may be needed instead of reaming if the steel is short. If it is long some connections may have to be removed and replaced, or reaming may have to be done before replacing the connection angles or welded plates.

With derrick or climbing crane erection, plumbing up should be done promptly so that permanent fastening can proceed before the rig is jumped, since the permanent members supporting the rig and the members between the bottom ends of the guys of a guy derrick and the foot of the mast must be thoroughly secured in order to transfer the stresses produced in working the derrick or crane. With mobile crane erection, plumbing up should also proceed promptly so that the permanent fastenings can be installed to hold the structure safely and correctly in place.

6.13 Bolting

In the case of high-strength bolting, the gangs frequently perform the fitting-up operation as part of their regular work. In that case they will usually be equipped with a pneumatically or electrically powered impact wrench as well as a powered reaming tool. A pattern of holes is selected, starting near the center of a connection (if there are many holes) and working out to the edges, tightening enough scattered bolts to bring the various faces of the different plies of steel into close contact. Drift pins are driven through the remaining open holes, and permanent bolts are then installed in those holes. If the fitting-up bolts used in bolting were some of the permanent bolts, they are then tightened. If not, they are removed and permanent bolts inserted in their place and tightened.

An approved procedure should be used with high-strength bolting. This can consist in tightening each bolt to a snug position, that is, hand tight, or with an impact wrench used when the socket ceases spinning and starts to be impacted. Then an additional one-half, three-quarter, or one full turn of the nut or head of the bolt (depending on which is being turned) is completed. The amount also depends on the diameter and length, in accordance with the current specification of the Research Council on Riveted and Bolted Structural Joints of the Engineering Foundation, or as required by the contract specifications. By marking the socket at various equidistant points, the correct amount of turn is easily checked.

An alternate procedure permitted in the Research Council specification involves the use of a testing device, usually once or twice a day, to tighten a bolt of the same type and diameter as those to be installed in the structure. The impact wrench to be used later is used in the test, and a control valve with a gauge in the line is installed to control the pressure of the compressed air at the wrench. This pressure is reduced until the tension in the bolt in the testing machine is that which is specified for that diameter and length of bolt when the wrench stalls. By means of the gauge, bolts of several different diameters can be tested, the gauge reading being noted for each diameter at the time. Then, when actually installing a particular diameter, the valve must be set to give the correct air-pressure gauge reading.

If an impact wrench with an automatic cutoff valve is tested, the cutoff point is adjusted for the wrench to produce the required tension in the bolt in the testing device. (This type of wrench is known as a "calibrated wrench.") The objection to either of these calibrating methods is the need to change the pressure or the cutoff point for each different diameter or length of bolt.

The "turn-of-the-nut" method is preferable since the amount of turn is specified and the machine does not need to be tested each day, nor the air pressure changed repeatedly. Furthermore, it permits using the full air pressure which usually results in better impacting of the bolts. The owner's or customer's inspector usually tests a certain percentage of the tightened bolts to verify that they have been tightened to the required tension.

As a check on the work done by the high-strength bolting gangs, a good foreman will test some of the bolts himself by means of either a hand torque wrench or an impact wrench he has previously calibrated and set to the required pressure for the bolt diameter and length. By backing off the nut or head, whichever was torqued, he then brings it back to the point where it was, reading the hand torque wrench or checking that the impact wrench he is using will just bring the nut or head back to that point. Occasionally a good man is able to use a long-handled open end wrench and by feel (which comes with experience) check the tension.

Another method uses specially designed washers called "direct tension indicators" with a group of protrusions pressed out of one of the two flat surfaces. When the tension in the bolt has reached the desired value, these protrusions will flatten somewhat and should leave a small gap under the head or nut, depending on which has been tightened against the washer. Inspection is then made by means of a thin feeler of specified thickness, which should just be able to penetrate the gap.

This method permits a visual inspection of the tightened connection, eliminates the need for the foreman or the inspector to use special wrenches in checking the tightened bolts, and helps the bolter-up know when the desired tension has been achieved.

By assigning identifying symbols to each gang and having them mark the points they have bolted with their symbol (using a marking crayon or paint-stick), the foreman or the inspector can identify which gang is responsible if for any reason the bolts in a particular point have not been properly tensioned. Whenever feasible the gang (usually two men) should be assigned to tightening the same type of joint. Thus, they can become more proficient in that particular operation and the joints will be properly tightened.

The men should be trained to store their impact wrenches in protective toolboxes or in a shanty overnight and weekends to protect them from the weather and to discourage theft. They should also be taught to blow out their ½-in. leader hose (tail hose) from the ¾-in. main air hose to the wrench itself before connecting it to the wrench. The tail hose should be kept as short as can be used efficiently to help reduce air-friction losses. The men should be shown how to inspect and clean the air inlet screens when needed and to lubricate where required. When a wrench starts to produce less power than normal, it should be returned to the toolhouse or the manufacturer for proper servicing, rather than having the men who use the wrench try to repair it.

The foreman should check to see that his high-strength bolting gangs are getting an adequate supply of compressed air. If the supply line between the compressor and the work is long, an air receiver should be set up close to the work. This will help keep the pressure up to that which is required for good workmanship. The size of the compressor and of the receiver depend on the sizes and number of wrenches and any other pneumatic tools being used.

The wrench and hammer sizes in turn depend on the diameter of the bolts involved. A good foreman will arrange to mark the steel at each point before the gangs start there, giving the type, diameter, and length of bolts to be installed, and noting if washers are to be used under the head or nut or both (depending on the slope of the steel, on whether the head or nut is to be torqued, and on the material of the bolt). He should indicate when it is important to have the head on one side or the other of a connection and when the head should be turned instead of the nut.

He should organize the distribution of bolts to the bolters, to the fitters, and to the raising gangs when the connectors can use the permanent bolts in setting and connecting the steel. This saves replacing erection bolts later with permanent bolts, and avoids the possibility of leaving the wrong type of bolt in a connection. He should educate the inspector as to what method will be used in tightening, as this often does much to prevent the inspector from making unreasonable demands in questioning whether or not the proper tension has been achieved. Arrangements should be made for inspection before a scaffold or float is moved, to avoid an inspector demanding that a scaffold or float be rehung for his inspection after a gang has completed a point and moved away.

Normally a bolting gang will require the services of two structural iron-workers since the man operating the wrench cannot generally reach the other side of a connection to install the bolt and hold the head if the bolt should start to turn as he tightens the nut, or if the nut turns while he is tightening the head. In addition, if he is driving drift pins through holes that are only slightly misaligned, his partner can catch the pin if it comes through too quickly and thus prevent it from flying out or falling, thereby causing an accident. Ordinarily the nut is tightened, but occasionally the job requires the head of the bolt to show on the only side from which the bolt can be tightened, due to architectural or other features of the connection.

When the bolt will be exposed to view after the structure is completed, some engineers or architects insist on the heads of the bolts being placed on the exposed side. In some instances the nut and part of the bolt shank could extend through the concrete or other covering, and then the head must be on the side of the connection that is to be concreted in order to reduce the height on that side. This requires tightening the head on that side if the other side is inaccessible for the wrench to be operated.

With a high-strength bolting gang, if one of the two men does not report for work, a new man can be trained in a matter of hours, and production, although it may be slowed temporarily, is not completely lost.

Bolting gangs usually work from floats or needle-beam scaffolds. Only one pair of needle beams is generally needed at a point to support the planks laid across them if they are slung in such a way that there is support on both sides of the connection for men to work on each side. With floats, two are usually required for safe operation, one hung on one side and the other hung on the opposite side of the connection. Floats are not always needed for bolting since the men can often sit on the steel itself, install the bolts, and operate their wrenches.

A good foreman will check his gangs to make certain that they know how to hang their needle-beam scaffolds or their floats safely, using proper knots, and he will see that the men inspect their ropes frequently. The rope lines must be adequate for the loads, with some extra reserve strength in case a rope is cut or damaged inadvertently while the men are on the scaffold or float.

For compressed air use, ½-in. tail hoses should be readied, and there should be enough ¾-in. main hose on hand to reach all parts of the work from the receiver or directly from the compressor. Impact wrenches should be serviced ahead of time and checked to make sure they are properly lubricated, the correct size chucks are on hand and correctly secured, and the men know how to use the wrenches properly.

On a tall building where the compressor must be left below, a 2-in.-diameter air hose or pipe is used between the compressor and the receiver on the floor where the work is being done. Additional sections of hose or pipe are added as the receiver is moved up to each new floor. The compressor itself should be located where it is protected from falling objects, welding sparks, or burning (or cutting) slag, or some form of protection should be provided over it.

6.14 Welding

Welders are normally tested to qualify for overhead, vertical, and horizontal welding. If a welder does not report to work, he probably cannot be replaced until a new man is tested and hired. The testing is usually done at the job site and the test specimens sent to an accredited testing laboratory for bending, breaking, etc., as required by the test generally described in the job specifications. The usual test is the one prescribed by the American Welding Society specifications.

If no test is stated as part of the contract, it would be advisable for the erector to establish certain minimum requirements of his own to be sure the men are qualified to make adequate weldments. When alloy or high-strength steel is involved a welding procedure should be set up well in advance, possibly welding test strips in the erector's toolhouse. The men must follow the qualification procedure found to be satisfactory for the particular joint and steel metallurgy.

In some cities local ordinances require welding to be done by men who have been certified by an authority set up for the purpose. Welders who pass the specified tests are given certificates authorizing them to perform welding on structures in that city. Some erectors give certificates to welders who have passed tests on one of their jobs, indicating that they have passed tests satisfactory to the erector. Then, for a specified time these men are not required to pass new tests on future jobs for that erector. As in the case of bolting, welding can often be done without scaffolding, the welder simply working from the steelwork itself. Otherwise, floats or special platforms can be hung or set up. Because the men doing the actual welding are so highly skilled, it is usually more economical to have a gang swing the necessary floats or platforms in advance for them to work from instead of delaying the welders to do this work. These floats or platforms are then removed by this auxiliary gang.

For best production the welding generators, transformers, or rectifiers should be as close as feasible to the work being done. They should be stationed on good, solid planks, raised high enough off the ground by blocking to prevent a heavy rain or water from melted snow or ice from reaching them, which could create a dangerous electrical ground. Men have been severely shocked by standing on wet, grounded planks and touching some part of a generator or electrical contact point.

A good ground cable must be secured from the machine to the steelwork and protected if there is any chance of its being dislodged or damaged. The lead cable should be of the correct size for the current being used, and it should be well insulated to prevent accidental short-circuiting to the steel itself. The welding foreman should watch that the correct type of electrodes and the corresponding correct polarity are used. Electrodes must be kept dry, and if they are of the low-hydrogen variety, electric heating ovens are advisable to ensure dryness. Only enough electrodes for immediate use should be taken from the oven each time. Small, portable ovens are available so that the men can bring a small supply of electrodes close to their work.

A constant check should be made by the foreman or an individual assigned to inspect to make sure the men follow the qualified procedure and sequence, using the predetermined number of passes and in the correct direction. Electrode holders must be maintained in safe condition so that a man cannot make a contact with himself or with the steel if he lays the holder down incorrectly.

It is best to inspect while the men are actually welding, but if this is not possible, it should be done as soon as possible after completion of the weldment and before a scaffold, float, or platform (if used) is moved away. When radiography or other form of testing or inspection is required, this should also be done promptly to save the cost of bringing men, equipment, and scaffolding back to correct a defective weld.

A good foreman will have the necessary accessory equipment on hand, such as wire brushes, peening hammers, welding shields, hoods, or helmets. He will also have the correct shade of flash goggles for men assisting the welders or men working close by to prevent eye flash injuries. If possible, the welder's shield or hood should be of the type that can be secured to his safety hat by a hinged connection (Fig. 3.7.70).

Some form of container should be used for the electrode stubs in order to discourage the men from dropping them indiscriminately, especially where someone can be hit or could step on such a stub and have it roll underfoot, possibly causing an accident. By assigning an identifying symbol to each welder, which he is to mark on the point he has completed, it is easy to check on men making good or bad welds.

6.15 Lintels

Loose lintels should be kept separate from the main steel. It is customary to deliver them to the particular locations desired by the subcontractor who will ultimately set them in place. It is advisable to have the general contractor check them and indicate in some way, preferably by a receipt, that all required loose lintels have been delivered. They should be piled safely on adequate blocking so that they cannot be dislodged accidentally and cause injury to someone. Hung lintels to be secured to the steel but shipped separately are bolted in place just before the spandrels to which they connect are raised off the working floor.

A follow-up gang should be used to align hung lintels within the required tolerances after the steel has been plumbed and permanently fastened. When pouring concrete floors may cause them to be deflected or moved out of place, it is sometimes necessary for the erector to return to the site later to align and adjust such hung lintels. It is well to arrange with the customer that lintels will be adjusted by the erector only once at no extra cost. Too often they are out of line later and the erector must return repeatedly for different floors to readjust and duplicate his previous work. To align hung lintels a transit in the street at a definite distance off the building line can be used for the alignment with respect to the building line. A level set up on the floor is used to align them vertically. An alternate method of alignment uses a heavy piano wire stretched across the faces of the columns in the outer row, but this is liable to sag and the deflection would prevent accurate vertical setting.

The steel details should provide some way to adjust horizontally and vertically by means of shims or slotted holes or both. On a tall, heavy building the shortening of lower tier columns under the load of the upper tiers of steel must always be taken into account when setting hung lintels to elevation before the balance of the steel has been erected. If the adjustment requires reaming, drilling, burning, or welding, adequate scaffolds or floats should be hung to protect the men doing the work. Safety belts with life lines are advisable for this work.

6.16 Wall-Bearing Steel

If it can be accomplished, when some of the steel is wall-bearing the bearing walls should be in place when needed to erect these members as part of the floors involved. If the walls are not ready the steel can be landed on falsework. The falsework shoring and supports should be installed so that no part can be dislodged accidentally by others after the erector has left the site, thus permitting the steel landed there to fall and injure someone or become damaged. The falsework must always be strong enough to support not only the steel, but any possible loads that other trades may place on such falsework-supported steelwork, since the erector will have no control of this after he leaves. Any bolts or electrodes that will be needed when the steel is finally erected must be carefully stored so that they will be available when needed.

It is usually better to land the steel directly below its final position on blocking that is high enough to permit pouring the floor with pockets around the blocking. Then later, when the walls are built, the men return and roll the pieces off the blocking on rollers or timber dollies on the completed floor. A set of hand falls can be hung from above, or a light gin-pole or dutchman can be used to erect the steel on the walls. The pieces can also be left on the steel erected at their final level, and rolled out later, using a lightweight pole or dutchman on the floor below or a set of falls from above to hold the free end until it is landed on the wall.

Another method requires a jigger stick with a set of hand falls at its outer end for lifting the various pieces into place. This is costly since the jigger stick must be moved repeatedly and lashed down to the steel above for each position of the pieces being erected by it.

As an alternative—to eliminate extra trips back to the site—the wall-bearing members can be connected at one end to the structure, while the other end is held up by a wire-rope guy and turnbuckle to a beam or column above. The free end should be left just high enough for the masonry subcontractor to build the wall to the proper elevation with a small gap under the ends of the beams. By slacking the turnbuckles he can land the steel on the wall. The guys and turnbuckles are then removed and shipped back to the erector's toolhouse by the general contractor.

Some erectors prefer to erect all the self-supporting steel that is possible, returning repeatedly as each floor is ready for the wall-bearing members. Other erectors wait until all the walls for a floor are in place and then erect that entire floor, returning for each successive floor to erect both the self-supporting and the wall-bearing material.

With a derrick-erected structure, some erectors prefer to erect as much as possible with the derrick and then leave it in place on the roof steel. This permits them to start up the derrick for the wall-bearing steel on each floor, reaching down from the roof to set the wall-bearing steel on each of the lower floors as the walls for that floor are completed. Openings must be left for the lead lines to the boom and load falls of the derrick through the building from the hoist below.

6.17 Open-Panel Construction

With open-panel construction, the erection scheme should be followed carefully to be sure that the proper supports for the floor planks (usually of extra thickness and length because of the greater than normal supporting panel) are used where intended. The steel should be unloaded where predetermined to be sure that the floor planks are not overloaded and that the load is transmitted to skids directly over permanent or temporary supporting steelwork. When wire ropes and turnbuckles are to be used stretched across the structure to support the floor planks, skids should be laid directly over the steel framing into the columns, and the loads of steel should be unloaded diagonally in order to be supported by the skids and beams and not by the wire ropes. Some slack should be left in the ropes to reduce the stresses in them. This type of floor-plank support gives a very springy floor that most erectors dislike (Sec. 5.21).

6.18 Dismantling—Demolition

When dismantling a structure, a safe procedure and sequence should be set up in advance. All men involved should be called to a safety meeting for a briefing on the procedure and safety precautions to be followed. When bolts, rivets, or weldments are to be removed from old members in place, it should be clearly established how many erection bolts are to be placed temporarily in the connection, how many bolts or rivets are to be left in place, or how much welding is to be left until ready to remove the piece. When a piece is to be burned off, the amount of metal to be left until the piece is ready to be taken down must be specified. Preferably this should be marked on the piece to make sure that it will be able to support itself, any men who may walk on it, or material that may be landed on it until the final cut is made.

The final cut or removal of the bolts, rivets, or welding should not be made until a sling is in place and a very slight strain taken on the load line. Too much strain will cause the piece to spring up or swing when released and may cause an accident. All loose material should be removed before a piece is cut loose to prevent such material from falling on men working below.

The space below the dismantling or demolition should be roped off, and guarded against anyone passing through the area. Where burning or cutting with an oxyacetylene or other type of torch is required, fire protection should be arranged below with adequate extinguishers or water hoses, etc., on hand. A protective covering should be placed in advance over flammable material below that could be set on fire.

The safety of remaining portions of the structure must be checked as well as the effect of removing individual pieces. If shoring is inadvisable, material may have to be hung temporarily from above until the time for its removal.

6.19   Joint Operations

If two derricks, two cranes, or a crane and a derrick will pick a girder or truss together, it should be definite which of the two foremen in charge of the rigs is to be in charge of the operation. In addition, it should be clearly established who is to give the signals to the signalman for each machine, to transmit to his hoist or crane operator, so that both rigs will raise, lower, or swing together or as required.

Before beginning a lift the man in charge should make sure the lifting slings or other devices are correctly placed and hooked on at the points where each rig will lift the correct weight for its capacity. By locating the two hitches at different distances from the center of gravity of the piece, one rig will pick a relatively heavier or lighter load, depending on the location of its sling or hitch (Figs. 5.25.7 and 5.25.8). Before lifting, the man in charge, preferably the superintendent himself, should check to see that the cranes are adequately supported and blocked, outriggers in place if needed, or derricks in good working order, and the hoists themselves in proper condition with brakes adjusted for the load.

When it can be avoided, a piece should not be carried by two (or more) cranes operating jointly because there is a risk of one crane moving faster than the other. This will tend to make the piece start swinging, and the lifting slings or hitches will no longer be vertical and steady, with the chance that the piece may get out of hand and tip one or both cranes. Such heavy pieces should be delivered as close as possible under their final location, and the cranes picking jointly should be spotted to swing but not travel. To move a piece the cranes can always swing, land it, move to new positions, and repeat this operation until the piece is in position to be lifted directly into place.

If a pole is to be used with a crane or derrick, it should be set up in the proper location, adequately supported and guyed, and the lifting falls correctly reeved, with the lead line in place to an auxiliary hoist or to a drum on the main hoist, while the other end of the piece is controlled by the derrick or crane.

The raising gangs should be briefed on the operation so that every man knows what he is to do and there will be no floundering or confusion at a critical moment in the operation.

6.20 Responsibilities

Although the superintendent is in charge of labor and the resident engineer or field engineer is in charge of engineering, there must be complete cooperation between the two for the satisfactory running of the project. The superintendent should observe that all phases of the work—raising, plumbing, fitting, bolting, welding, handline, tool maintenance, etc.,—are under control and are being processed safely, economically, efficiently, and expeditiously. Nevertheless, the engineer should also be observing all of these operations, and when he notices deficiencies, either visually or through an inspection of the records of progress and production, he must inform the superintendent and confer with him. There should be a rapport between the two men so that they can work together in the best interests of the job.

Both the superintendent and the resident engineer should be alert to all operations keeping up their necessary relative progress, so that plumbing-up, fitting-up, and fastening operations keep close behind the actual raising of the steel. They should see that other trades are not encroaching on the areas needed by the erector.

The stability of a structure may depend not only on the work of the erector but on the work of the trades following, for example, the pouring of concrete floors should keep up with the steel erector in a tier building. When the safety or stability of a structure is endangered by the steel erector, or by other trades not keeping up with him, the superintendent and the engineer are both responsible for seeing that steps are taken promptly to correct such a condition, either by actions under their control or by informing the proper persons controlling the other trades. The office representative of the steel erector can often accomplish more by taking care of informing the customer or owner of the conditions under their control that need remedial action.

The disposition of the gangs should be such that their work will aid the general contractor in taking over areas for completing his work, provided it does not interfere with a logical erection sequence and does not increase erection costs. Both the superintendent and the engineer should be alert to anything being done by the erector's forces that may violate local ordinances, codes, or contract obligations or specifications. If the engineer discovers such conditions, he should notify the superintendent promptly to permit the latter to have these matters rectified. When errors in the steelwork are discovered, whether caused by drawing-room mistakes, improper fabrication, or incorrect erection procedure, the engineer is usually the one best equipped to check the errors and supervise the corrections, but the superintendent is still the man to select and assign the men to do the actual corrective work.

The two men should watch the job constantly to keep everything running smoothly. Equipment should be lined up to be on hand when needed for dismantling and loading out erecting equipment and tools. The men should be observed to see that they are wearing safe clothing (work trousers with no cuffs, shirts to protect them from the sun, and work gloves to protect their hands), that they are using safety hats (hard hats), safety goggles when performing operations requiring them, and safety belts with good rope properly tied off where needed. Tools in use should be noted to be sure they are in safe condition, that open-end wrench jaws are not worn or sprung, that tips on connecting bars are safe for use, that bolt bags are supplied to the men who need them, and that welders are using safety hats with hoods or shields attached. In general, the entire job must be watched to see that tools and equipment are in safe condition and are being used safely.

If an accident occurs, a thorough investigation, analysis, and report should be made immediately to pinpoint the cause and to aim at preventing a recurrence. Safety meetings should be held regularly and the men impressed with the real aim of the superintendent and management—to have the men work safely. Safety briefings before any unusual or particularly hazardous operation will pay in better work since the men will feel that they are part of management by being informed about what the work will be and how it should be done, and by feeling that they are being depended on for the safe completion of the operation.

When a safety engineer is on the job he should cooperate fully with the superintendent. If there is no safety engineer one of the men should be chosen as a safety observer, changing the appointment each week. On a large job, it may even be worthwhile to appoint several men as a safety committee, permitting them to make a report to the superintendent of what the committee considers unsafe conditions or conditions that could be made safer. By permitting them to report at the next safety meeting, together with what the superintendent has agreed to have done to improve conditions, the committee receives prestige and recognition and the job will generally proceed with greater cooperation and safety on the part of all the men. After all, the safer the work, the quicker the job will proceed, the better the men will work. The final production costs will generally show far greater savings than the cost of safer tools, safer equipment, and proper, safe workmanship in the actual progress of the job.

Instead of assigning a safety engineer to a single job, a roving safety engineer visiting a number of jobs at unexpected times can frequently see a dangerous condition far more quickly than the men who are working close to an unsafe trap on the job. He can help educate the men to work safely and to eliminate old-fashioned superstitions that they can do nothing to prevent accidents, that "when their number is up, that's it," or that a job must not be started on a Friday. Accidents do *not* have to happen—they can be prevented. A no-accident record results from carefulness and planning. Carelessness cannot fail to result in accidents. It is a matter of luck when an accident is only a minor one.

7

The Guy Derrick
and Accessory Equipment

7.1 Construction

The guy derrick is the type of equipment that is used most for erecting tall structural steel-framed tier buildings. The basic components are a mast, boom, load falls, boom falls, and guys (Fig. 7.1.1). The guys hold the mast upright. The mast in turn supports the lower end of the boom. The wire-rope boom falls (topping lift), between the top of the mast and the head of the boom, is used to raise and lower the boom. The boom tip supports the load falls (main falls) to lift and lower loads. This derrick is called a guy derrick because of the guys holding the mast in place.

Many years ago the mast and boom were made of heavy timbers, round, square, or rectangular in cross section. Because of the physical limitations of delivering long timbers to the construction site through city streets and around corners, and because of the burden of handling these heavy, cumbersome pieces, steel replaced wood. Long members of steel could be made of shorter sections to be bolted together after delivery. This resulted in easier and safer handling in shipment from an erector's toolhouse, easier handling at the job site, and easy assembling and later dismantling for removal. The four main members of the boom and mast are usually made of angles or pipes at the corners of a square or rectangular cross section. These are braced by a series of diagonal angles, bars, or pipe lacing on all four sides between the four main members. The various parts are welded, riveted or high-strength bolted together. Batten plates are used at the ends of each section to stiffen and help hold the entire member in shape. Either splice plates, splice angles, butted splice castings, or heavy, pinned connections are used to connect the various sections together (Fig. 7.1.2).

200

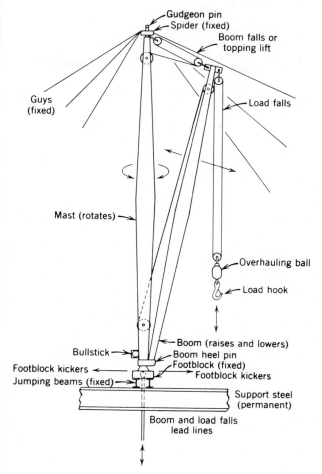

Gudgeon pin
Spider (fixed)
Boom falls or
topping lift
Guys
(fixed)
Load falls
Mast (rotates)
Overhauling ball
Load hook
Boom (raises and lowers)
Bullstick
Boom heel pin
Footblock kickers
Footblock (fixed)
Jumping beams (fixed)
Footblock kickers
Support steel
(permanent)
Boom and load falls
lead lines

FIG. 7.1.1. Guy derrick.

The foot of the mast is supported by a casting with a curved (convex) surface at its lower end that rests on a similarly curved (concave) casting to permit the mast to rotate and move slightly out of plumb in any direction while being completely supported. The lower casting in turn rests on roller bearings, ball bearings, or on a bronze or other type of antifriction plate to permit rotation that is as frictionless as possible. The roller bearings, ball bearings, or bronze plate in turn are supported in a casting fastened to a pair of steel beams. The entire assembly of supporting casting, bearings, support, and beams is known as the footblock.

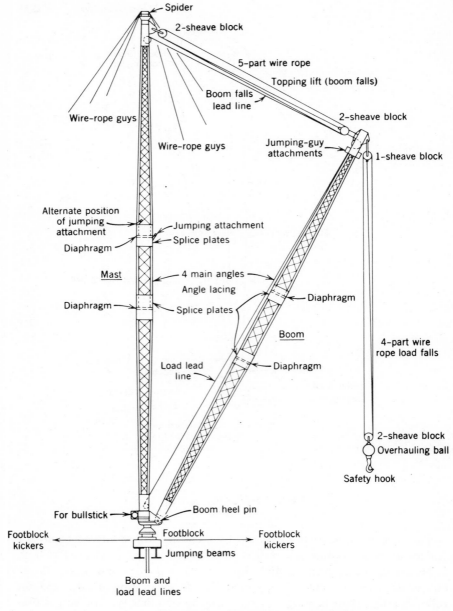

Spider

2-sheave block

5-part wire rope

Topping lift (boom falls)

Boom falls
lead line

2-sheave block

Jumping-guy
attachments

1-sheave block

Wire-rope guys

Wire-rope guys

Alternate position
of jumping
attachment

Jumping attachment

Splice plates

Diaphragm

Mast

4 main angles

Angle lacing

Diaphragm

Diaphragm

Splice plates

Boom

4-part wire
rope load falls

Load lead
line

Diaphragm

2-sheave block

Overhauling ball

Safety hook

For bullstick

Boom heel pin

Footblock
kickers

Footblock

Footblock
kickers

Jumping beams

Boom and
load lead lines

FIG. 7.1.2. Guy-derrick construction.

202

The castings have a vertical hole through the center for the passage of separate lead lines through the bottom of the mast to raise and lower the boom by means of the topping lift or boom falls; to raise and lower the main load by means of the load falls or main falls; and sometimes one for raising and lowering lighter loads by a whip or runner of a single part or two parts of wire rope.

Two or three vertical sheaves, usually on the same shaft or pin, in the footblock beams, are used to guide the lead lines into the mast. If the hoist is located on the same floor or ground level as the foot of the derrick, these sheaves permit the lead lines from the hoist drums to change direction from nearly horizontal to vertical through the foot of the mast. When the hoist is located on a lower level, as the derrick is turned for actual operations 180° in either direction, the lead lines will tend to make a half turn in the direction the mast has been rotated. The leads are kept aligned through the mast by means of these sheaves in the footblock. In the ends of the two beams of the footblock, or at some intermediate points in them, reinforced holes are provided for attaching anchoring guys (footblock kickers) in four directions to hold the footblock in position against the horizontal thrust of the boom against the mast.

A short distance up the mast, idler sheaves are positioned to divert the main load line and the runner line to the head of the boom. Near the top of the mast there is a sheave to permit the topping lift lead line to change direction and to actuate the topping lift. At the top of the mast a heavy steel pin (gudgeon pin) is secured to permit the mast to rotate in a horizontal casting (spider) separated from the top of the mast by a flat bronze or brass washer or antifriction bearing. This permits the mast to rotate as freely as possible in the spider. (In some designs, the spider is tipped slightly from the horizontal.)

The spider is held in place by guys, preferably eight, but sometimes only six or seven. The top ends of the guys are looped over projections on the top of the spider, or connected to vertical pins in the spider, or shackled to appropriate holes. The lower ends of the guys are secured through turn-buckles and slings to anchorages. Preferably there should be two guys on each of the four sides of the derrick, since the stress in the guys can be distributed best by using eight guys. At the connections of the guys to the spider, there are keeper plates, shackles, or similar devices to prevent the guys from jumping out of their connections.

A bracket type of support for the bottom end of the boom to rotate about a horizontal steel pin (boom heel pin) is located at the foot of the mast. A connection is provided on the upper side of the head of the boom for the topping-lift falls block, one on the underside for the upper load falls block, and a lead sheave for the main load line. This line runs from the lead sheaves near the foot of the mast to the main load falls over the idler sheave part way up the mast. An additional lead sheave is usually located on a separate pin in the head of the boom for the runner or whip line. In some cities two parts are required since a single-part whip or runner line is prohibited. If there are two parts it is common practice to reduce the diameter (and strength) of the wire-rope line so that the two parts have sufficient strength for the maximum load to be handled. The lighter the wire rope, the easier it is to handle and to reeve. The load will move half as fast with two parts for the same lead-line speed as it would with a single part. This is safer and gives better control. Heavier loads are handled by the main load falls with three or more parts.

The mast must be longer than the boom, usually by about 10 ft, so that the top of the boom can be turned under the sloping guys when the boom is in its uppermost position against the mast. Idler rollers or timbers are installed on some long booms, on the upper face, to prevent the lead lines from rubbing on the lacing and abrading the wire rope or damaging the lacing. It is important to check that the lines are feeding over the idler and lead sheaves provided, and have not jumped out of place to jam or rub against the sides of these sheaves or against parts of the derrick itself.

At the lower end of the main load falls, the lower load falls block will usually be provided with an attachment to which to shackle an overhauling weight (ball), which, in turn, has an attachment on its lower end for a hook—preferably a safety hook.

Ordinarily, a guy derrick, especially one of light capacity, will have two U-shaped straps provided on the back face of the mast near the foot, into which to insert a bullstick. The bullstick is a long pipe section about 6 to 8 in. in diameter and about 10 to 20 ft long, depending on the force needed to turn the derrick. The bullstick man can turn the derrick by means of the bullstick to spot the boom head (and load falls) directly over a load being lifted and then swing it to its position directly over the final location of the piece being erected.

Four sets of manila-rope falls are advisable to prevent the mast from rotating uncontrolled, using two in opposite directions at each end of the bullstick. The other ends are secured to the four columns around the derrick or to four anchorages when the derrick is set up at ground level. By proper manipulation, using only two falls at a time at either end of the bullstick, the bullstick man will be able to hold the derrick steady even under windy conditions, taking in on one set of falls and releasing the other; and when necessary, taking a turn of the free end of the rope around the bullstick itself.

When a column is being upended, the boom must be turned to keep the load falls directly over the hitch. The manila falls aid the bullstick man to control the boom and prevent it from whipping or drifting back as the column is finally lifted upright from the floor or ground. If the derrick turns easily, a single part of rope can be used in place of each of the falls. An area around the derrick must be kept clear of steel, toolboxes, or other encumbrances to permit the bullstick man to work in turning the derrick. Planks or a platform should be laid, if needed, on which he can work safely and efficiently.

Frequently a bullwheel is used for turning the derrick, especially with a derrick that does not turn easily because of its greater capacity and correspondingly greater weight. The bullwheel consists of a circular frame of channels or angles, about 10 to 16 ft in diameter, secured to the mast by horizontal angles or other bracing. It is held up rigidly in place to the mast by diagonal bracing in a vertical plane. In some designs, the boom heel pin is supported on framing between the mast and the bullwheel itself instead of directly on the mast. This bracket must be properly designed and reinforced for the purpose. It permits the head of a bellied design boom to be brought up closer to the mast to clear under the sloping guys when the mast and boom are being turned. The mast and boom should always turn as a unit. The area in which the bullwheel turns must be kept clear of obstructions.

To turn the derrick in either direction, the central portion of a wire rope is secured to the bullwheel and the ends wound around it in opposite directions and then off to two separate powered drums on a single shaft of a swinger engine. The ends of the wire rope are wound in opposite directions on their respective drums so that as the shaft is rotated one rope will be wound on its drum and the other will be paid out at the same rate of speed. As the drums are rotated in either direction, the bullwheel and derrick will be turned. If the bullwheel is controlled by a hand-operated winch, the operator must be cautioned against the danger of the crank handle whipping out of control if the boom drifts unexpectedly and the derrick turns.

Frequently, but principally on a heavy derrick, wire ropes with turnbuckle adjustment are secured between the bullwheel and a point part way up the boom. Then, as the bullwheel turns, the ropes help to swing the boom around with the turning of the bullwheel and mast instead of depending on the boom heel pin alone to bring the boom around with the mast.

Whether a bullstick or a bullwheel is used, it is important to use a telltale rope of clothesline or bell cord type, one end secured to the mast, the other end to the footblock, and just long enough to reach when the mast is turned 180° in either direction. This serves as a warning against having more than a half turn or twist in the lead lines. This telltale must be watched continually in order to know when the mast has been turned half-way around since it must not be turned further. The mast must be swung back and turned around in the opposite direction to prevent undue rubbing and chafing of the lead lines.

The advantage of the guy derrick over other types of derricks lies in its ability to rotate through 360°. The disadvantage is the need to boom it up each time it is moved from an area between any two guys to an area between any other pair of guys. The ease with which it can be jumped vertically from floor to floor makes it a very important piece of equipment for erecting tall structures. A fairly new piece of equipment in this country that can also do this is the climbing-tower crane. While the current models have capacities comparable to some guy derricks, they are more cumbersome, limited in scope, and have certain drawbacks such as the possibility of fouling surrounding structures when swung around.

7.2   Initial Setup

The derrick will generally be shipped to the job completely knocked down (from the erector's toolhouse if owned, from a vendor's storehouse if rented, or from the manufacturer if purchased new). It should be checked immediately upon delivery for any damage in transit, which should be reported promptly to the delivering carrier. Any damaged material should be repaired or replaced before using to prevent possible failure of the equipment.

If a mobile crane or other derrick is available, it should be used to unload all parts of the derrick. If no crane or other powered equipment is available, the sections must be unloaded by hand. The hoist to be used to power the derrick should be on hand in advance or delivered at the same time the derrick arrives. By means of the hoist, sections can be "snaked" off a truck, down skids laid between the truck and the ground, or off a railroad car with skids similarly laid between the end of the car and the ground. The section should be jacked up enough to slide planks or skids and timber or steel rollers underneath. Sections can then be rolled by hand or with a set of manila rope falls powered by a spool on the hoist. The hoist should be securely anchored, even if only temporarily, so that as power is applied it will not move.

The boom and mast are assembled after unloading all sections. Blocking should be laid on the ground at such heights that when the various sections of the boom and mast are completely assembled the centerline between the top and bottom of each will be a straight line. All splice bolts are installed and tightened or, if pin-connected, all pins are placed and secured with all corresponding faces of each section on the same side. The hoist should be located in its working position while the mast and boom are being assembled and the wire ropes wound on their respective drums. The jumping beams should be placed where the derrick will be set up, with the footblock on them in its predetermined position for supporting the mast, and it should be blocked or guyed against the pull of the hoist lines. A boom shoe (jumping shoe) is also set either on the jumping beams or on suitable blocking across them. Gantlines are secured to the mast at this time. (Gantlines are manila ropes reeved through single sheave blocks secured to fittings near the top of the mast.)

The load falls should be reeved to the upper load block secured to the head of the boom at the proper place, leading the line through the bottom section of the mast and through the footblock. The topping lift can be reeved at this time, securing the lower block to the head of the boom on the side opposite the upper load block. The upper topping lift block is secured to the head of the mast, with plenty of slack in the falls. Four boom-jumping guys are now fastened to the four fittings provided for this purpose near the head of the boom. Four sets of manila falls, each usually a four-part falls, are now made ready. One end of each set is hooked onto four equally spaced anchors, or to slings through or around suitable anchorages, and the free ends hooked into the ends of the boom-jumping guys. The hooks of both blocks of each set of falls should be moused to prevent them from coming loose.

All anchorages for the various guys should be checked, whether split-end eye anchors in rock, hairpin anchors, or lashing embedded in footings or foundations, to be sure they were installed at the correct angle, that is, pointing toward the head of the mast in its working position. They must be deep enough to hold but not so deep that the shackles or slings for the guys cannot be secured to them. They should have sufficient weight in the anchoring material to resist uplift and sliding from the pull of the guys, and they must be in the locations indicated on the erection-scheme drawing.

With a crane or other derrick available the assembled boom is now picked by suitable slings at a point above its center of gravity and the bottom end set in the boom shoe. As soon as the boom is upright the four manila falls are taken in to make the boom guys taut and are tied off. The boom should now be standing with the load block between the boom and the mast assembled on the ground. The load block is now secured to the mast by adequate slings, preferably hooking onto fittings provided on the mast for jumping. If such fittings are not provided, the slings should be placed around the mast at a point above its center of gràvity and at a splice plate or other suitably reinforced area to prevent crushing or distorting the mast section.

The spider should be in place on the gudgeon pin and all mast guys secured to the spider before picking the mast. Any keeper plates or similar devices should be in place to hold the guys on the spider. The guy turnbuckles, opened almost fully, should all be in place, their lower ends secured to slings around their respective anchorages or by shackles to anchors embedded in the foundations or footings. Wire-rope clips to be used on the guys and wrenches to fit them should be on hand.

The mast is picked by going ahead on the load line, upended and landed upright in its proper position on the footblock. The mast guys are now threaded through the spools held in place by the clevises and pins on the free ends of the turnbuckles. One by one the free end of each guy is pulled up toward the head of the mast by using the gantlines. As each guy is pulled snug, wire-rope clips are installed to hold the loop and tightened, to be retightened after the derrick is working. Spools or thimbles should be used wherever loops are formed in wire-rope guys to avoid crushing the individual wires in the loops. The gantlines are unfastened from the ends of the guys and secured to the lacing on the mast until needed again. A similar procedure using the gantlines is followed when the derrick is jumped or moved to a position requiring a change in the length of the guys.

By tightening or loosening individual turnbuckles, the mast can be plumbed vertically. As soon as all the mast guys have been secured and the mast plumbed, the topping-lift lead line is taken in until there is a very slight strain on the top of the boom. The boom guys are released and the topping lift picks the boom off its temporary support. The boom shoe is removed, the boom rotated, the heel forced into its proper place at the foot of the mast, and the boom heel pin inserted.

If it is within the capacity of the crane or other derrick at the reach involved, the mast can be picked first, the mast guys tightened, and the boom then picked by using the topping lift falls and set into position. If only a short-boom crane or derrick is available, unable to pick the boom or the mast above their centers of gravity, weights can be fastened temporarily to the bottom end of the boom or mast, as the case may be, to lower the center of gravity. This will then permit the short-boom rig to pick the member safely, provided the total weight is now within the capacity at the reach involved. These added weights can be overhauling balls, plumbing cables, or the like.

The foundation contractor will often permit the use of his derrick or crane to assemble and set up the steel erector's derrick at the start of the job. This would be in exchange for dismantling and loading out the foundation contractor's rig after the steel erector's derrick is in operation, without additional cost to either.

If no powered equipment is available to set up the derrick, the sections of boom and mast are snaked into position on blocking by means of rollers and skids, by manila-rope hand falls, or by the power of the hoist whenever possible. The sections are assembled on blocking as described above, and the load falls and topping lift reeved as before. In this case it is important to have the load fall side of the boom uppermost. The foot of the boom is guyed back to anchorages to resist the horizontal force that will occur in tripping the boom upright.

An anchorage well away from the foot of the boom is now used, slacking the main load falls sufficiently to connect to this special anchorage, either directly or through slings or pendants. (A pendant, often called a pennant, is a length of wire rope intended to be used as a straight piece, with an eye spliced in each end. It is never used as a choked sling or wrapped around anything.) By going ahead on the load-line drum the boom should start upward (Fig. 7.2.1). In this case, the falls on the boom-jumping guys must be used to keep the boom rising in a vertical plane, slacking off the back guys and taking in on the front guys.

It may be necessary to jack up the head of the boom at first to tip the boom up enough for the load falls to start the boom upward, or it may be necessary to use a small pole, 6 to 8 ft high, properly guyed, with a set of hand falls to pick the head of the boom high enough for the main load falls to take effect. Once the boom is upright on its protective boom shoe, with the four boom-jumping guys secured, the mast is picked by the main load falls and the derrick made operative, as in the procedure described for a crane or derrick picking the boom or mast.

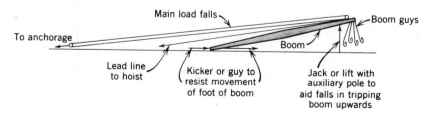

FIG. 7.2.1. Tripping a boom without auxiliary powered equipment.

If an area has been assigned by the general contractor for placing the hoists in the excavation, a check should be made to be sure the leads from the drums to the footblock in the derrick's first position will be satisfactory. They should not be too long nor in an area where others have to pass through. Otherwise, arrangements should be made to have these leads barricaded to avoid accidental encroachment by anyone. After the hoists have been moved up to the street-level steelwork, the leads to the fiddle blocks should be barricaded to prevent accidents, unless they are in an area that will not be used by others.

The hoist will normally have two drums: one for the boom and one for the load line. A third drum may sometimes be needed for a runner line. A swinger, if a bullwheel is used, will usually be a separate double-drum hoist on the working floor or at the main hoist. A pair of drums on a single shaft geared to the main hoist mechanism can also be used.

7.3   Jumping

As much as possible of the tier above the working floor (or ground level on the initial setup) is erected to permit jumping. A second pair of jumping beams is landed on the supporting steel of the uppermost permanent beams to which the derrick is to be jumped. The jumping beams are spread apart just enough to permit the derrick and footblock to pass between them and clear when being raised to the new level. If a bullwheel is used, this is partially or completely dismantled for the jump, and the wire ropes actuating it are removed. The dismantled members and the hoist or swinger that powered the bullwheel are landed on the new working floor.

The connections of the permanent steel on the newly erected upper floors should be fitted up and, if possible, permanently bolted or welded before jumping the derrick. Even if only the top level of steel can be permanently connected, this should be done; then the men can go back to the floor below to complete the permanent fastenings at that level after the derrick has jumped and the upper floor has been planked over. Fastening the connections of the permanent beams that support the jumping beams under the derrick must be completed before the derrick jumps, or else they must be completely fitted up. Fastening all connections involved in bringing the stresses from the guys back to the foot of the derrick should be completed or fitted up thoroughly before the derrick operates again at the new level.

The four kickers on the footblock are released by loosening the turnbuckles about two turns, and the shackles securing the turnbuckles to slings around the four columns to which they were fastened (or anchorages on the initial setup) are released. The kickers, shackles, and anchor slings are landed on the new working floor above. When landing the derrick above, it is important to spot the footblock in nearly the same relative position as below. Then the kickers will fit by merely tightening the turnbuckles by the same amount as they had been released below. When first setting up the derrick at ground level, the kickers may have to be anchored at some distance away, depending upon what anchorages were available. After jumping the first time, the kickers should be anchored to the four nearest columns surrounding the derrick. If the derrick is spotted as described, these can be used repeatedly without changing their lengths. The exception occurs when the derrick must be moved sideways from its former position below because of changes in location of adequate supporting steel or changes in the location of columns above.

The boom heel pin is then removed ("pulled") and the topping-lift falls picks the boom slightly (Fig. 7.3.1). As a safety precaution, a light, short sling should be fastened to the end of the pin, the other end secured to the boom. This will serve to prevent the pin from falling if it should be dropped inadvertently by the men while removing it or replacing it later. The boom shoe is positioned close by on a pair of short beams or heavy timbers on the jumping beams supporting the derrick. The boom is rotated 180° by means of the bullstick and lowered to land on the boom shoe, inserting the boom heel pin through the boom heel and the holes provided for it in the boom shoe. (With a heavy boom it may be necessary to use a manila-rope falls to pull the bottom of the boom out if it cannot be "breasted" to its new, temporary position manually.) This now puts the load falls between the boom and the mast and puts a twist in the topping lift. This also causes the load line to pass halfway around the boom instead of following its usual path on the upper face of the boom. Although this will cause some rubbing of the wire rope it is usually not serious. (Some erectors have come up with special devices that make it unnecessary to rotate the boom. These devices save some of the wear on the load line, but since they are patented they will not be described.)

The lower ends of the four or more boom-jumping guys are now secured by manila-rope falls of four or six parts (as needed) to four or more columns at the new, upper working-floor level. They are spaced as nearly equally around the boom as feasible. The upper ends of the guys were previously secured to fittings provided on the four corners (or the four sides) of the boom near its head when the boom was initially assembled. In addition, it is advisable to use an extra guy from the head of the

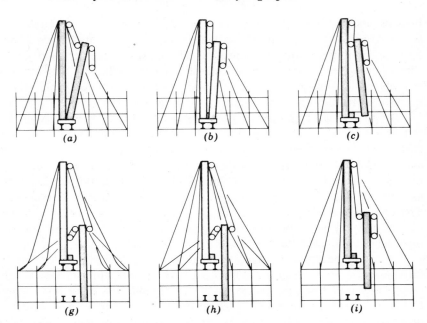

FIG. 7.3.1. Jumping a guy derrick; schematic arrangement of sequences. (a) Derrick erecting steel. (b) Derrick completes erection of tier (two floors) directly above working floor. (c) Boom is separated from mast. (d) Boom is turned 180°, anchored in a boom shoe and guyed to serve as a temporary mast or gin-pole. (e) Load falls is hooked on to mast. (f) Mast guys are cut loose and moved up to top level of steel. Boom starts to pick mast. (g) Boom lifts mast to

boom directly opposite the mast. This is needed to resist the force of the load falls picking the mast. The topping lift is released and slacked as soon as the guys have been secured to columns at the new working-floor level above so that the boom can be used as a gin-pole to lift the mast.

It is advisable to remove the overhauling ball in order to reduce the weight of the mast, footblock, and guys that the load falls will lift. The lower load block is now shackled to a jumping plate or other attachment provided on one of the mast sections. If none has been provided, a sling should be secured around the mast part way up, preferably at a point where there are splice plates or reinforcements, to prevent crushing or distorting the mast. This point should be above its center of gravity and yet low enough for the drift required to raise the mast to the new level.

Slings and turnbuckles are installed to hold the footblock to the foot of the mast, and a slight strain is taken on the load line so that the boom is

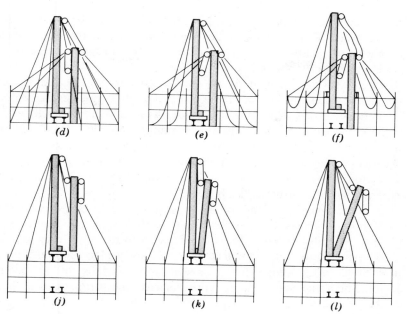

new position two floors above. (**h**) Mast is landed on jumping beams at new level and guys are adjusted. (**i**) Temporary boom guys are cut loose. Topping lift (boom falls) starts to pick boom up to mast. (**j**) Boom has been lifted up to new position of mast. (**k**) Boom heel is again pinned to foot of mast. (**l**) Derrick is again ready to set steel on new level.

now supporting the mast. The mast-guy turnbuckles are loosened about two turns and disconnected from their column anchor slings. A snub line should be used between the end of each guy and its anchor column since the guy will tend to whip inward when the turnbuckle is released from its anchor sling. A line should also have been secured to one of the loops of the eye-and-eye column anchor sling since it will also tend to whip or drop when the turnbuckle is unshackled from it. Since the sling is partly below the floor level, this line will aid in lifting it up to the floor when released.

When first setting up a guy derrick in the hole, the guys and turnbuckles will usually be shackled to anchors at ground level. As a result, the guy lengths will generally have to be changed after jumping to the top of the first tier steel. This is done by loosening the wire-rope clips that hold the bottom ends of the guys to their turnbuckles, again using the gantlines to pull the ends of the guys upward and retightening the clips. They should be retightened after the derrick has been operating in the new position. These wire-rope clips must be checked frequently to prevent the wire rope from slipping through them.

Each guy should be secured to a column anchor sling placed around a column and under the beams framing in from either side of that column. This is done so that these beams, together with the column itself, will resist the uplift from the guy. The beam framing into the column, at right angles to the two framing into the sides of the column around which the sling was placed, will resist the horizontal force from the guy toward the foot of the derrick. If there is no beam in this direction at that point, a temporary shore or strut should be used.

The column anchor slings for the mast guys should be installed at the new working floor level, preferably around columns at the same locations as before. With the turnbuckles still connected to the guys, they are lifted to the upper floors and reconnected to the newly placed column anchor slings. Single manila rope lines can usually be used for lifting the turnbuckles, but if the derrick is of large capacity, the turnbuckles and guys may be so heavy that, for safety, the turnbuckles should be lifted separately after disconnecting them from the guys. The ends of the guys are then similarly lifted and reconnected to the turnbuckles after the latter have been secured to the anchor slings in their positions at the new working floor level. When the turnbuckles and guys are too heavy to lift, even separately with a single line, the gantlines on the mast can be used with a snub line to hold back the turnbuckles and then the guys against the pull of the gantlines toward the mast.

A snub line for a light mast, or a set of manila rope falls for a heavy mast, should be in place to help hold the mast away from the boom as it moves up to its new level. This is hooked onto the mast at a point opposite the new working floor level before the mast is lifted.

By going ahead on the load line, the mast will be lifted by the boom acting as a pole guyed in place by the boom-jumping guys. As the mast is raised it is important to make sure the topping lift is loose and being paid out as the mast rises since it is still connected between the head of the mast and the head of the boom. There must be no strain exerted by it to resist the boom lifting the mast. The load line through the footblock and then through the bottom of the mast helps to hold the mast upright as it moves upward.

The jumping beams at the new level are pushed together to their proper position on the supporting steel as soon as the bottom of the footblock is slightly above them. Planks should be laid under them on the permanent steel and on top of them under the steel footblock beams to prevent the sliding of steel on steel. As the derrick is jumped tier by tier to the upper floors, the supporting steel and beams framing into the columns used as anchorages may become shallower. In that case additional planks are added under the jumping beams as needed to save constantly changing the lengths of the guys. The jumping beams must be placed as shown on the erection-scheme drawings since the strength of the supporting structure was calculated for them to be used in those positions.

If a double layer of jumping beams is needed, the height of the bullstick above the working floor will be greater than before, and a temporary platform will have to be provided for the bullstick man to work safely. This will be about 12 or 14 in. higher above the working floor, depending upon the depth of the new set of beams. Although these beams are called jumping beams, they also function importantly to support the derrick when operating.

As the mast comes up, the guys will gradually change from their slack condition to tighten and help hold the top of the mast steady. It was necessary to loosen the turnbuckles at least two turns when removing them below, before the jump, since the mast and footblock must be lifted high enough to slide the jumping beams underneath them and into place.

When the supports are in place under the footblock, the load falls is slacked down to land the mast, and the mast guys are tightened to hold the mast upright in place again. The number of turns on the turnbuckles to tighten them should be the same as when loosening them previously, if the derrick has been spotted in the same relative position as below, and the guys are secured to the same columns at the new level as were used before.

All guys should be tightened so that they have the same tension. This can be measured by gauges, but it is usually done by shaking each guy to feel how tight each has been made. (This feeling of the guys comes only with experience.) There should be some slack in the guys so that as a heavy load is picked, as in unloading from trucks below, the mast will lean slightly out of plumb toward the load. At least two guys should take the strain from picking a heavy load opposite. As such a load is picked, the side guys should hold the mast plumb sideways.

In checking the plumbness of a mast by eye or by spirit level, it is important to observe if the design is such that the gudgeon pin is directly over the mast foot casting. Some designs call for an offset between the two, and the detail drawings of the derrick should be checked carefully to see if this is so. Ordinarily, a straight intermediate section will be exactly vertical when the derrick is properly plumbed.

Once the mast has been secured by the guys, the load line is slacked off, the load block is unfastened from the slings or shackles securing it to the mast, and a slight strain is taken on the topping lift. The boom guys can then be released. The topping lift is now used to pick the head of the boom up toward the head of the mast. The boom will rotate back the 180° it had been turned previously. Extreme caution must be observed at this time and a snub line used to help hold back this rotation to prevent the boom from whipping out of control and injuring men about it. As soon as the foot of the boom is high enough, the boom shoe is removed and the foot of the boom is pushed into its place at the foot of the mast, replacing the boom heel pin.

The footblock kickers are reinstalled and tightened, and the boom guys are secured out of the way to the boom. If a bullwheel and swinger or hoist are being used these are reinstalled. The bullwheel is reassembled to the mast, the wire ropes reinstalled around the bullwheel and reconnected to the swinger or hoist. If the latter is powered by compressed air or electricity, the necessary hose or electrical connections are made, the swinging mechanism properly tied down and anchored, and the derrick is again ready to operate. When picking the first load after the jump, the guys should be checked for tightness and the wire-rope clips carefully inspected to be sure they are tight enough to hold the guys in place.

As soon as the derrick is operating fully the steel in the area directly below it is erected. This steel usually has to be omitted to permit jumping the derrick. This frequently requires one part of a pair of spreaders consisting of a long eye-and-eye sling with a hook spliced into one end, or a long pendant with a hook shackled to one end, the other end in either case being shackled to the overhauling ball or secured to the load hook. The pendant, or spreader part, is used to lower the hook to the lower floor without having to lower the main load block or the overhauling ball below the level of the new working floor. Otherwise, the block and ball would have to be threaded between the jumping beams and would be liable to foul in raising the pieces being erected directly below the derrick.

7.4   Planking

Planks are spread on as much of the new working floor area as possible before the jump is started. The best method to follow is to start planking while the derrick is still erecting steel above its working floor. The planks from the previous working floor, one tier below, are gathered into piles two wide and about ten or twelve high, piling them at the edge of the building and extending out a short distance. They should be landed on blocking to make it easier to place slings around the bundle. Then the load falls is lowered outside the structure and a pair of choked slings is used to permit the load to hang level, placed around a pile of planks with the free ends of the slings hooked up to the load hook. The planks are then raised and spread out to form the new working floor. This leap-frogging operation is repeated for each tier as soon as the planks can be freed and gathered up from below. But these planks should not be moved until all work below the derrick working-floor level and above the planks has been completed since they were left behind when the derrick jumped previously to protect men working on the open floors above the planks.

Occasionally the lower floor planks cannot be released until after the derrick has jumped to the new level. Work should be expedited below to release them as soon as possible so that they can be raised to the new level working floor and spread out over the entire floor before the new steel is taken in. If they cannot be safely released in time, additional planks should be delivered to the job for that purpose.

Skids consisting of two or three planks one on top of the other, or timbers such as 4-by-4s, should be laid on the floor planks in such a way that when bundles of steel are laid on them they will help to carry the load into the permanent steel structure under the floor planks, or at least help to spread the load over a number of planks.

Floor planks should be of a size and weight so that two men can pick them up, carry them, and lay them down easily and safely. Ordinarily, 2 × 12-in. select Douglas fir structural grade planks, 22 to 24 ft long, will serve the purpose best. If heavier planks (such as 3 × 12 in.) are required because of the span they may have to cover, a plank gang of four men, instead of two men, should be used to handle them, with two men at each end.

7.5 Signals

The signaling system for informing the hoist operator what operation is wanted can be by hand-operated bells, but this introduces the hazard, especially on a tall building, of men below brushing against the long lines used to operate the bells and inadvertently ringing a bell with possible disastrous results. Accordingly, unless the lines can be boxed in at each floor high enough to prevent this, it is advisable to use an electrical system. With such a system there must be a safety device that will give a positive signal to the hoist operator if the current has failed or the wires have been cut or broken, or have become disconnected.

With a hand-operated or electrically actuated system, the accepted signals are as follows: If standing still, one bell or one flash on a light means "go ahead." If moving, one bell or one flash means "stop." If stopped, two bells or two light flashes mean "lower." If the derrick is to be tied up for some time, it is customary to ring the bells or flash the lights three or four times alternately on two of the bells or lights. This permits the operator to dog off his drums instead of being forced to hold them with his brakes for an appreciable time. To start up this same signal alerts him to release the drums and wait for an operating signal.

Some systems are devised to have two lights, one for raising and one for lowering respectively, for each drum, that is, boom, load, and whip. This is much safer because the proper light is kept turned on during the time it is desired to raise or to lower, and when a light goes out because of a signal, or a cut or broken wire, or a power failure in the signaling system, the operator stops the motion that is being performed.

As soon as the derrick is being readied to jump, additional lengths of bell cord must be tied on that are long enough to reach the next level working floor plus enough to permit the bellman to move about. In the case of the electrical system additional sections of electric wire are plugged into the line and securely clamped to help prevent inadvertent disconnecting.

Another system that is sometimes used is the voice system; here, the signalman speaks into a microphone or telephone, and the hoist operator receives the signal either from a loudspeaker near the hoist or through earphones. Many operators object to wearing earphones for an entire day. Here again, there must be a positive safety device such as a light, buzzer, or horn, actuated by relays on a separate battery, that will go into action in case the current fails or the wires become broken or disconnected. The power for the electrical system can be the regular 110-volt current usually available, stepped down through a transformer to a safe voltage, or a battery can be used.

Some erectors use a walkie-talkie, but this, as well as the voice system, can be dangerous in case the power in the system should fail after a hoisting or lowering signal has been given, unless the signalman keeps up a continuous conversation and the hoist operator stops immediately when the conversation stops. A walkie-talkie is prohibited if blasting is being done in the vicinity, since the electronic circuit has been known to set off nearby blasting caps prematurely. When work starts in the hole, the signalman is usually easily visible to the hoist operator, in which case hand signals can be used (Table 12, Appendix E).

One advantage of the bell cord and bell system when utilized by a skillful signalman consists in his pulling the cord just enough to lift the clapper on the bell, but not enough to cause the release of the clapper, which would ring the bell. When he does this once it is a signal for the operator to go ahead a very small amount. Then when he releases the cord, the operator stops immediately. If he rings the bell once and then pulls the cord slightly, but not enough to ring the bell a second time (the signal to lower), it is a signal to slack down a very little bit. When he releases the cord, the operator stops immediately. When connectors are trying to connect a piece and are just short of being able to pin a hole in a connection, this signal is of great value since too often the operator will raise or lower the load too much, and the connectors then have difficulty in spudding the holes as they line up. In addition, a good signalman can jiggle the bell cord without letting the clapper strike the bell, after he has given a signal, indicating "raise quickly" or "lower quickly," as when lowering the empty load hook to the street to pick up a load.

The load-falls lead line will have been marked with white paint or chalk to give the hoist operator a warning when the load block is nearing its lowest level. He then slows his drum in anticipation of a "stop" signal. Similarly, as soon as the derrick has jumped, he will re-mark the load-falls and the boom-falls lead lines to indicate the maximum high position of the boom and the new lowest position of the load block. Additional marks are often placed ahead of the final mark at each jump, as a warning to slow down.

7.6   Removal of Derrick

When all of the steel has been erected the derrick must be taken down and shipped away. There are several methods of doing this safely. The simplest, when the derrick must take itself down, involves jumping the derrick down one tier, dismantling the sections, and loading them out with a jinniwink, a Chicago boom, a jigger stick, another derrick, or a crane.

When there are several derricks erecting steel, if they can reach one another, an excellent method is for each derrick in turn, except the last one in this operation, to dismantle the one next to it and load out the sections. This requires the steel between the derricks to be left down in that tier until later. The lower load block and the overhauling ball are lowered and landed on top of the working-floor level.

The load falls is unreeved, after first taking the precaution of snubbing the load line below the footblock so that it will not slip down uncontrolled toward the hoist below. This is done by means of manila rope with a rolling hitch around the wire-rope lead line, tying off the manila rope to hold the lead line securely, or by clamping a wire-rope sling to the lead line and securing the other end of the sling at the floor level. The wire rope is gradually taken in at the hoist below, snubbing the manila line as the rope descends so that it will always be under control.

If a wire-rope sling was used, a manila line is used to lower the sling fastened to the running rope; or the running rope can be pulled up by hand and coiled on the floor below the derrick after disconnecting the end from the hoist drum. If there is a whip or runner a similar procedure is followed.

With the boom held upright by the topping lift, the second derrick now reaches over and hooks onto the boom. The topping lift lead line is snubbed as was done with the load lead line. The topping lift is unreeved and the rope lowered; or, as before, the rope is pulled up by hand and coiled on the floor below the derrick, to be removed with the coiled load line. If an auxiliary piece of equipment is to be used instead of the next derrick, the load and topping lift lines are neither lowered nor pulled up. Instead, they are snubbed to be held until needed to power the other equipment, and only the falls are unreeved at the proper time.

The boom heel pin is now pulled and the second derrick lands the boom on top of the steel. The splices are disconnected and the sections are prepared for loading out by the second derrick. In this case, the steel between the second derrick and one side of the building must also have been left down in that tier to permit it to pick the dismantled sections, swing around, and lower them to the street.

If the second derrick has the capacity at the reach involved to lift the mast bodily, it hooks onto the mast of the first derrick, the mast guys are cut loose from the spider and lowered one by one by means of the gantlines. The mast is then picked and laid on top of the steel (as was done with the boom), the splices are disconnected, and the sections are made ready for lowering and removal from the job. The footblock, jumping beams, bullwheel and swinger (if used), guys that have been coiled and tied with manila rope or mousing, and all other parts of the first derrick are picked and swung to be lowered to the ground for removal from the site. The second derrick then fills in the steel between it and the first derrick, and between it and the side of the building, which had been left down in that tier in order for it to reach the other derrick and for loading out.

If the second derrick does not have sufficient capacity at the reach, the boom and mast can be dismantled in place by it and the sections removed one by one. In this case additional plumbing guys should be secured to the columns around the derrick being removed so that guying or blocking the boom and then the mast will not introduce stresses for which the columns were not designed. (Buildings have been wrecked and men injured or even killed by the failure to provide such additional guys.) The boom is guyed or blocked to the structure temporarily near the lowest section, and the second derrick hooks onto the top section. The splice at the lower end is disconnected and the individual section is landed on the steel. This is repeated with each section of the boom. The mast is similarly guyed or blocked, the mast guys are removed, and the splices are successively disconnected as each piece is hooked onto and removed.

Finally the last derrick of those dismantling and removing the others must be dismantled by itself, unless some other means for removing it have been provided. When only one derrick is used on a job, the simplest method is to jump the derrick down one tier. Jumping beams are placed one tier below the last derrick working floor, and a working area is planked over for working at that level.

The boom heel is pulled and the mast lowers the boom one tier by means of the topping lift, landing it on its boom shoe on timbers or steel beams, as for jumping upward. The boom is now guyed with the boom-jumping guys after turning it 180° so that the load falls is between the boom and the mast. The load falls is hooked onto the mast at the same point as earlier when it jumped the mast to the upper working floor. The mast guys are removed, cutting them loose from the spider and lowering them one by one by means of the gantlines, to be tied in coils for loading out. The topping lift lead line is snubbed to prevent it from going down under its own weight, and the topping lift is unreeved. This line is handled as described above, either lowering it to the hoist, pulling it up to the new working floor, or holding it snubbed until needed.

Temporary timbers or beams are landed one or two floors below the place where the boom is standing, and the boom lowers the mast far enough to land it and still be able to hook onto the top section of the mast. A snub line or a set of falls is used again to hold the mast away fom the boom as it is lowered. The mast is temporarily guyed by means of manila-rope falls, or blocked near the lowest section, or lashed to the boom before the top splice is disconnected. The load falls then lowers the top section to the floor where the boom is standing. Some erectors do not use timbers or beams to land the mast, but instead insert the bullstick, if long and strong enough, through the lacing of the lowest mast section.

The ends of the extra-heavy-pipe bullstick rest on the permanent steel to support the mast while it is being dismantled. The boom proceeds to handle each section of the mast as the splices are disconnected, finally lifting the bottom section up to the new working floor. Then the temporary support beams are lifted from where they had been placed one or two floors below.

After snubbing the lead line to prevent it from going down under its own weight, the load falls is now unreeved and rereeved to a set of blocks hung from the permanent structure above, as nearly over the boom as possible. The surrounding columns should have additional plumbing or other guys installed temporarily to hold the structure against the stresses from this new set of falls handling the boom sections. The bottom boom section should now be guyed or blocked, as was done with the mast before dismantling. The new set of falls is hooked onto the uppermost boom section. The splice is disconnected and the section is lowered to the floor. The remaining boom sections are similarly dismantled, the falls unreeved, and the lead line lowered to the hoist below, coiled on the floor, or left snubbed until used again to power other equipment.

Planks are laid so that the various parts of the dismantled derrick can be rolled out to the edge of the building and lowered to the ground, preferably landing them directly on trucks or trailers, in cars, or on other carriers for removal. Timber or steel rollers, or a timber dolly, can be used for bringing the pieces out to the edge of the building.

7.7   Jigger Stick

For this lowering operation, several devices can be used. One of these is the jigger stick. This is a heavy timber or steel beam, or a pair of beams or channels fastened together with diaphragms or timber blocking using long bolts or rods through them. It is cantilevered out beyond the edge of the steelwork, on a floor above the new working floor where the various sections have been dismantled. The inner end of the jigger stick is lashed securely or tied down to the permanent steelwork. The permanent structure must have the strength and stability to withstand the stresses to be imposed. A set of blocks is secured at the outer cantilevered end and preferably reeved with the original derrick load-falls rope. In this case the load line was snubbed and held until needed. By means of snatch blocks this line then leads to the upper block on the jigger stick. The hook on the lower block or on the overhauling ball shackled to the lower block is used to lower each section and the other parts of the derrick. At least one tagline is used, but preferably two, one at each end of the piece, to control the pieces on the way down. This prevents the pieces from turning, which could twist the new load falls or let the pieces strike the structure.

7.8   Chicago Boom

A Chicago boom can be set up if a jigger stick is not used. This should be done by the derrick before it fills in the steel between itself and the edge of the building where the Chicago boom is to be used. A derrick boom, or a boom designed and built specifically for the purpose, is used. A boom seat is secured two or three floors below the roof to an outside column (Fig. 7.8.1). This seat consists of steel framing that can be clamped to

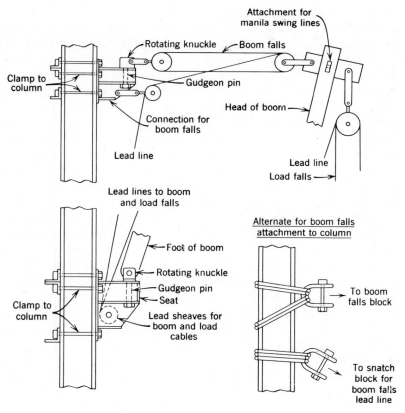

FIG. 7.8.1.   Chicago boom.

the column or to a beam if of sufficient size and strength to withstand the horizontal and vertical reactions of the boom. A knuckle in this seat permits horizontal rotation by means of a vertical pin through the knuckle, and vertical rotation of the boom by means of a horizontal pin through the knuckle, or a vertical pin is used with a horizontal pin hole in the head.

The seat is secured to take both vertical and horizontal stresses, and it should be installed just above a column splice if possible, the splice plates preventing it from slipping down the column. The clamps generally use long rods with threaded ends to bolt the seat on the outside face of the column to temporary angles, etc., on the inside face. This provides horizontal resistance when the boom is swung to either side. The boom is pinned to the knuckle by means of the horizontal pin, permitting the head of the boom to be raised and lowered, while the vertical pin permits the boom to be rotated with the knuckle in a horizontal arc.

Meanwhile a heavy shackle is lashed two or three floors above on the same column. A set of blocks is connected to the shackle to be reeved with the original derrick boom falls line for raising and lowering the Chicago boom, or a structural component is clamped to the column above with a provision for reeving the boom line to the new boom falls and for attaching the upper boom falls block. This must be done in such a way that the block can swivel to permit the boom falls to move with the boom raised or lowered or swung to either side. Since the boom is generally longer than the vertical distance between the seat and the upper attachment, provision must also be made to prevent the upper attachment from sliding upward on the column when the boom is in its highest position.

When the Chicago boom is used, neither the load line nor the boom line originally powering the derrick is lowered or coiled. Instead, the original derrick boom line is used, through snatch blocks, to power the new boom falls on the Chicago boom, and the load line is similarly used to power a load falls secured to the head of the Chicago boom. Long manila lines (point lines) are secured on each side of the head of the boom to be used to swing it manually in one direction or the other. Extra guys should be installed from the point where the topping lift blocks are secured to the column, into the structure away from the boom, and to either side to resist the pull of the boom falls at that point. The new load falls is now used to hook onto and lower the various parts of the dismantled derrick and the floor planks both from the working floor where the derrick was dismantled and the previous working floor one tier above.

7.9 Jinniwink

Instead of a jigger stick or a Chicago boom, a jinniwink is often used. The jinniwink is made up of five main members fastened together, plus a boom, boom falls, and load falls (Fig. 7.9.1). Normally each of the main members is light enough to be handled safely by two men. The members are secured to each other by steel pins, with threaded ends for nuts to hold them in place, or by long bolts. Straight steel pins can be used instead, with cotter holes, using cotter bolts and nuts rather than cotter pins, which often come out of place or break off.

Two upright sloping members are connected to a horizontal front sill of timber or built-up structural steel member, which must be securely tied down at its ends. The sloping pieces in turn are connected at the top to form a vertical A-frame that acts as a mast. These members, are also made of timber or built-up structural steel or beam sections. They should be guyed temporarily in place when setting up the jinniwink and later when dismantling the rig. A horizontal member (back sill) is connected to the center of the front sill, preferably at right angles to the front sill. The rear end of the back sill must also be adequately tied down. A backleg connects the rear end of the back sill to the back of the top of the A-frame.

A set of falls is connected to the other side of the top of the A-frame to be actuated by the boom rope from the dismantled derrick. A set of wire-rope falls can be used that is operated by a hand crab or winch secured

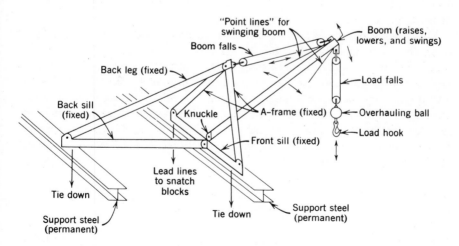

FIG. 7.9.1. Jinniwink.

waist-high to one of the A-frame legs. A knuckle is located at the center of the front sill, with a vertical pin permitting horizontal rotation and a horizontal pin to which the bottom end of the boom is secured, permitting the boom to be raised and lowered; or a vertical pin is used with a horizontal pin hole in its head. The head of the boom is held by the falls that is secured to the top of the A-frame. A load falls from the head of the boom is usually powered by the load line that originally operated the derrick load falls, leading through snatch blocks to the hoist below.

The jinniwink can be set up, except for reeving the boom and load falls, while the derrick is still operating; it is placed on the topmost steel, to which the front and back sills are lashed or otherwise tied down. The front sill is usually set at right angles to the edge of the building since the boom can only swing 180° from one of the A-frame legs to the other. In this way the boom can reach into the structure, lower its load falls through the open steelwork, pick a section of derrick or a load of planks, etc., lift it clear of the top, and swing it out over the edge, well clear of the structure below, to lower it to the ground. Again, taglines must be used for safe operating, to hold the load being lowered from turning and thus twisting the load falls, and to keep the load clear of the structure.

When the derrick has been completely dismantled and removed from the structure, the jinniwink in turn is dismantled. The falls are unreeved, the boom removed, and the boom splices disconnected in order to handle the individual sections easily. The A-frame is guyed as when setting up the rig, the backleg removed, and the various other members disconnected and lowered to the ground by falls either through an open shaft or on a construction hoist elevator.

If the building is low and a crane is available with a boom that can reach the working floor, this may be an easy way to load out the various derrick sections instead of setting up a jigger stick, Chicago boom, or jinniwink. The ideal situation would be to use the steel erector's derrick to set up another contractor's derrick (usually a stiffleg derrick) on the topmost steel before closing in. Then this new derrick is used to dismantle the steel erector's rig and load out the various parts. This is frequently done at no cost to the steel erector if the steel erector had set up the other derrick at no cost to its user.

7.10   Moving a Guy Derrick

It is sometimes necessary to move the derrick as steel framing steps back where setbacks occur on a tall building, or for other reasons. Extra jumping beams are laid on the steelwork in the direction of the move, the ends abutting the ends of the jumping beams on which the footblock is resting. The boom is lowered and laid down on rollers, or on a rolling dolly, either in the direction of travel or in the opposite direction. This is done to distribute the weight to be moved and to reduce the concentration of weight under the footblock. The lower ends of the guys are disconnected one by one after their lower ends have been secured by manila falls, and the turnbuckles are loosened. With the mast now stayed by the manila tackle on the guys, the column anchor slings and turnbuckles are moved to their new locations where they will be used again to hold the mast in the derrick's new position.

The guys in the direction in which the derrick is to be moved are now transferred, one by one, securing the manila falls to the column anchor slings or other slings at the new forward locations. The footblock is jacked up enough to slide rollers of heavy pipe or solid steel underneath it, using planks between the bottom of the footblock and the rollers. Planks are laid on the jumping beams for these rollers for the full distance the derrick is to be moved. Two sets of falls are attached to the footblock, one in the direction of travel for moving the derrick forward and one in the opposite direction to help hold back and thus control its movement and prevent the derrick from rolling out of control. As the forward tackle is taken in, the rear tackle is paid out. At the same time, the falls on the forward guys are taken in, the rear guys falls are paid out, and the falls on the side guys are likewise controlled so that at all instants the mast is kept plumb in all directions.

If the move is a long one, it may be necessary to stop and move the falls that hold the side guys, one by one, to the new locations of these guys. Once the derrick is in its new location the guys are again connected, one by one, to their respective turnbuckles, which were moved ahead earlier, and the turnbuckles are retightened to hold the mast. (The reason for stressing the changing of the guys one by one is that enough guys must always be acting to hold the mast from falling over in any direction.) Occasionally a stiff strut is installed between the mast and the boom after it has been laid down. This helps to prevent the mast from going over in that direction.

Some erectors prefer to grease the tops of the beams or skids on which the derrick is to be moved in order to avoid jacking and installing rollers. Although the skidding can be done in this manner, it introduces an unnecessary hazard of men slipping on the greased beams or skids, or of getting grease on their shoes, clothes, or gloves. Any of these can result in an accident through slipping and falling, or a man may be unable to hold tools securely in greasy gloved hands. The grease must be thoroughly cleaned off the beams before they can be used again.

7.11   General

When a guy derrick, or for that matter any type of derrick, is first assembled, all parts requiring lubrication must be checked. Most important are those parts inaccessible after assembly, such as the casting on the bottom of the mast that fits into its opposite-shaped casting in the footblock; the bronze or other bearing plate, washer, or roller-bearing or a ball-bearing race in the footblock; and the brass or other antifriction plate washer between the spider and the head of the mast. Old caked grease or oil must be cleaned out and fresh lubricant applied.

All points that can be reached after the derrick is operating should be lubricated whenever required, and all such points should be inspected regularly. Wire-rope clips on the guys should be inspected frequently and the nuts tightened as needed. After every jump or move the entire derrick should be checked carefully to make sure everything is in proper working order. The wire ropes must also be inspected for excessive wear or abrasions, broken wires, and proper seating in the various sheaves over which they operate. Blocks should be examined for unusual wear of sheaves, which is usually a warning of wire rope leading improperly over them, or even of a rope having jumped a sheave probably resulting in damage to both the rope and the sheave. In such a case, it must be decided promptly whether or not to replace the rope. Wire-rope clips on becketed ends of boom, load, and runner lines must not be overlooked. Load hooks should be inspected for evidence of damage, distortion, or defects.

8

The Stiffleg Derrick and Travelers

8.1 Construction

Many years ago stiffleg derricks were frequently used to erect structural-steel tier building frames. Their use for this work was unsatisfactory because of the difficulties encountered in jumping them upward and their usually excessive weight. Guy derricks were found to be much more efficient for erecting tall tier buildings. The stiffleg derrick was found to be excellent for erecting certain other types of structures.

The stiffleg derrick consists of six "stiff" members: a boom, a vertical mast held upright by two sloping back legs, and two horizontal sills, plus the wire-rope topping lift or boom falls, the main load falls, and usually a runner or whip line (Fig. 8.1.1). Just as the derrick with its mast held in place by guys is known as a guy derrick, the derrick with its mast held in place by stiff back legs is called a stiffleg derrick. Originally the stiff members were made of timbers—round, square, or rectangular in cross-section—but with the improvement of steel and, for reasons outlined in Sec. 7.1, timber construction was generally replaced by steel construction. However, some small, light-capacity stiffleg derricks are still made of timber.

The steel boom is usually constructed in spliced sections and fabricated from plates and angles, or pipe sections, using angle or pipe sections for the four corner main members, and angle or pipe sections as lacing on the four sides. It is secured to the foot of the mast or to the bullwheel by a boom heel pin, permitting it to rotate vertically on this pin, being raised and lowered by means of the topping lift or boom falls, similar to the operation of the guy-derrick boom. The mast is fabricated from plates and angles or channels, and unless it is short enough for normal shipment on cars or by truck, it will also be spliced part way up (Fig. 8.1.2).

229

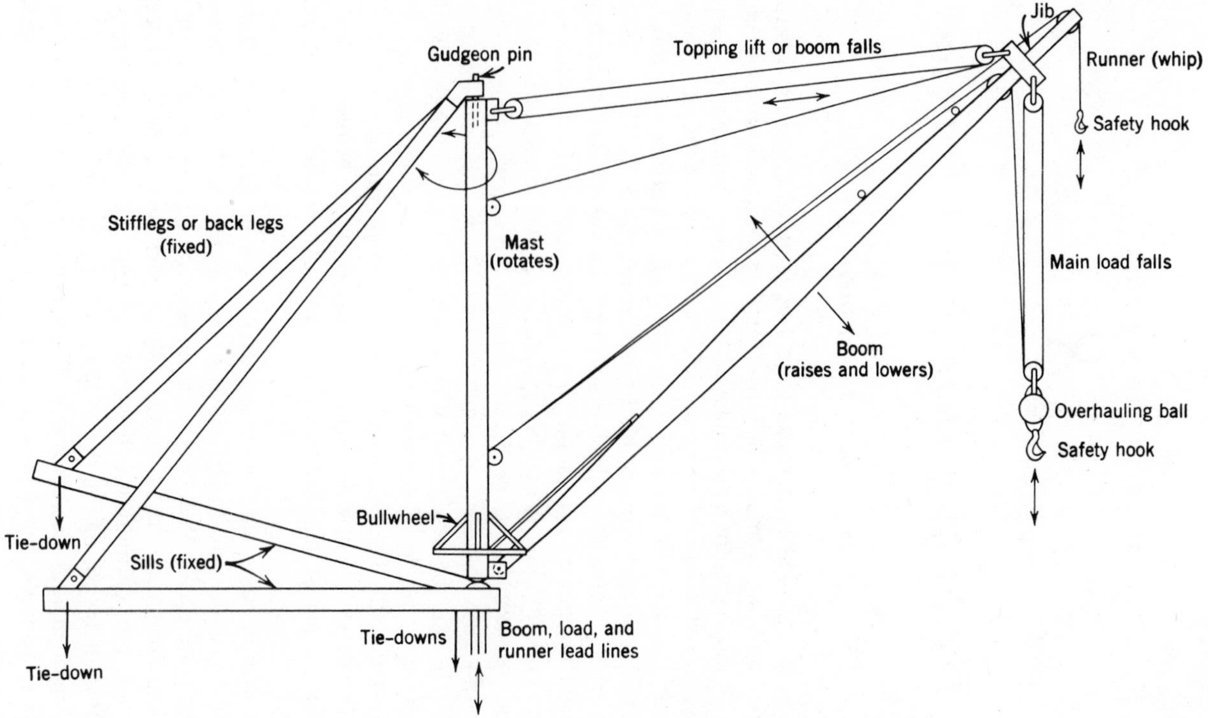

FIG. 8.1.1. Stiffleg derrick.

Labels in figure:
Runner (whip)
Safety hook
Main load falls
Overhauling ball
Safety hook
Jib
Boom (raises and lowers)
Topping lift or boom falls
Mast (rotates)
Gudgeon pin
Stifflegs or back legs (fixed)
Bullwheel
Boom, load, and runner lead lines
Tie-downs
Sills (fixed)
Tie-down
Tie-down

Unlike the guy derrick, where the bottom support is always in compression, at times the bottom of the stiffleg mast can have either tension or compression. Accordingly, the ball-like (convex) casting at its foot is supported by a casting with a suitable antifriction plate washer or bearing on which the bottom casting on the mast itself can rotate, and it is generally held down in place by means of a split, concave casting shaped like the mast foot casting, but acting as a locking ring to prevent the mast from lifting up. These castings have a vertical hole through the center for the lead lines as described for the guy derrick foot castings (Sec. 7.1). A heavy, vertical steel (gudgeon) pin is fastened to the head of the mast, long enough to hold a casting in which it rotates. There is a separate casting above this one to which the bails or straps will be fastened for the topping lift or boom falls. The back legs are fastened to the casting in which the gudgeon pin rotates.

Because the boom is customarily longer than the mast, there will be an upward pull by the topping lift on the upper casting on the gudgeon pin in certain positions of the boom. This will produce tension in the mast and back legs. There must be some form of keeper to prevent the casting from being lifted off the gudgeon pin. Some designs will have the pin for the straps holding the topping lift upper blocks fastened directly through the gudgeon pin to eliminate the upper casting. There will also be an antifriction plate washer, bearing, or bushing between the castings and the surface of the head of the mast directly below the castings.

The castings at the foot of the mast are supported on a built-up support to which the front ends of the two horizontal sills connect. The sills can be spread apart at various angles usually up to 90° with provision to permit angular spreads generally in multiples of 15° from 45° to 90°. This is done by means of plates with appropriate holes that connect the sills to the built-up support or footblock. The sills are made of sections fabricated from plates and angles or channels. If the sills are excessively long, they are spliced for ease in handling and shipping.

The back legs are also made of built-up sections, spliced if necessary. They are pinned at the bottom ends to the rear ends of their respective sills. At their upper ends they are pinned to the "ears" on the casting in which the gudgeon pin will rotate. This casting must permit the same angular spread as the plates that connect the sills. The purpose of the sills and back legs is to support the mast in a vertical position and to transmit the stresses induced by the boom falls (from the boom and the load being handled), through the back legs to the sills, and through the sills back to the foot of the mast.

There can be tension as well as compression in these back legs, depending upon the direction of the boom and the height to which it is raised. For this reason, the back ends of the sills must be tied down adequately or counterweighted to resist the uplift developed at these points. Similarly, the front end of the sills at the foot of the mast must be tied down or counterweighted to resist the uplift that might be developed in certain positions of the boom.

The boom and mast of a guy derrick can be rotated 360° as a unit. The boom and mast of a stiffleg derrick can only rotate as a unit from the position of the boom against one back leg to the position of the boom against the other back leg. Normally the legs are at 90° to each other, which gives the best stability to the derrick and yet provides a reasonable working area for the boom to swing and handle loads. If the boom needs to be rotated more than the 270° thus provided, the angle between both the legs and sills must be reduced to 70°, 60°, or 45° as described above. This permits an additional rotation of the mast and boom. There may be cases in which there is no support for the back ends of the sills if they and the legs are spaced at 90°. Then the sills and legs are connected closer together as noted. In a few cases the boom must be installed between the back legs, that is facing toward the rear. The procedure for setting up the derrick in such a situation is similar to that used when the boom is operating normally.

A topping lift or boom falls is secured between the bails or straps from the casting at the top of the mast and bails or straps are connected to the upper side of the head of the boom. The boom lead line runs over a sheave located well up on the mast and leads down through the mast over lead or idler sheaves in the footblock. A set of main load falls is reeved to the head of the boom on its underside. The load line is then passed over a lead sheave in the boom head to a lead or idler sheave part way up the mast and then down to a second sheave in the footblock, next to and on the same shaft as the boom falls lead line sheave. Rollers or timbers are usually mounted at several points on the upper side of the boom to prevent the wire rope from chafing or rubbing on the steel lacing of the boom, which might damage both the rope and the boom.

An additional sheave is located in the head of the boom for a whip or runner line of one or two parts. This line then leads to a sheave also part way up the mast, usually placed on the same shaft as the main load lead line sheave. This runner line then passes down through the mast to a third sheave in the footblock, on the same shaft as the topping lift and main load lead line sheaves. Often a jib is added to the head of the boom and cantilevered beyond as much as 10 to 20 ft. In this case the runner line leads directly from the end of the jib. The jib is usually made up of a pair of channels tied together by diaphragms or plates and secured to the boom by steel pins or by a bolted connection.

Since the stiffleg derrick is usually a high-capacity rig, a bullwheel is used to turn the mast (and boom). This is similar in construction ·to the one described in Sec. 7.1 and is secured to the mast in the same way.

The three lead lines for boom, load, and runner (or whip), after passing through the foot of the mast and around the three sheaves in the footblock under the mast, turn to lead nearly horizontally to the hoist drums. The hoist is located in the area between the sills and back legs (Fig. 8.1.3) or at a reasonable distance further back but in the same general direction. A timber or built-up structural member is usually installed on top of the sills, but not too close to their intersection under the mast. Two sheaves or two single-sheave wire-rope blocks laid horizontally on their sides are mounted on top of this timber or steel support. These serve as leads for the two ends of the wire rope wound around the bullwheel, the two ends then going to two drums mounted on a single shaft geared to the hoisting engine. These lines are wound in opposite directions on their drums to operate as described in Sec. 7.1. Also, a separate swinger can be used, usually operated by the operator of the main hoist. Occasionally it may be necessary to use a luff tackle to increase the power of the swinger drums in rotating the mast and boom.

8.2 Initial Setup

The various parts of stiffleg derrick are usually shipped completely dismantled. Since the sections are generally heavy, it is best to have an auxiliary piece of equipment on hand, such as a crawler or truck crane, to unload and assemble the derrick. If such a crane is available, the various parts are unloaded by it from the delivering carriers. The sections should be checked for possible damage in transit, which should be reported to the carrier's agent promptly for a possible claim later and to permit him to inspect the damage. Any damaged material should be repaired or replaced if there is any possibility of equipment failure as a result of the damage.

The sills are assembled, if spliced and are laid in place as level as possible where the derrick is to operate initially. They are spread apart to the prearranged angle and fastened at their front ends to the gusset plates on the footblock that is to support the mast. In assembling the sills, back legs, and mast, if they are spliced, extreme care must be taken in each case to have the blocking on which the sections are laid at the correct height so

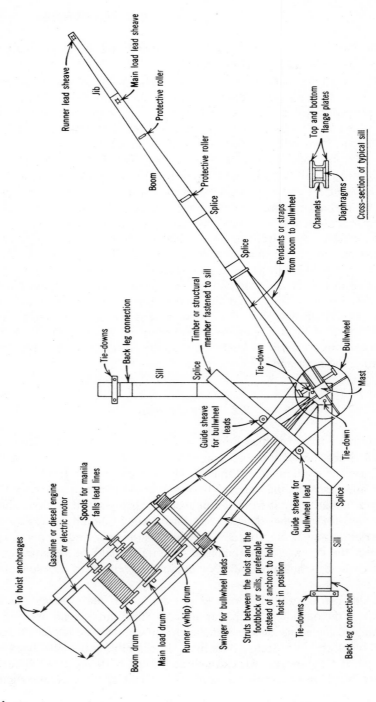

Runner lead sheave

Main load lead sheave

Jib

Protective roller

Boom

Protective roller

Splice

Top and bottom flange plates

Splice

Channels

Diaphragms

Cross-section of typical sill

Pendants or straps from boom to bullwheel

Tie–downs

Back leg connection

Timber or structural member fastened to sill

Sill

Splice

Tie–down

Bullwheel

Mast

Guide sheave for bullwheel leads

Tie–down

Gasoline or diesel engine or electric motor

Spools for manila falls lead lines

Guide sheave for bullwheel lead

To hoist anchorages

Splice

Sill

Boom drum

Main load drum

Runner (whip) drum

Swinger for bullwheel leads

Struts between the hoist and the footblock or sills, preferable instead of anchors to hold hoist in position

Tie–downs

Back leg connection

FIG. 8.1.3. Layout of stiffleg derrick and hoist.

234

that the sections line up exactly with each other. In addition, the sections must be assembled in their correct positions relative to each other. All splice bolts should be installed and properly tightened.

The hoist should be on hand at this time and can be unloaded from its carrier to the ground by means of skids supported on blocking. This can be done while the derrick members are being assembled. Rollers are inserted under the frame of the hoist by jacking. The machine is rolled down the skids and over planks or timbers, laid for the purpose on the ground, into place where it is to operate. Here it is anchored to resist the pull of the wire ropes to the derrick or to falls used in setting up the derrick. Instead, struts can be installed between the hoist and the derrick to resist these forces.

A set of falls must be used to hold the hoist back as it is rolled down the skids to prevent it from coming down too quickly and "getting away," which can cause injury to the men or damage to the hoist. If it rolls easily, crow bars can be used to move it. If not, manila falls can be used to start it onto the skids and move it on the ground.

If a crane of suitable capacity is on hand, it can be used to unload the hoist and place it in its proper location. The ends of the boom, load, and runner ropes are secured to their drums by means of wire-rope clips, clamps, or other devices suitable for those drums. The ropes are wound on their drums, generally with the boom on the top or rear drum, the main load on the center drum, and the runner on the front or bottom drum, to be ready for use when needed to reeve the different sets of falls to operate the derrick.

The crane assembles the mast if it is spliced. The head casting for the back legs is placed on the gudgeon pin, making sure the gudgeon pin is properly fastened inside the mast. Any old congealed lubricant on the gudgeon pin or on the mast foot castings, which will be inaccessible later, should be cleaned off and fresh lubricant applied where required. As described for the guy derrick, but applicable to any type of erecting equipment, care must be exercised to assemble all parts correctly in order to prevent overstressing any part beyond the designed stress, for example, assembling boom, mast sills and back legs in correct alignment and with the sections rotated to match.

If the bails for the topping lift are pinned to a separate collar that fits over the gudgeon pin, this collar should be in place before setting the mast upright. Instead of three lead sheaves on the face of the mast, some masts have an offset cap known as a "rooster" on top of the gudgeon pin. This has three separate sheaves on one shaft for the two lead lines from the head of the boom for the main load and runner, and the one lead for the topping lift. These three pass over the sheaves in the rooster, down through a vertical hole in the center of the gudgeon pin and the mast, to the three sheaves under the mast in the footblock, and then to the hoist drums. In this case the rooster should also be in place before picking the mast, and the sheaves should be checked and lubricated. The back legs, if spliced, are assembled next on blocking.

At this time it must be decided whether to reeve the topping lift before or after setting up the mast. Some erectors prefer to send men aloft on the top of the mast later, laboriously pulling the wire rope up to form the topping lift, threading the rope through the proper sheaves, and overhauling the rope to lower the end to be threaded through a sheave in the boom head block. The end is then pulled up again by a manila rope and the process is repeated over and over, since many stiffleg derricks have as many as 30 or more parts in the topping lift. If it is decided to reeve the topping lift before setting up the mast, then the lead line from the boom drum is fed through its sheave in the footblock, through the bottom mast castings, through the mast, over the lead sheaves on the mast or in the rooster to the lower block of the topping lift system and over the correct sheaves in the upper block now held by the bails to the head of the mast, the mast being still on the blocking on the ground on which it was assembled.

At this time the lower load block does not need to be secured to the boom, which can be assembled later. The reeving for the topping lift should be made as shown on the erection-scheme drawing since an efficient reeving can be somewhat complicated; one of 30 or more parts just mentioned generally requires tandem blocks of as many as 16 sheaves each. The outer ones will have a larger diameter than the inner ones so that the parts will not foul each other. The parts must be reeved to balance so that the blocks will stay in line as the boom is raised and lowered.

Four sets of manila falls to act as temporary guys are secured near the head of the mast, the mast is now picked and set upright in its place on the support, fastening the guys to reasonably equally spaced temporary anchorages to hold the mast safely by itself. The back legs are lifted next and fastened by pins at their upper ends to the ears on the mast head casting and pinned at their lower ends to the sills. The temporary guys on the mast can now be removed.

The boom is assembled next, preferably with the bottom end pinned to the mast by its boom heel pin. Blocking is laid, either level or sloped, so that when the boom sections are all in place and the splices are completely bolted or pinned (if pin-connected), the centerline from head to foot will be in a straight line. The lower topping-lift block is secured to the bails from the upper side of the head of the boom. If the topping lift was not reeved earlier, it should be done at this time, sending one or two men aloft, up one of the back legs to the head of the mast, to pull up the lead line and reeve the blocks as noted above. Ladders are installed on some large masts for men to reach the top so that they do not have to climb up a back leg.

While the boom is being assembled, tie-downs should be installed at the outer ends of the sills and at the inner ends as close to the mast as the construction of the derrick permits. On some derricks, holes for these inner tie-downs are provided in the gusset plates that connect the front ends of the sills together and to the built-up supporting footblock. Counterweights must be laid on the sills if tie-downs cannot be used because the support under the derrick does not permit this simple method of resisting uplift. A small platform is then generally needed over the sills, on which to place counterweights of metal or concrete blocks; or a bin can be built into which to load sand or other material to act as a weight.

With the topping lift in place, if a jib is to be used, it should be installed at this time, unless the erection scheme requires it to be left off temporarily because the added weight of the jib may overload the boom in handling near-capacity loads. In this case the jib is installed after the heavy loads have been erected. A jib to lift a light load is generally bolted or pinned directly to the head of the boom, either in line with the centerline of the boom as an extension to give additional height, or at an angled offset so that the runner single-part or two-part falls will not foul the main load falls. But this does not appreciably increase the height or reach for handling loads.

When the jib is an extension of the boom, if a substantial load is to be picked by the jib falls, or if the jib is of light construction, it may be hog-rodded to the boom (Fig. 8.1.2). This is done by means of a strut (known as a "horse") on top of the head of the boom, with straps, angles, or wire-rope pendants from the end of the jib to the top of the strut, and from the top of the strut to a point on the upper side of the boom. This strengthens the jib by forming a simple vertical truss with the jib and the boom forming the bottom chord of the truss.

With the jib in place, the boom is lifted by means of the topping lift just high enough to install the upper main load block. The lower load block should be directly below it on the ground or on blocking. The main load-falls lead line is now reeved through the sheaves under the mast, through the lead sheaves on the face of the mast or through the rooster, and through the lead sheave near or at the head of the boom, and to the blocks. The reeving should again be as shown on the erection-scheme drawings, based on a balanced reeving so that the lower block will tend to hang level with or without a load, and the falls will resist twisting.

The overhauling ball is now shackled to the lower load block and the derrick is ready to operate. Some erectors prefer to use a heavily weighted lower load falls block, eliminating the overhauling ball. Although this decreases the drift required in setting steel, it increases the weight of the block so appreciably as to make it difficult to maneuver it into place for reeving. If a jib has been installed, the runner line should also be reeved at this time. Since most stiffleg derricks are intended to handle heavy loads and thus have the main load falls reeved with many parts, most erectors use two parts in the runner reeving so that the slower, many-parts main load falls need not be used for moderate or light loads that can be handled faster on the runner with only two parts of wire rope.

If no crane or auxiliary powered equipment is available to set up the derrick, the parts must be unloaded by hand. In this case it is advisable to deliver the hoist first, unloading, positioning, and anchoring it for use. Manila-rope falls are usually ample to snake the sections of sills, mast, back legs, and boom from their carriers by means of skids, blocking, and rollers or dollies. The sections are assembled on blocking by means of these falls, and planks or skids are laid on other blocking at various heights powering the lead line of the rope falls by using a spool usually provided on the hoist. A pneumatic or hand-powered winch can also be used.

The sills are placed at their predetermined angle and in their correct location and elevation. The boom is assembled next on blocking, but with only enough sections in place for the load falls, when reeved, to reach above the top of the mast. The boom is assembled with its foot near where the mast will be set up and with its normal underside uppermost so that the load falls will be on top of the assembled short boom. The splices are bolted or pinned (if pin-connected) and the load falls is reeved directly from the hoists to the head of the boom, leading through a snatch block anchored near the foot of the boom. There should be only as many parts in this temporary load falls as will be needed to pick the mast.

At least four sets of manila falls are fastened, one end of each to the head of the boom to act as temporary guys, and the other ends to four equally spaced anchorages around the foot of the mast. Timbers are placed under the foot of the boom to protect it as the boom is rotated upright. Lashing or slings are secured from the foot to temporary anchorages to act as kickers to resist the horizontal forces produced in tripping the boom upright. The boom is stood up by means of the main load falls (Fig. 7.2.1), using the four manila falls to steady it and keep it rising in a vertical plane. Once the boom is upright on the timber protecting the foot, the four manila falls guys are tied off to their temporary anchorages to hold the boom in place.

With the boom acting as a gin-pole (Sec. 9.11), the load falls is used to set the mast and back legs in place. The boom is then lowered back to the ground by reversing the previous operation, that is, the load falls is anchored again, the lead lines to the temporary guys of manila falls are untied, and as the load falls is slacked off, the manila lines are taken in or paid out (as needed) to keep lowering the boom. in a vertical plane. As soon as the boom is down the load falls is unreeved and the splices making the short boom are disconnected. The boom is now assembled in its proper length, with the load-falls side underneath and the topping lift side on top. The topping lift, main load, and runner are now reeved and the boom is installed in its working position.

8.3   Dismantling

There are many ways to dismantle the derrick. If a crawler or truck crane or other derrick is available, it may be satisfactory reversing the sequence described for setting up the derrick. The boom is lowered, the falls unreeved, and the boom sections dismantled. The mast is again guyed temporarily as before, and the back legs are removed to be dismantled. The temporary mast guys are removed after hooking onto the mast with adequate slings. The mast is lowered after releasing the locking ring castings at its foot, laid down, and the splices disconnected. The tie-downs or counterweights are removed and the sill splices disconnected.

If the stiffleg derrick has erected steel over itself satisfactory for the purpose, this steelwork might be used to dismantle the rig, following the procedure described for a similar situation in dismantling a guy derrick (Sec. 7.6); or a Chicago boom can be set up on the face of the permanent structure for the purpose (Sec. 7.8).

It may be necessary to dismantle the derrick by using its own boom as a gin-pole if there are no auxiliary powered aids. The boom is lowered onto blocking by the topping lift. The load falls, runner, and the topping lift are unreeved and the ropes wound on the hoist drums. The jib is unfastened next from the boom, and the boom splices are disconnected. The boom is unpinned from the foot of the mast and is reassembled by hand to a shorter length—just enough to reach the top of the mast when the boom is stood up to be used as a gin-pole, with the load block side of the boom now uppermost. The load-falls blocks are reeved directly from the hoist through a snatch block to the head of the newly assembled short boom. Using the same process described for tripping up the boom to assemble the derrick, installing foot kickers, and using temporary manila falls for guys, the boom is stood up alongside the mast, the foot of the boom on timbers to protect it.

Temporary guys are secured to the mast as was done in standing it up. After hooking onto the back legs with adequate slings, each leg is unpinned from the ears on the masthead casting and from the sills and lowered onto blocking for disconnecting the splices, using the boom as a pole to do this. The mast is hooked onto with proper-size slings, the temporary mast guys are removed, the locking ring castings at the foot are disconnected, and the mast is lowered onto blocking for disconnecting the splices. The tie-downs or counterweights are removed and the sills are next disconnected from the footblock and their splices disconnected.

The load falls is again secured to the special anchorage so that the boom can be slacked down to the ground safely. The four manila falls acting as guys are used to keep the boom in a vertical plane as the load falls is slacked off and the boom rotated back onto the blocking on the ground. This new load falls is unreeved, and the wire rope is wound on the hoist drum to be shipped away on the hoist with the topping lift and runner ropes on the other drums; or the ropes are unwound from the hoist drums onto wooden reels for shipment elsewhere. The boom is disconnected, and the kickers used to hold its foot in place are removed.

Rollers or dollies are used to good advantage, jacking the various pieces enough to slide the rollers or dollies underneath and on top of planks or timbers, depending on the need due to the weights to be handled. By means of manila falls powered by the spool on the hoist, or a pneumatic or hand-powered winch, the various sections are then skidded or rolled over these planks or timbers on blocking to carriers for shipment away from the site.

8.4   Travelers

Frequently a long structure can best be erected by one or two rigs traveling its length. This is especially true when it is a low, heavy-membered building. If cranes of sufficient capacity are available, it may be preferable to use them rather than derricks. However, there are some situations when cranes cannot be used because of ground conditions or other reasons. Then, instead of using derricks in place of cranes, travelers may be best, eliminating the need to move a few derricks, or avoiding the need for a great number of such rigs to cover the area. Another example is a building with an excavation so deep that a crane cannot be placed in it easily or removed later, and with more than the usual amount of heavy steel in the area below the steel level. This type of structure lends itself admirably to traveler erection.

If one or more derricks are set up in the deep excavation, steel must be fed in by auxiliary equipment at street level. Instead, a traveler may be safer, more economical and more efficient. Buildings over underground operating railroad tracks are of this type. Where a crane or derrick working in the hole would interfere with the railroad operations, a roadway for a crane to work on would have to be built on top of the steel as erected, for the crane to advance and continue erecting and for trucks to feed steel to the crane.

In this type of situation, a platform to support a derrick should be constructed at the edge of the excavation. The various members of the platform or traveler are unloaded and assembled, preferably by auxiliary powered equipment. The parts are bolted together for ease in dismantling and for economy in installation. The derrick, usually of the stiffleg type, is then unloaded and set up on this platform (Fig. 8.4.1).

A guy derrick can be used if the platform is large enough so that the guys, fastened to the platform, will be far enough away from the mast to act effectively for the loads to be handled. They must be at least far enough away to hold the guy derrick safely when the platform is moved, and suitable anchorages further out must be available after each move. This procedure is not desirable because it takes longer, costs more, and introduces unnecessary hazards in disconnecting the guys from the platform to move them out far enough to make it safe for the derrick to handle the loads after the traveler completes each move.

A stiffleg derrick on a platform eliminates the hazards of changing guys. It is ready to work immediately after each move is completed and the platform is tied down. Adequate counterweights can be used on the derrick or platform, with the derrick sills connected to the platform. The derrick is erected on the platform with the hoist located on the platform (Sec. 8.2). Counterweights will be required to start the operation of the stiffleg derrick erecting one or two bays ahead of itself at the edge of the excavation. The platform, if properly designed, will have such a span that it can be rolled or skidded over two lines of permanent steel. The particular lines of beams or girders on which it will rest when operating, and on which it will be moved, should have been thoroughly checked in advance for their strength and stability to support the traveler, to permit it to move over them and to provide sufficient weight, if possible, for tie-downs to be placed around them. If these supporting members of the permanent structure are otherwise satisfactory, counterweights can be used instead of tie-downs, but this increases the load that must be carried, makes moving more difficult, and is usually slower because of moving the additional weight.

If no auxiliary powered equipment is available to assemble the platform and erect the derrick on it, anchorages will have to be found so that the derrick boom can be used as a gin-pole. It is stood up as described earlier for assembling the stiffleg derrick when no powered equipment is available, and is used to assemble the platform and then the derrick on the platform. The procedure for using the boom as a gin-pole is described in Sec. 9.11.

After the derrick erects one or two panels ahead of itself and to either side, the tie-downs are released, the platform is moved ahead on this newly erected steelwork, tied down again, and the derrick repeats the operation until reaching the other side of the structure. The platform is moved each time as the steel ahead is erected, fitted, and bolted (preferably with permanent bolts). With welded construction the traveler may have to be tied up while the permanent connections are welded sufficiently for moving the traveler ahead safely. On reaching the other side of the structure, after erecting and moving the traveler repeatedly, the derrick is dismantled, the platform members disassembled, and a crawler or truck crane can be used to good advantage to load out platform hoists, and derrick.

Rails are usually laid on the steelwork when travelers are moved. Heavy plates that can take the abuse of moving are installed on the bottom of the platform members to slide on the rails, or wheels may be installed on the underside of the platform to roll on the rails. If the supporting steel is suitable, the wheels can roll directly on the top flanges, taking care to keep the wheels on the steel by means of vertical guards or guides. The wheels must not be permitted to roll off to either side. When wheels are used the platform must be jacked up off the wheels and blocked after each move before the tie-downs, if used, are tightened to secure the platform in place before picking any loads. If counterweights are used, the platform must be jacked up to avoid overloading the wheels, after which adequate blocking is placed to transmit the loads directly from the platform to the supporting steelwork. Grease can be applied to the rails for sliding, but if this is done, extreme care must be taken to clean the grease off completely.

When moving a traveler two sets of falls should be used: one to move the traveler ahead, and one to hold it back so it does not move too fast or too far. These falls can be manila falls of enough parts for the weight to be moved. The lead from the forward falls can be actuated by a spool on the hoist, the rear falls line being snubbed and paid out as needed. Wire-rope falls can be used, reeved up to a drum on the hoist, if available, or even to a separate small hoist installed for the purpose on the platform of the traveler. Before every move a traveler should be checked carefully to make sure the trucks or parts involved in the move are in safe, satisfactory condition and are lubricated as required.

In erecting a tall building the steel below street level is often heavy enough to warrant the use of a high-capacity stiffleg derrick, while the framework above ground level is so much lighter that guy derricks are more appropriate. In such a situation the traveler erects as described, and at predetermined points sets up guy derricks on the steel erected. A tall building that covers a large area, over a deep excavation, is most adaptable to this procedure. The last guy derrick set up by the traveler will be used to dismantle and load it out instead of bringing in a crane to do this.

With a long, low building, a stiffleg derrick on a traveler platform can sometimes be used to erect the entire structure from ground to roof, backing away as each bay is erected. But this derrick must have a boom that is long enough to reach over the steel erected on the near side of each panel to erect the steel on the far side of the panels. In addition, the traveler counterweight and the capacity of the boom must be sufficient to handle the heaviest loads, usually at minimum radius, and the lighter loads at the extreme reaches. Here again, a crane may appear to be better suited for such an operation until a check is made and the ground found to be unsatisfactory. With a traveler, timbers may be able to distribute the load under the platform satisfactorily to the ground, whereas the concentrated load on the smaller area occupied by a crane may be too much for the soil to support. Furthermore, highway restrictions may prevent delivering a heavy crane to the site, whereas the lighter, individual members of a traveler platform and the parts of a derrick may be permitted.

There have been instances of long, fairly high buildings that were so wide that two stiffleg derricks on separate platforms traveling on the ground were used to erect the steelwork two or four floors above as they moved back away from the erected steel. (These travelers were assembled as described above for starting a traveler with a stiffleg derrick mounted on it.) Then at the end of the building each one sets up a guy derrick on top of the erected steel at that point. These guy derricks then picked the travelers bodily, turned them around, and set them on top of the steelwork at the new upper level. The travelers then proceeded to take down the guy derricks and continued to erect, moving back toward their starting points, where they again set up the guy derricks. These guy derricks then dismantled and loaded out the stiffleg derricks and traveler platforms, finally dismantling themselves after filling in the steel around them that was left down by the travelers to permit this final operation of dismantling and loading out.

A gin-pole can be used when a crane is not available to assemble a stiffleg derrick or a traveler platform. Although it will be slower and less efficient, the gin-pole may prove to be more economical than shipping a crane to the site. A small guy derrick can be set up just to assemble the traveler and derrick on the platform, leaving the guy derrick in position until all the erecting has been completed. The traveler is then moved back to be dismantled and the guy derrick is dismantled. The sequence when using a pole or a guy derrick to assemble a platform or to set up a stiffleg derrick is the same as when using a crane, except that the method of setting up and reeving a pole (Sec. 9.11) is different.

A special kind of traveler, called a "creeper" traveler, can be used on a very tall building with severely tapered faces. Erecting rigs placed within the limits of the structure would not be able to use the booms that were long enough to reach out to unload steel delivered in the street below.

A stiffleg derrick is mounted on a horizontal, framed platform. The outer edge of the platform is supported by diagonal braces to steel below. Two vertical lines of running rails of structural shapes are fastened to the structure as each tier is erected above the traveler. The platform and the lower ends of the diagonal supports are fastened to these rails after each jump. Lifting tackle is reeved between fittings at the top of each section of rails and the platform when a tier has been erected and the rig is ready to jump.

The derrick and platform then travel on the rails to the new working floor level and are refastened to the rails. The lifting tackle is removed and reinstalled on the tops of the new rail sections when the new tier has been erected and the rig is to be jumped.

8.5   A-Frame Travelers

Occasionally an A-frame derrick (Fig. 8.5.1) is used instead of a guy derrick or a stiffleg derrick on a traveler platform. This type of derrick has very limited areas of operation, and unless material can be delivered either directly in front of it, or on either side ahead of it, the boom will not be able to swing around to unload material to be erected.

Essentially, an A-frame derrick has two vertically sloping front legs fastened by means of pins or bolts to connection plates on the two ends of a front sill to form a vertical triangle or A-frame. From the two ends of

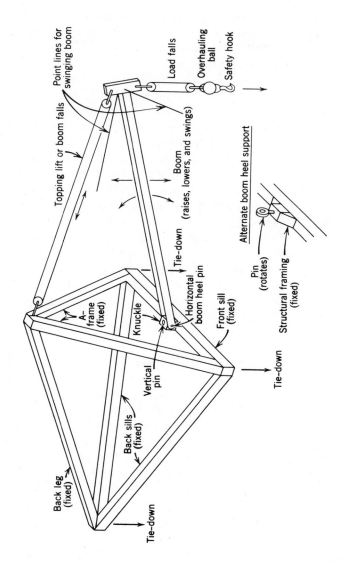

Point lines for swinging boom

Load falls

Overhauling ball

Safety hook

Topping lift or boom falls

Boom (raises, lowers, and swings)

A-frame (fixed)

Knuckle

Tie-down

Horizontal boom heel pin

Front sill (fixed)

Vertical pin

Tie-down

Back leg (fixed)

Back sills (fixed)

Tie-down

Alternate boom heel support

Pin (rotates)

Structural framing (fixed)

FIG. 8.5.1. A-frame derrick.

245

the front sill two back sills are bolted or pinned, and they extend to connect together at their back ends to form a horizontal triangle. A stiff back leg extends from the top of the intersection of the front legs to the intersection of the back ends of the back sills. The front vertical triangle acts as a mast, and a topping lift is fastened to the front of the intersection of the front legs to support a boom hinged to a special knuckle at the center of the front sill. This knuckle is similar to a jinniwink knuckle (Sec. 7.9). The knuckle permits the boom to rotate vertically and swing from against one leg 180° to the other front leg. The swinging of the boom is done by means of "point lines," or falls, to each side. These limitations in swinging and the area in which the boom can operate normally make the A-frame derrick of too little value in building erection. It has its place in erecting a long, low structure where the ground conditions are suitable for either skidding it along or mounting it as a traveler. A bay is erected, the derrick or traveler is moved back to erect another bay, and it continues moving and erecting. As in the case of a stiffleg derrick or a stiffleg derrick on a traveler platform, tie-downs or counterweights must be used. The process of setting up and dismantling is similar to that described for the stiffleg derrick (Sec. 8.2) and the jinniwink (Sec. 7.9), taking care to guy the front legs temporarily as they are set in place and again when removing the back leg during dismantling.

8.6   Two-Boom Travelers

A variation of the stiffleg on a traveler platform is the two-boom traveler. This is the equivalent of a pair of stiffleg derricks mounted on a platform, with the masts near the two front corners of the platform and one leg of each derrick crossing at the front of the platform. A common sill is used across the front of the platform. The back legs are generally placed at 90° to the front crossed legs on a rectangular platform, with separate hoists operating each of the two booms and their respective load falls. If they are located properly on the platform, the two hoists serve as part of the needed counterweight to reduce the size of the tie-down or counterweight requirements. With this type of traveler a heavy girder or truss or other member that is too heavy for one boom can be picked by the two booms jointly, and the balance of the steel can be erected by each boom working on its half of the structure. Two lighter capacity rigs can then be used instead of a rig with one boom capable of handling the excessive weight pieces alone. This in turn speeds up erection since a light boom rig can always work faster than a rig that has a high capacity for only a small part of the work but is too heavy for erecting the lighter steel speedily and efficiently.

The process of setting up and later dismantling a two-boom traveler is similar to that described from the stiffleg derrick traveler. The main difference is that one of the front crossed legs is continuous, while the other is split to connect the two pieces together where they cross the one-piece leg. The platform should be assembled in place on the ground or other structure. The hoists should be placed on the platform and anchored to it. The mast to which the continuous front leg connects is set up and temporarily guyed. Its two legs are installed, connecting between the sills anchored to the platform and the casting at the top of the mast. The temporary guys are removed.

The second mast is then set up and guyed. Its back leg is connected to the remaining sill and to the top of its mast, and then the two parts of its front leg are joined together over the other mast's front leg. The guys are removed and the two booms are connected to their respective masts. Separate bullwheels are put in place to turn each mast and its boom. All reeving is installed as shown on the appropriate erection-scheme drawing. Dismantling will proceed in the reverse order. When powered equipment is available for setting up and later dismantling, this should be used. Otherwise, one of the booms can be used as a gin-pole, but this is usually slow and costly. The cost of bringing in a crane to do the work is usually far less than the cost of assembling and dismantling by a gin-pole, and so the crane is recommended where feasible.

Train sheds lend themselves to traveler erection since permanent tracks can be laid on ties on the permanent roadbed at ground level and used for rolling the traveler. Standard railroad car trucks are placed on the tracks and the platforms bolted to these trucks. In this case tie-downs cannot generally be used, and so counterweights must be laid on the platform or trucks to offset the uplift produced by the derrick picking a load. A tower can be built on the platform and a much smaller stiffleg derrick mounted on top of the tower, high enough to reach the topmost steel framing, but of sufficient capacity to handle the heaviest loads. This rig must be tied down to the tower, which in turn must be secured to the trucks and adequately counterweighted.

Hangars are also adaptable to erection by a stiffleg derrick on top of a wide tower supported by a platform moving on skids, rails, or other supports. As sections of the roof truss or roof arches are erected without setting a complete truss or arch in place, these sections can be landed on blocking on the tower at the correct final elevation. The remaining sections are then connected to these tower-supported portions. In this way the tower acts as falsework for the truss or arch erection and as a support for a smaller derrick to handle all the steelwork from ground to roof. The tower and platform can be made wide enough, if the hangar has a very wide span, to support two derricks; or two separate travelers can be used. In this case the entire truss or arch can sometimes be preassembled and permanently connected on the ground and then picked by the two rigs to be set in place as one piece. This procedure eliminates the need for temporary falsework in case the traveler is not to be used for that purpose. It also eliminates the need for sending men aloft, after connecting separate pieces, to install permanent fastenings in the air. Since no scaffolding or floats have to be placed because no bolting or welding is being done aloft, there is far less hazard.

9

Cranes and Miscellaneous Equipment

9.1 Crawler-Crane Construction

The basic crawler crane for use in steel erection is equipped with a body containing the internal combustion engine that operates the drums for a main load falls, a single or two-part runner or whip line, and a topping lift or boom falls for raising and lowering the boom (Fig. 9.1.1). In addition, the engine is used for rotating the crane body through 360° as well as for propelling the crawler treads. A few models are driven by electric motors instead of internal combustion engines.

The boom is secured through boom heel pins to the front of the crane body, and the topping lift or boom falls is secured to the back end of the crane body. On the heavier capacity cranes the boom falls is secured to a gantry fastened to the back end of the crane-body frame. This gantry is attached in such a way that it can be lowered with an empty boom to reduce the vertical clearance needed for traveling under obstructions when moved on a carrier or shipped from job to job. The entire body rotates on a crawler tread underframe, held either by a central, heavy kingpin, or by means of round or tapered rollers on a large-diameter forged or cast supporting ring or turntable mounted on the underframe.

Hooked rollers are sometimes also provided to bear against the underside of the top flange of the supporting ring. In some models these hooked rollers also act as the supporting rollers riding on the bottom flange of a channel-shaped ring.

The front hooked rollers must be strong enough to take the uplift stresses imposed by the counterweight when the crane is handling no load. The rear hooked rollers (opposite the boom) must resist the uplift forces when the crane is lifting a load. All the upper rollers, and the hooked rollers when they act in both capacities, must take the compressive stresses

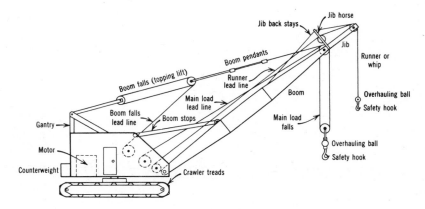

FIG. 9.1.1. Crawler crane.

due to the weight of the crane body itself as well as the additional stresses when the boom is handling loads.

For greater stability the crawler tread frames on the higher capacity cranes are extended outward when in use. They can be retracted to a narrower width as required for loading on a standard railroad flat car or on a carrier for transporting over highways. Most states prohibit a width "over the road" as great as is needed for stability in picking loads of 100 to 200 tons. In fact on some crawler cranes it is even necessary to remove the entire crawler tread mountings, after blocking the crane on its carrier, truck, or railroad car, in order to reduce the width to that required for shipping clearance.

Crawler cranes are available with capacities as high as 500 tons on a basic 80-ft boom. Some with a lower maximum capacity can use a boom and jib to reach almost 500 ft above the ground. These high-capacity, long-boom cranes often weigh considerably more than 400 tons exclusive of the load. The counterweight alone (detachable when necessary to reduce the weight for transporting the crane) often weighs more than 70 tons.

When using extremely long-boom cranes, exceptional care must be exercised to observe dangerous wind velocities and pressures, because a much greater area of the boom can be affected when compared to the shorter booms. Also, when swinging, it must be done slowly enough to keep the load falls vertical; otherwise, if the load were to swing, it might exceed the rated capacity.

Some crane manufacturers are offering crawler and truck cranes that have been modified to permit handling loads far in excess of their normal capacities. They use a basic crawler or truck crane and add a guy-derrick mast and boom mounted on the crane body. Guys are added from the head of the mast to surrounding anchorages. This reduces its mobility since some guys must be slackened and others taken in when the crane must be moved. The guys must then be adjusted for the new position of the crane (Fig. 9.1.2). With this arrangement, such a crawler crane with guy-derrick attachment and guys can lift as much as 600 tons on a 130-ft boom and 150-ft mast at minimum radius; or it can lift more than 60 tons on a 390-ft boom and a 270-ft mast. Without the guys, but with the guy-derrick attachment, it can lift more than 300 tons on a 100-ft boom at minimum radius, or more than 40 tons on a 350-ft boom.

Another type of crane, called a "Sky-Horse®,"* using a mast and boom, leans the mast back until the top is directly over an added counterweight on the back of the crane body, to which the mast is secured by falls and pendants (Fig. 9.1.3). With the Sky-Horse, the mast can be leaned further back until the head is over a counterweight that is equipped with swivel wheels permitting the crane to rotate with the trailing counterweight following on its own wheels on the ground (Fig. 9.1.4); it is able to use a 350-ft boom and 150-ft jib to reach 500 ft.

One manufacturer arranges a circular support around a crawler crane (called a "Ringer®"** crane) with an outrigger fastened to the crane body. This outrigger rides on the supporting ring. In this way the crane, which normally uses no outriggers, is given additional distance beyond the edges of the crawler treads to provide greater moment to resist overturning. With a 60-ft diameter ring, this type of crane is able to lift 600 tons with a 140-ft boom on a 130-ft mast (Fig. 9.1.5), or more than 100 tons on a 240-ft boom at maximum radius.

However, this arrangement limits the mobility of the crane since it must remain in the fixed position where its circular support has been blocked, unless it is raised off the circular supporting ring and the ring is removed, and then the crane uses its treads to move to a new location. Here the ring is again installed and blocked, and the crane is able to operate.

A circular support ring is also used with a truck crane. A special outrigger is added to the truck crane body to ride on the ring. This outrigger is considerably longer than the usual ones on the crane and permits heavier loads to be handled.

* American Hoist & Derrick Co.
** Manitowoc Engineering Co.

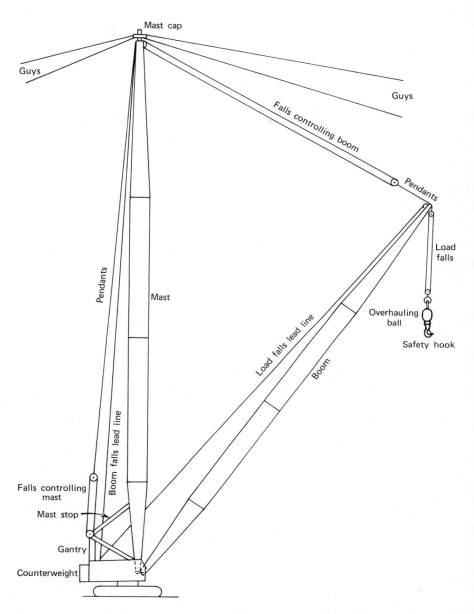

FIG. 9.1.2. Guy-derrick crane (crawler crane with guy-derrick attachments). (Truck crane with guy-derrick attachments is similar.)

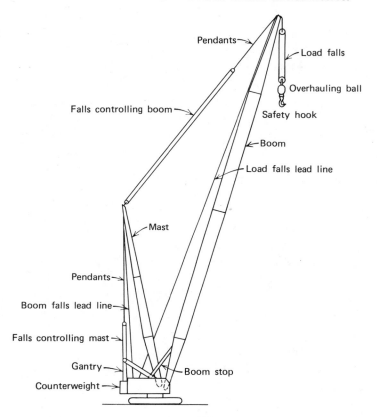

FIG. 9.1.3.   "Sky-Horse®" crane with fixed counterweight. ® American Hoist & Derrick Co.

9.2   Truck-Crane Construction

Truck cranes consist of a truck mounting with a cab at the front end for operating the traveling mechanism. This is separate from the main crane body, which is mounted toward the rear end of the truck frame in such a way as to permit rotation of the crane body through 360° (Fig. 9.2.1). The portion of the truck frame under the crane body is equipped with one or more axles, each with two or four pneumatic, rubber-tired wheels. The front or steering wheels are generally on one, two, or even three axles on the higher capacity cranes at the front end of the truck frame. There may also be one or even two additional axles near the midpoint of the truck

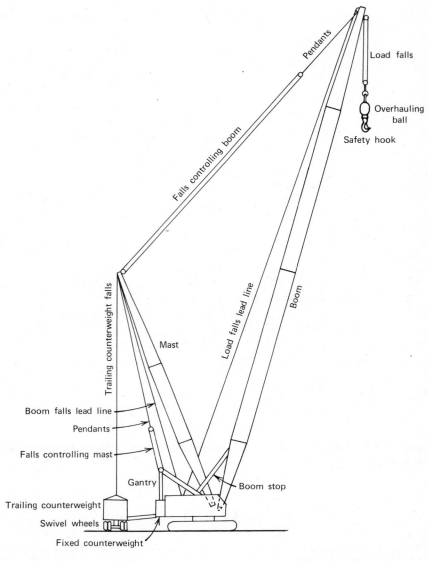

FIG. 9.1.4. "Sky-Horse®" crane with trailing counterweight. ® American Hoist & Derrick Co.

254

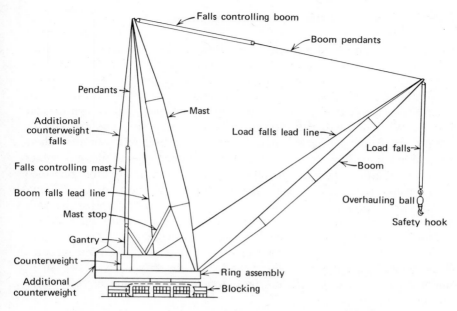

FIG. 9.1.5. "Ringer®" crane. ® Manitowac Engineering Co.

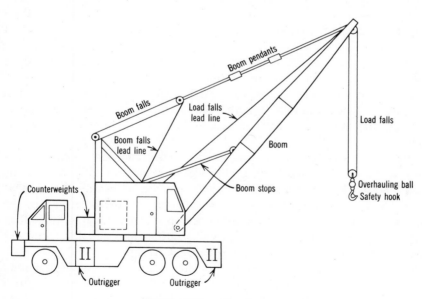

FIG. 9.2.1. Truck crane.

frame, fitted with single or double rubber-tired wheels at each end of each axle to help support the load of the crane on the ground.

Outriggers on each side of the crane support are used to raise the truck-supporting frame high enough so that no load is transmitted to the wheels when lifting heavy loads. These outriggers are normally withdrawn in their frames. When needed, they are pulled out to each side and blocked on the ground. Some outriggers must be rolled out by hand, whereas others are equipped with hydraulic systems that extend them outward under power. In addition, the hydraulically operated outriggers are often provided with an additional device that extends supports at the ends of the outriggers, under pressure, down to the ground. Some systems pivot the ends of the outriggers vertically directly to the ground under hydraulic pressure.

Others without power have a screw jack incorporated in their ends to help support the outriggers on timber blocking placed on the ground. Some lighter capacity rigs require wedges to be driven from two sides, under the outer end of each outrigger, on top of blocking placed on the ground.

The outriggers are needed to prevent any load from being carried into the tires from the lifting operation. Some truck-crane models have the outriggers hinged to lie flat against the sides of the supporting truck frame. They are swung out at right angles when needed, to be wedged or otherwise blocked to lift the truck frame, to support the crane body clear of the wheels.

As in the crawler-crane construction, the boom is mounted on the front end of the crane body by means of boom heel pins. A jib can be secured to the end of the boom to give additional reach or height for light loads. A topping lift or boom falls from the head of the boom to the crane body, or to a gantry at the rear of the crane body, controls the raising and lowering of the boom.

The internal combustion engine in the truck-crane body controls only the rotation of the crane body itself and the raising and lowering of the main load block, the runner or whip line, and the boom falls. The traveling mechanism is controlled by a separate engine at the front end of the truck frame. The crane body rotates either on a center kingpin, on rollers (round, tapered, or hooked), or on ball bearings in a suitable race on a turntable ring mounted on the supporting truck frame.

Some truck cranes have capacities of 300 tons on a basic 70-ft boom, and are able to handle light loads on a 330-ft boom and 100-ft jib to a height of more than 400 ft above the ground. The crane in full operating condition, with no load being lifted, may weigh almost 200 tons. The weight can be reduced to about 100 tons by removing the boom, outrigger devices, and any detachable counterweights, so that the crane can travel over the highway within legal weight limits, at speeds of as much as 45 mph.

As in the case of the crawler crane, some manufacturers provide facilities for adding a guy-derrick mast, boom, and guys to increase the capacity to 300 tons on a 100-ft boom and 120-ft mast (see Fig. 9.1.2).

9.3   Wagon-Crane Construction

The wagon crane is a variation of the truck crane (Fig. 9.3.1). There is no cab at the front end of the truck frame support. Instead, the movement of the crane is controlled by the operator in the crane cab. The engine in the crane body controls the movement of the wheels as well as the motions of raising and lowering the load, runner, and boom, and the rotation of the crane body. The wheels are mounted on two, three, or four axles, and the crane often has the same capacity to the front as to the rear. With a truck crane the greatest capacity is to the rear, with little or no lifting being done over the front since this would endanger the driver of the crane and the crane-supporting frame is usually not designed for it; however, with a wagon crane the load can be lifted in all directions.

Wagon cranes are normally of considerably lower capacity than the standard truck cranes, although some have a lifting capacity of as much

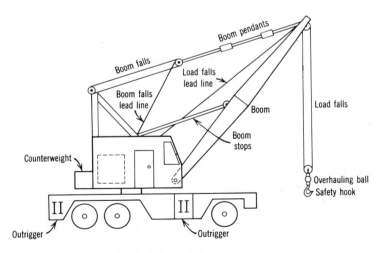

FIG. 9.3.1.   Wagon crane.

as 35 tons. Some are made with a boom and jib as long as 140 ft, but then they are rated to lift only very light loads on that length of boom and jib, even at minimum radius.

9.4   Locomotive-Crane Construction

The locomotive crane is similar in construction to the crawler crane, except that it is supported by a railroad flatcar type of frame. This in turn is supported at each end on railroad trucks of four or six wheels (Fig. 9.4.1). The crane is equipped with standard railroad couplers and air brakes, which can be connected in the usual manner to the railroad train couplers and air-brake lines while in transit. While the crane is erecting, the brakes are controlled by the operator in the crane body. Controls in the crane itself operate the boom falls, the main load falls, the runner, the swinging of the crane body, as well as the propeling mechanism to the railroad trucks under the supporting frame. Some locomotive cranes are still operated by coal-fired or oil-fired steam engines, but most of them use internal combustion engines.

Some locomotive cranes are available with a capacity to lift as much as 200 tons on a basic boom, and with a boom and jib available that can reach as high as 200 ft above the rails but with considerably less capacity.

9.5   Setting Up Crawler Cranes

A basic crawler crane can be delivered to the job site on two kinds of truck or tractor-drawn carriers or on a railroad flatcar. It can be hauled on a low-boy carrier if the crane is not too heavy or too wide to be hauled over the highways with its crawler treads in place. The platform of such a carrier is close to the ground but high enough for the treads to clear the roadway when the crane has been loaded on the carrier. To unload the crane wedge-shaped blocking is placed against planks or timbers laid on the ground just back of the treads. The crane is walked up the wedge-shaped blocking onto the planks clear of the carrier, which is then pulled away. The crane is moved off the planks down to the ground. Loading on such a carrier is done in a similar but reverse order.

Another type of carrier for a crane that is not too heavy uses two horizontal steel beams pivoted at the front ends and supported at the rear ends by a set of wheels on a detachable axle. The crane is carried on these beams. To unload, the detachable axle and wheels are disconnected

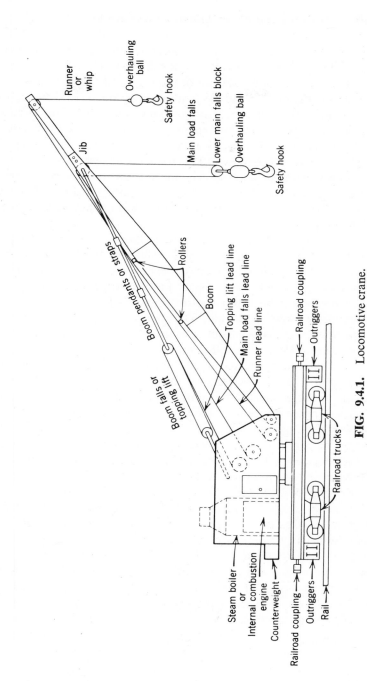

FIG. 9.4.1. Locomotive crane.

Runner or whip

Overhauling ball

Safety hook

Main load falls

Lower main falls block

Overhauling ball

Safety hook

Jib

Rollers

Boom pendants or straps

Topping lift lead line

Boom

Main load falls lead line

Runner lead line

Boom falls or topping lift

Railroad coupling

Outriggers

Railroad trucks

Steam boiler or Internal combustion engine

Counterweight

Railroad coupling

Outriggers

Rail

259

from the beams. A pair of jacks is used to lift the rear ends of the beams with the crane well forward on the beams. The detachable axles and wheels are then rolled back out of the way, and the jacks lower the beams enough for the crane treads to land on the ground. The crane is thus able to walk back clear of the carrier, the jacks raise the empty beams, and the axle and wheels are rolled into place to be refastened to the beams.

The crane is loaded in a similar fashion. It is walked over the beams after they are jacked down. The beams are then jacked up to pick the crane on the beams, and the axle and wheels are fastened in place so that the carrier can haul the crane over the highways. Either of these types of carrier may be pulled by a truck or a tractor, or they may be integral with the hauling mechanism.

If the crane is delivered on a railroad flatcar, as is often necessary when it is too heavy to be hauled over the highway or when the treads have to be removed because of delivery clearance, the problem of unloading may be somewhat complicated. Some cranes have facilities for retracting the tread frames to a narrower width than when in use, and those cranes can usually be walked off the car. The treads are extended outward after the crane has been unloaded onto the ground. One of several methods must be used to walk the crane off the car.

If the car can be backed to an unloading platform that can support the crane, and if there is a ramp to the ground from the platform, which is usually at flatcar height, the crane can be walked off the car onto the platform and down the ramp. When there is no such platform and ramp, the car should be spotted so that blocking can be placed on the ground at the end of the car. A temporary ramp or skids should be laid on and secured to the blocking between the car floor and the ground. The ramp should not be at too steep an angle. The crane is then walked off the car, down the ramp or skids, to the ground.

When it is not feasible to build such a ramp at the end of the car, blocking is usually required alongside the car to the height of the car floor. Sufficient timbers or planks are laid on this blocking for the crane to turn and walk off the car. In this case it is very important to have blocked under the floor of the car on both sides to prevent the car from tipping as the crane turns and moves. Once on the temporary platform, blocking and skids must be laid for the crane to walk down to the ground, either alongside the car or at right angles to it, depending on the terrain. Once the crane is off the car and all loose blocking or equipment is removed, the car should be released to the railroad to save unnecessary demurrage on the car.

When the crawler tread frames have been completely removed for shipment, the bottom section of boom can be left in place, with the topping lift reeved and temporarily secured to a fitting at the upper end of this lowest boom section. A temporary load falls can then be reeved to a fitting on the upper end of this boom section. In some instances only a sling is used, secured to the end of the boom section instead of a set of falls. By means of this sling and the boom falls or the temporary load falls and the boom falls, the treads are picked and landed in their proper places on blocking built up alongside each side of the car. Here again, the flatcar must be blocked to prevent it from tipping as the crane swings to either side to place the tread frames in place. With the treads in position, secured, and the traveling mechanism connected, the crane can be walked off the car in one of the ways described above. Adequate blocking and timbers must be in place for it to walk on.

Good practice requires the balance of the boom, the jib, detachable counterweights, etc., to be shipped in a car just ahead of the crane-body car. Any blocking and tie-downs used to hold the crane body in place for shipment must be removed, taking care that the crane cannot tip over backward when the blocking under the rear end is removed. Some cranes require the complete boom to be in place to prevent this. In such a case it is imperative that the rest of the boom be assembled and shipped in a connected car in such a way that the bottom section in place can be lowered and aligned with the balance of the boom. The remaining sections are then pulled to the bottom section by falls or a line from the crane load drum, and the splice is fastened. The topping lift must then be removed from its temporary attachment and connected by its pendants to the head of the complete boom.

As an alternative, the boom can be completely assembled for shipment and placed in the next car ahead of the crane body. The bottom section is then not in place on the crane body. The boom is skidded on timbers, or rollers are used to drag it to the crane body, securing the boom heel at its proper place on the crane. The topping lift is reeved and attached by pendants directly to the head of the boom.

When the bottom boom section is in place for shipment and the balance of the boom is shipped completely dismantled, the section is used as a form of working boom. A temporary falls on the end of the section is used to pick the various loose boom sections and assemble them on blocking on the ground.

When another crane is available it is more economical and faster to use it to assemble the boom and place it in position, to assemble the crawler tread frames if they were detached, and to build the ramp if one was needed for walking the crane off the car. To load a crane when its use is completed, the unloading process is reversed, again being careful to block under both sides of the flatcar before running the crane on the car. The boom can be detached and pulled ahead into a gondola car or onto a flatcar ahead of the one on which the crane body is being loaded.

The railroad inspector should be contacted before loading to determine what the railroad requirements will be. These will affect the position of the crane on the car and trucks for safe loading. All blocking and tie-downs that will be required should be secured and installed, both for the crane body and for the boom, jib, and other fittings in a separate car ahead of the crane car.

Once the crane is on the ground, whether delivered by railroad car or truck-drawn carrier, the gantry must be erected in its high position for proper use (if the particular model of crane is equipped with a gantry). Some cranes have their gantrys pivoted so that it is a simple matter to rotate them into position and secure them in place, usually by pinned connections. Some models require a small pole to be lashed to the rear of the crane, with a set of falls going from the top of the pole to the gantry previously laid down on top of the crane. Some cranes are designed so that the boom can be raised with the gantry in a lowered position. Then the gantry can be connected in raised position after the boom and gantry have been lifted to the point where the connection holes in the gantry line up with the corresponding holes in the crane body.

When an auxiliary rig is available it is best to use it to set up the gantry, assemble any sections of boom that had been dismantled, and install any necessary counterweight that may have been removed for shipment.

If the size and weight of a high-capacity crane are so great that it has to be completely disassembled for delivery as separate parts (such as crane body, crawler assemblies, counterweights, boom and jib sections, etc.), then truck or tractor-driven carriers or railroad flatcars cannot be used for delivery. Instead, it is then necessary to use separate trucks or railroad cars and auxiliary equipment such as a truck or crawler crane, ginpole, or available derrick to unload and assemble the crawler frames and treads, mount the body in place, assemble and install the boom and add the counterweights, and later to disassemble them when ready to ship the crane away.

Whichever method of delivery has been used, as soon as all the dismantled parts have been assembled in place the main load falls and runner or whip are reeved and secured in their proper places. The boom stops are installed between the boom and the crane body. The crane should then be thoroughly inspected to make sure that it is in safe working condition, that no damage has been done in transit, in unloading, in the previous loading, or by the previous user. All parts requiring it should be checked and lubricated.

This inspection and check should be made whether the crane has come from the erector's toolhouse, from a rental vendor, or directly from a manufacturer on rental or purchase. Any damaged members of the boom or jib should be repaired or replaced before use. The wire ropes, pendants, blocks, sheaves, pins, and all parts that are subject to wear must be checked before using the crane and regularly thereafter to be sure they are lubricated when needed and that there has been no damage. When the crane can be delivered under its own power over the highway, without its treads causing any damage, this may be advisable since it can usually be delivered completely assembled, except that its gantry may have to be lowered because of limited clearances en route. In this case it is only necessary to land the boom temporarily on blocking, raise the gantry, connect it, and raise the boom to its working position.

9.6   Setting Up Truck Cranes

Ordinarily a truck crane is delivered over the highway, either completely assembled or with the boom and/or counterweight removed. When delivered over the road completely assembled, it should be inspected as described for the crawler crane. In addition, the tires, outriggers, and the traveling engine and mechanism should be checked. If the highway is used, any necessary licenses or permits should be arranged for in advance. Some states even require license plates on the truck crane in order to operate on the job site. This should be checked in advance so that there is no delay in securing them.

When the crane is so large and heavy that it cannot be moved over the road completely assembled or even with the boom, counterweights, and outrigger devices removed, then it is necessary to deliver the disassembled parts such as the counterweights, boom, jib, tower (in the case of a tower crane), or tower and mast (in the case of a guy derrick, "Sky Horse®,"* or "Ringer®"** crane) on a number of trucks or railroad cars. Auxiliary equipment such as an available mobile crane derrick, or gin-pole must be used to unload the various parts and assemble them at the site.

At times it is more economical to deliver a truck crane on a railroad flatcar than to run it a great distance over the highways. If the gantry was

---

* American Hoist & Derrick Co.
** Manitowoc Engineering Co.

lowered because of clearances, it must be raised back into place. This is done either by a mechanism provided by the manufacturer as a permanent part of the crane or by an auxiliary rig. Without these aids a small pole should be used as described for the crawler crane. If delivered on a car with the boom dismantled, it will generally be advisable to install the bottom section of boom in place using the boom falls to a fitting at the upper end of this section. This section is then used as a form of boom to unload the remaining sections and assemble them on blocking at the correct heights so that the centerline of the assembled boom will be a straight line from top to bottom. The truck crane then moves into position so that the bottom section can be slid into its place in line with the rest of the boom. The temporary boom falls and the temporary load falls or lifting slings are then disconnected, and the assembled balance of the boom is pulled into place on the end of the bottom section. The boom falls and load falls are then reeved and secured in place. If a jib is to be used, it should be installed, and the runner or whip line reeved over the proper sheaves. Any detachable counterweights are installed, the boom stops are fastened in their place between the boom and the crane body, and the crane is ready to be inspected to be sure it is in safe working order.

9.7   Setting Up Locomotive Cranes

Locomotive cranes are customarily shipped with the rear end of the crane body tied so that it cannot rotate in transit, with the boom completely assembled and loaded in a gondola car ahead of the crane, with the propelling mechanism disconnected, and with steel "keeper" plates or blocks removed from between the trucks and the underside of the supporting frame. The ties at the rear must be removed and stored for reuse when the crane is shipped away later, the propelling mechanism connected, and the plates or blocks installed between the tops of the trucks and the underside of the car-supporting frame. The boom is assembled into its proper place and the various falls reeved, all as described above for the crawler and truck cranes.

In addition to inspecting the crane itself, the wire ropes, boom, jib, sheaves, pins, etc., the trucks and wheel bearings should be checked. The air-brake system must be in safe working order before the crane is moved or put into erection service. If rail clamps are provided they should be adjusted to be slightly loose when secured to the rails so that they act as a warning if the crane is overloaded and the rear end starts to pick. They should not be depended upon to utilize the rails and ties as extra counterweight; they are merely a safety warning to stop before the crane can tip over.

9.8   Mobile Tower Cranes

In recent years, a new type of crane has been developed that uses a tower to support an erecting boom. There are several forms of tower cranes: tower mounted on a crawler-crane body (crawler-mounted tower crane) (Fig. 9.8.1 *left*), tower mounted on a truck-crane body (truck-mounted tower crane) (Fig. 9.8.1 *right*), tower mounted on a fixed foundation

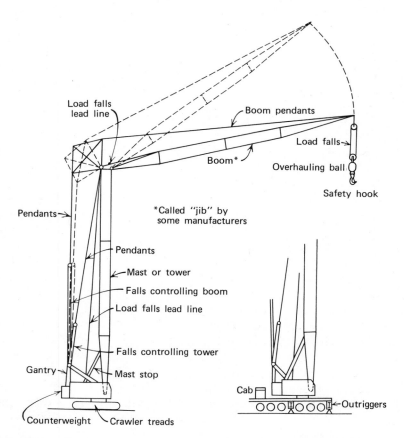

FIG. 9.8.1. (Left) Mobile tower crane crawler mounted. **(Right)** Mobile tower crane truck mounted.

(fixed, static, or stationary tower crane) (Fig. 9.8.2), to be mounted to climb vertically in a structure (climbing tower crane) (Fig. 9.8.3 and 9.8.4), or mounted directly on trucks to slide or roll on rails laid on the ground (traveling tower crane) (Fig. 9.8.5).

With the tower mounted on a truck-crane or crawler-crane type of support, the crane-operating mechanism is installed in a crane body rotating on, and supported by, a truck-type frame with wheels, etc., as in the case of the truck crane, or on the treads or crawlers in the case of the crawler crane. On some crawler-mounted tower cranes, a tower as high as 250 ft with a 200-ft boom can be used to lift more than 10 tons at a radius of 200 ft. A 100-ft jib can be added to the boom for additional height and reach to handle light loads. In the case of the truck-mounted crane, as much as a 200-ft tower with a 150-ft boom can be used to handle about 5 tons at maximum radius. In both types the tower is mounted on the front end of the crane body. A boom is hinged at the top or near the top of the tower, depending on the particular manufacturer's design. The tower and the boom are made of chord angles or tubular or pipe sections, with angle, plate, or pipe lacing.

By means of falls and a gantry on the rear end of the crane body, the tower is raised from the ground into its upright working position, being rotated on pinned connections at the front end of the crane body (Fig. 9.8.6). This set of falls is powered by a lead line from a drum in the crane-body mechanism.

The boom is pinned in place and the boom falls reeved before the tower is raised. The boom is then lifted from a position where it hangs from a pinned connection at the top of the tower to its working position above. The boom can then be rotated to move up or down for erecting, between an approximately horizontal position for maximum radius to its highest position for maximum vertical reach.

The lead line to a load falls secured to the head of the boom is operated by another drum in the crane body. The drums controlling the lines that raise and lower the tower, boom, and load lines are usually actuated by an internal combustion engine in the crane body, and in the case of the truck-mounted crane, there is a separate cab at the front end of the truck frame for controlling the movement of the truck-supporting frame with its separate engine. Some models are powered electrically. The entire crane body supporting the tower, boom, and load falls must be rotated to swing the boom and load around. This rotating movement is controlled by the operator in the crane cab.

For dismantling, the boom is lowered, rotating it about its pinned connections at the top of the tower, until it hangs flat against the tower. Then

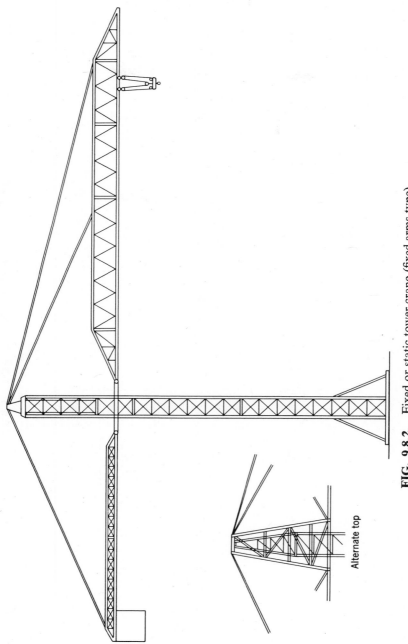

FIG. 9.8.2. Fixed or static tower crane (fixed arms type).

Alternate top

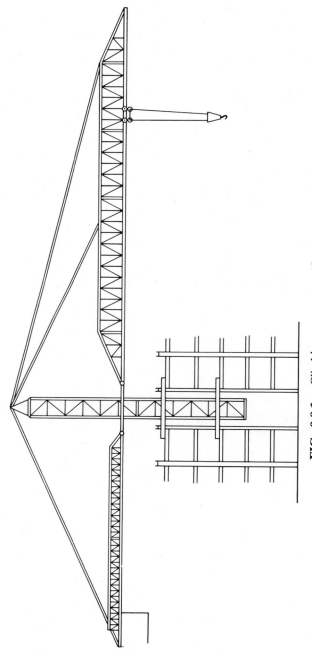

FIG. 9.8.3. Climbing-tower crane (fixed arms type).

268

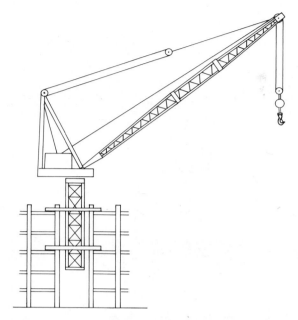

FIG. 9.8.4.   Climbing-tower crane (luffing boom type).

the tower, with the boom secured to it temporarily, can be lowered by rotating the tower (and boom) to the ground for disconnecting them from the crane body. Or the tower (and boom) can be rotated to land on a truck or other carrier and disconnected from the crane body.

Several manufacturers design the head of the boom to fit on a small rolling carriage, or a pair of wheels on an axle is fitted directly to the head of the boom. In either case when being dismantled, the boom is lowered to hang loose against the tower. Then, as the tower is lowered by rotating it on its pins at the front of the crane body, the boom hangs vertically from the top of the tower. When the tower has been rotated far enough, the head of the boom is landed on the rolling carriage on the ground, or the wheels are secured to the head of the boom. This permits the head of the boom to ride forward away from the tower on the ground or on planks laid for the purpose. As the tower is lowered to the ground, the head of the boom rolls forward and the boom pivots on the boom heel pins at the top of the tower.

When the tower is finally level, near the ground, it is landed on blocking. The falls used to erect or lower the tower are moved and connected

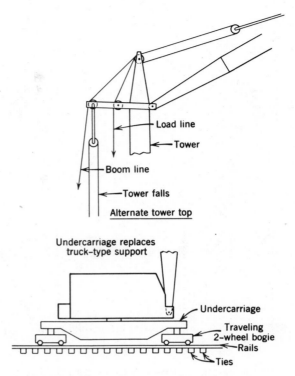

Load line

Tower

Boom line

Tower falls

Alternate tower top

Undercarriage replaces
truck–type support

Undercarriage

Traveling
2-wheel bogie

Rails

Ties

FIG. 9.8.5.   Tower crane: rail mounting.

to a special connection at the top of the first section of the tower. The splice just above this connection on the tower upper face is now unfastened, and the bottom section of tower is used as a form of boom. The load falls is moved, or a new set of temporary load falls is reeved and fastened to another special connection on the underside of the upper end of this bottom tower section.

This falls is then used either to load the boom and the rest of the tower on trucks and trailers or other carriers; or the balance of the tower and the boom section splices are disconnected and the various sections are loaded individually by means of the short, bottom tower section used as a boom. As an alternative, the tower can be disconnected from the crane body completely assembled, to be landed on trucks and trailers for shipment. In this case, the boom is generally similarly loaded completely assembled, merely disconnecting it from the tower.

Some manufacturers have combined the truck crane or the crawler crane and the tower crane into one machine. The crane with high capacity is fitted with a boom in its usual place on the crane body so that the crane can be used in normal fashion as a truck crane or crawler crane. The boom is then detached and replaced by a tower to be raised to a vertical position on the front end of the crane body. The boom is reinstalled at the top of the tower to be used as a tower-crane boom.

This is an ideal combination since the crane as a truck crane (or crawler crane) can unload equipment at the start of the job, including the tower to be used later and used to load out equipment on completion of the work. In addition, with a greater available load capacity on the boom in its normal position, heavy steel near the ground can be erected by the crane and the tower can be set up later. The boom is then moved up to the top of the tower for use in erecting the upper, lighter steel.

9.9   Fixed and Climbing Tower Cranes

A variation of the mobile tower crane is the static or fixed tower crane anchored to a foundation. This type is best suited to be used inside a building frame. The tower is erected to the maximum height required for the first few tiers of the structure. It is then extended upward as the building is erected installing bracing at various floors between the tower and the permanent structure as erection proceeds. The static type (on fixed foundations) is sometimes set up just clear of the permanent structure but braced to it at various floors as they are erected. This eliminates the need to leave open panels in the structure for the tower, which is usually necessary when the tower is inside the building.

One type of static or fixed tower crane consists of an outer, generally square tower of built-up angles, pipe, or tubular main members with angle, plate, or pipe lacing. Inside this tower is a second, shorter telescoping tower, near the top of which the erecting equipment is mounted. This equipment consists of a long arm, usually of built-up members, generally triangular in cross section, and named by different manufacturers a "boom" or a "jib." This boom is pinned to the tower and extends out horizontally from the tower in one direction. A lifting tackle mechanism or trolley rolls on the underside of this erection arm. A second, shorter arm called a "counterweight boom," "counterweight bridge," or "counterweight jib" extends horizontally from the tower in the opposite direction. The counterweight is placed on the outer end of this counterweight arm, and in the high-capacity models, additional weights move between the tower and the fixed counterweight at the end of the arm to balance the

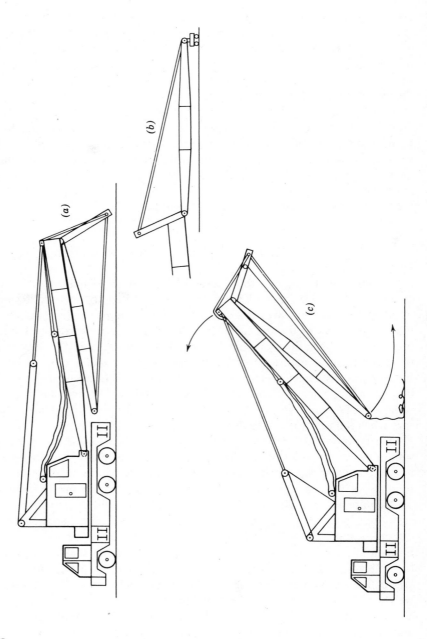

(a)

(b)

(c)

272

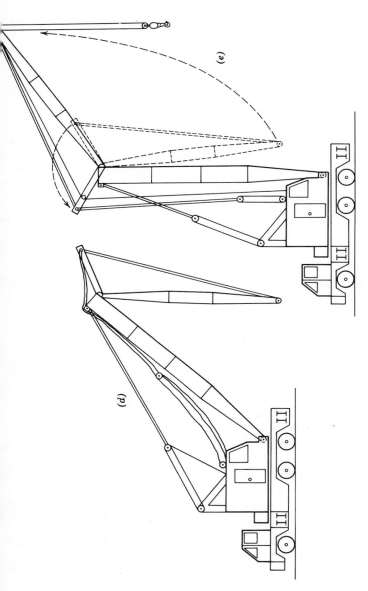

FIG. 9.8.6. Setting up truck-mounted tower crane. (**a**) Boom is pinned to connection on top of tower held by falls from gantry on crane body. Boom can be laid back against underside of tower or (**b**) laid ahead with tip on dolly to permit it to roll as lifted. (**c**) Tower is started up, letting boom swing out or roll on dolly. (**d**) Boom hangs down as tower is lifted toward its final position. (**e**) Tower upright, boom is rotated to working position by topping lift (boom falls).

273

weight of the load being lifted. (In some models the counterweight arm consists of a counterweight completely enclosed and extending from the tower to the end of the arm.) On other models the operating equipment is also placed on the counterweight arm. The two together are then heavy enough to counterbalance the heaviest load to be handled by the erecting falls at whatever radius it is rated to pick such a load on the long, front erecting arm. On some cranes of this type, only the counterweight is placed on the back arm, the operating mechanism being located below or on the tower itself. Still others have only the equivalent of the outer tower with a short, separate section of tower together with the two arms rotating on the top of the main tower.

Presently this type of tower crane has a capacity as high as 350 tons at 110-ft radius and 130 tons at about 270-ft radius, with the erecting hook more than 250 ft above the ground. The weight of such a crane may be more than 800 tons including the counterweight and ballast. Therefore, the foundation must be able to support this weight plus the load being lifted and any additional stresses from the wind.

However, this high capacity is ordinarily not required for the average steel-erection project. Instead, a similar crane is generally used, lighter in weight and able to lift only about 65 tons at minimum radius 160 ft above the ground, or more than 15 tons at 250-ft radius.

The great advantage of the static crane (or the climbing crane) is that it can lift a steel member (or a load) from the ground or from the working-floor level anywhere within the radius of the front erecting arm and land it or erect the steel member in place.

The two arms are supported by wire ropes, pendants, or links, depending on the manufacturer's design, from the top of the tower, which is erected high enough above their level to support them, or in some models the two arms are secured to the rotating tower top and cantilevered outward without the need for supporting ropes or pendants. A foundation must be provided under any type of stationary tower. The foundation must be strong enough to support the imposed vertical loads and have adequate anchorages to resist any uplift or horizontal forces that may be developed. The outer tower section must be braced against the permanent structure at various floors. The upper portion of the tower that holds the wire ropes, pendants, or links supporting the two arms is designed to permit horizontal rotation through 360°. The wind pressure on the exposed surfaces above the working floor must be allowed for in the construction of the tower and foundation. Some manufacturers attach the horizontal arms to a section of tower that rotates on top of and outside the main supporting tower instead of inside the fixed tower as most of the other manufacturers do.

As the building progresses upward, if the tower has not been erected initially to the maximum height required, additional sections are added to the outer tower. Hydraulic lifting jacks, a wire-rope lifting mechanism, or other forms of lifting devices involving yokes, jacks, jacking frames, etc., are used between definite levels of the outer, supporting tower and the bottom of the inner tower to lift the inner tower and its arms to a new level.

When the tower to be used is high enough the boom long enough, and the capacity sufficient to permit mounting the crane in one position to reach all the work, a foundation is prepared of concrete, timber, or thoroughly compacted ground. The tower is erected on this foundation. Heavy counterweights are added on a platform on the base of the tower to resist the overturning effect of a load being handled by the boom, or else anchorages are provided in a heavy foundation.

If adequate anchorages are used, some manufacturers claim that their cranes can use a tower and boom to lift a load over 1000 ft above the ground. The maximum capacity in this case is well under the capacity of the average static crane. Furthermore, the foundation must be adequate to support the weight of the crane, which can be more than 80 tons when handling as much as 65 tons at 85-ft radius, or 15 tons at about 250-ft radius. With this type of crane, the operating mechanism may be on the counterweight arm, near the top of the tower, or it can be in the base.

Still another form of this type of tower crane uses the original foundation for the outer tower and uses a hydraulic jacking mechanism to raise the inner, telescoping tower high enough to add a section about 10 ft high to the outer tower. This is done by adding four framed sides, one at a time, and connecting them at the four corners to each other to form a new section of tower. This is actually done with the main load trolley that runs on the underside of the boom (or jib). In the case of models using an inner fixed tower with the outer tower portion rotating, an auxiliary device is mounted to pick additional sections of the inner tower, 10 to 20 ft long, clear of the tower and high enough to swing them so that each can be dropped through an opening in the top of the rotating upper tower. The section is then lowered and secured to give added height to the fixed tower, and then the rotatable outer tower, with its erecting arm and counterweight arm is raised to the new top of the fixed tower to operate again.

For dismantling, the operation is reversed. The operating tower is lowered, the top section of the fixed tower is disconnected, picked up high enough to be swung clear, and lowered. This operation is repeated until the arms are down to the top of the structure for dismantling and loading out. The fixed tower is dismantled section by section by means of falls hung from the top of the permanent steel structure, after guying the structure in that area to resist any side pull from the operation.

The climbing crane is a variation of the static or fixed tower crane. There is only a short length of supporting tower. This is jumped by lifting the bottom end of the supporting tower by jacks or falls to a new level, or else a climbing frame is installed to lift it. In either case supporting beams are placed under the bottom of the tower, across the opening in which it is located, the ends resting on the floor members of the permanent structure under the bottom of the tower at the new level, after it has been jumped. This tower must be braced securely to the permanent structure at various floors.

Some of the cranes of this climbing type have an inner telescoping upper tower and are designed with auxiliary lifting frames, one placed under the bottom of the lower, outer tower, and the other placed on the permanent structure 20 or 30 ft above. A lifting mechanism is used between the two. This permits the entire crane to be lifted the 20 or 30 ft respectively, in one continuous jump. The lower frame is then used to support the tower on the permanent structure at the new level of the bottom of the tower. The upper frame is moved up the 20 or 30 ft (depending on the crane model), where it braces the tower to the structure. In all cases the area where the tower stands must be kept free until the rig has completed erection and has been dismantled.

In setting up these rigs, the tower is usually assembled on the ground on level blocking, or it may be delivered completely assembled. The counterweight arm is pinned to the tower at the proper point, and an auxiliary hoist is mounted as part of the base to be used to trip up the mast and the counterweight arm together. When the tower (or mast) is upright and has been stayed, the counterweight arm is lifted to a position at right angles to the tower by means of wire-rope falls. The front arm is then similarly pinned to the tower and lifted by separate falls from the top of the tower to a horizontal position. Links or pendants are then installed to hold the arms in place, and the falls that were used to lift the arms are disconnected.

For dismantling when the work has been completed, the process is reversed; the arms are lowered by rotating them about their pinned connections and dismantled, and then the tower is lowered to a point where it can be dismantled.

Those tower cranes that are electrically powered by 440 V 60-cycle ac current permit remote control of all operations. This in turn allows the operator's cab to be mounted successively at different levels of the tower so that he can be closer to the actual work being done, rather than being in the cab of a machine on the ground.

When the towers are jumped (climbing cranes), instead of having sections added (as in a static or fixed tower crane), the structure itself must be adequately braced to resist the forces developed at the floors where the crane or tower is braced against the permanent structure.

The details of the various operations will not be given because the intricacies of setting up, jumping, and dismantling vary so widely since each manufacturer has built his crane to a different design and has installed different jumping and dismantling mechanisms, so that one must follow the detailed instructions issued by the manufacturer for each particular climbing or fixed tower crane used.

Another form of the static (fixed foundation) or the climbing or traveling crane is one in which the boom (or jib) is neither horizontal nor fixed and there is no counterweight arm as such. It consists of a basic crawler-type body construction reduced to a minimum, containing the hoisting and slewing mechanisms, and is usually diesel powered (Fig. 9.8.4). It is cantilevered out from, and rotates on the top of, the tower, with a gantry at the outer end of the cab, from which a topping lift or boom falls raises and lowers the boom.

The boom is pinned at or near the front end of the crane body to allow the boom to be raised clear of any surrounding obstructions and to permit raising and lowering it to provide a flexible height and radius for operating the load falls at the end of the boom. The crane body has a fixed counterweight at the extreme outer end, with additional weights that can be moved back and forth under the body to help counterbalance the load being lifted.

This type, sometimes called a luffing, or tower gantry, type of crane (luffing refers to a boom falls raising and lowering the boom) permits heavier loads to be lifted with a high boom and a radius closer to the crane body or lighter loads as the boom is lowered—thus extending the radius.

As the need arises to increase the height of the supporting tower, this is done in a manner similar to that described for the crane with horizontal, fixed arms. This type was introduced to the United States some years ago; it was known as a "Kangaroo" crane since it was manufactured by an Australian firm. This type is now produced in the United States with capacities of about 115 tons at more than 20-ft radius or about 3 tons at 200-ft radius, when used as a fixed, climbing, or traveling tower crane; it is more useful for the average steel erection needs than higher capacity cranes with fixed arms.

Since the crane body is cantilevered outward, there can be excessive bending in the supporting tower, which must be resisted by enough anchorages as well as by bracing it to the permanent structure.

The tower on the static or fixed-crane type is theoretically capable of being extended to unlimited heights, but practical safety considerations limit the allowable height since these must take into account the stresses from wind, slewing, and handling loads on such an unguyed but braced tower. When the fixed tower is placed outside of, and braced to, the permanent structure, guys can be installed to permit higher levels of tower than free-standing towers. Similarly, when the length of the tower is kept constant and the tower is jumped upward as the erection of the building proceeds, with proper bracing the height of the top of the tower above the ground is unlimited. The extent of the tower above the uppermost floor is kept the same after each jump. The load under these cranes can be appreciable. One climbing crane model, for example, that can handle 10 tons at almost 100-ft radius to 3 tons at about 200 ft, may weigh more than 60 tons exclusive of the live load forces. Other models can lift 20 tons at minimum radius.

9.10   Traveling Tower Cranes

Still another form of tower crane of this same general type uses a tower similar to the one on the static or the climbing cranes, using either a horizontal, fixed erecting boom and counterweight arm, or the luffing type of boom that can be raised and lowered, which permits changing the height and radius at which the load falls can be operated. Instead of remaining in one location, the tower is mounted on a traveling frame. Rails are laid alongside the building for its entire length. These rails are spaced about 15 to 32 ft apart for the high-capacity machines, on ties on the ground after the ground has been leveled and compacted to form a good bed for the ties and rails. The supporting frame rides on four sets of wheels, or "bogies," on the rails. Heavy counterweights must be added on a platform on the traveling base frame to resist the overturning forces from handling loads on the main load falls as well as possible wind forces on the tower and arms or boom. The engine, drums, etc., are all mounted below in a frame on the trucks. In the case of the fixed arms type of crane, the engine is usually electrically driven, and in the luffing type it is usually diesel driven.

The tower is generally fixed in height. When first assembled and erected, it is built up to the maximum height needed for the boom to reach the required height and out to the most distant points needed. The capacities of the luffing type can amount to more than 100 tons at minimum radius on an 80-ft boom, or 5 tons at 200-ft radius on a 200-ft boom. One model of the fixed arms type can lift more than 20 tons at about 50-ft radius. The fixed arms type can be used to good advantage when a structure is long and not too deep for the crane boom and load to reach all points and when the steel members are not too heavy for the capacity of the machine. This type of crane, which is usually electrically powered, permits installation of the operator's cab at any level on the tower, since the crane can be operated by remote control. This enables the operator to have a better view of the work than with the truck crane, truck-mounted crane, crawler crane, crawler-mounted crane, or locomotive crane, where the operator is confined to the cab near ground level. With the luffing type, the operator is in the crane body and so he has the same advantage. The luffing type can also be used on a tower mounted on a traveling frame.

When all steel has been erected that can be reached by this type of traveling crane in any one position, the entire tower, boom, trucks, etc., are rolled on the rails or skidded (if not mounted on wheels) to a new position from which work continues. Such a crane mounted on a traveling frame may weigh more than 80 tons plus about 30 tons of ballast or counterweight in place.

9.11 Gin-Pole

A gin-pole is basically a form of boom, the bottom end of which is fitted with a shoe for support and for sliding. The pole is supported on the ground or on a suitable working platform, with the upper end held in place by guys. These guys must be easily adjustable and not of fixed length (as in the case of guys holding a guy derrick mast in place or those holding columns or trusses temporarily). This is done by attaching manila rope falls or tackle from adequate anchors to shorter lengths of wire ropes or short guys, secured to the head of the pole. The guys are adjustable to permit the pole to be leaned in various directions so that the load falls that is secured to the head of the pole can be spotted over a piece to be erected.

The gin-pole is composed of several spliced sections of heavy pipe, or it can be a spliced, built-up piece of pipe or angle main members, with angle, plates, or pipe lacing. If built up, it is usually of rectangular or square cross section. A derrick boom or mast can be used, revamping the foot to hold it in a shoe for support and sliding, and providing attachments at the head for the guys. Occasionally a timber of sufficient strength and length is used, with the necessary fittings installed at its head and foot.

When the pole is made of spliced sections it is assembled on the ground on level blocking so that it will be in one straight piece from top to bottom when the splices are in place and fastened. The head fitting, with provision for securing at least four guys and a load falls, is installed. On the foot, the fitting should provide for attaching a snatch block or a gate block or even an ordinary block for the load line from a hoist, and for two sets of falls on opposite sides. These falls will be used for moving the pole; they are also used as foot kickers while erecting.

In tripping the pole after it has been assembled, its foot is placed in approximately the first position for erecting. The hoist is located in line with the direction the pole will be moved as erection proceeds. The load falls is reeved to the head of the pole, with the lead line run through the block that is secured to the foot of the pole, and then to the hoist drum. The head guys are now secured by individual manila-rope falls to the first four equally spaced and suitable anchorages to be used for tripping the pole. (When only light loads are to be handled, a hand crab is often fastened to the pole at a height that is convenient for the men, rather than having a powered hoist actuate the load falls.)

The load falls is run out and secured, by means of a long pendant or a sling, to a temporary anchorage a reasonable distance away. Restraining kickers on the foot of the pole are placed to resist the pull of the load falls and to keep the foot of the pole in place.

If a powered hoist is not used, a set of manila-rope falls can be used to trip the pole, the rope being pulled by hand instead of by power. Also, the load falls of wire-rope tackle, powered by the hand crab, can be utilized. The shoe on the bottom of the pole is placed on planks or blocking to support the pole after it has been set up in working position and to distribute the load under the foot to the ground. By going ahead on the falls, the pole will be tripped up into position for use. (Jacking up the head of the pole onto temporary blocking before trying to trip it up will aid in starting the top of the pole upward.)

As the pole starts up, the manila falls on the two front guys are taken in, and the falls on the two back guys are paid out, keeping the pole rising steadily in a vertical plane. Once the pole is upright, the guys are secured, the load falls is disconnected from the pendant or sling used to trip it up, and the pole is ready to be used. This procedure is similar to the one described for tripping a derrick boom or mast when no other powered equipment is available to stand the boom or mast upright (Sec. 7.2). The guys should clear any rail, highway, or pedestrian traffic routes. They should be placed to avoid any possible contact with power, telephone, or trolley wires.

The steel to be erected can be skidded out of the car or off the truck delivering it by means of hand falls or by using power from the hoist that is to be used for actuating the load falls. Adequate anchorages are used to hold the hoist in place. The steel can be landed on dollies or rollers and moved into position under its final location in the structure.

The pole is leaned so that the load falls is directly over the piece to be erected. This falls is operated by the small powered hoist, or, as mentioned before, if the steel is not too heavy, a hand crab can be used or even an air hoist clamped to the pole. (Small pieces are usually raised into place by hand lines.)

The pole is moved by skidding over planks to the position of the next piece to be erected. If a powered hoist is used it is advisable for it to be located at the end of the area to be erected, in direct line with the direction the pole will be moved as erection proceeds. The foot of the pole must be tied back with a foot kicker in each position it is used for erecting, to resist the pull of the lead line from such a hoist (Fig. 9.11.1). This lead line is fed through the snatch, gate, or other block at the foot of the pole, up to the falls secured to the head of the pole. Even when no powered hoist is used, kickers should be installed to prevent the bottom of the pole from moving.

In erecting, the pole should be leaned to set the columns and bracing at one side of a panel, and then leaned to erect the columns and bracing at the other side of the panel. Then it should be brought back to erect the trusses, girders, or beams between those columns before it is moved to the next panel to repeat the operation and fill in between the two transverse members. The manila hand falls on the four guys are slacked off or taken in as needed to lean the pole in the different directions. If there is more than one aisle in the structure, as soon as an aisle is completed the pole and hoist are moved to the next aisle to continue erecting.

In moving the pole (Fig. 9.11.1) it is leaned in the direction it is to be moved, to be held by the two back guys by means of the manila falls between the ends of the wire-rope guys at the head of the pole and the anchorages. The lower ends of the two forward guys of wire rope and manila falls are now moved forward to the next anchorages to be used by

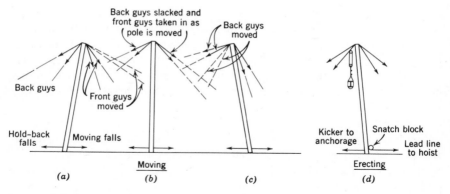

FIG. 9.11.1. Gin-pole. (a) Pole is leaned into front guys, held by back guys, to permit changing front guys to new anchorages ahead. (b) Falls on back guys are slacked; falls on front guys are taken in as pole is moved forward by falls at foot with holdback falls in opposite direction to keep pole under control. (c) Pole is leaned into back guys, held by front guys, to permit changing back guys to new anchorages. (d) Hold-back falls are replaced by kicker; lead line to hoist is led through snatch block at foot of pole. Pole is again ready to erect in new position.

them. One of the sets of falls secured to the foot of the pole is then used to pull the foot of the pole forward, the second set being in place to hold the foot back and keep the movement of the foot under control. Meanwhile, as the pole moves, the falls on the forward guys are taken in and the falls on the back guys are slacked, keeping the pole upright as it moves. When the pole is in its new position, it is leaned back so that the two forward guys already moved ahead can hold it, and the back guys are then moved forward and fastened to their new anchorages.

When a pole is used in conjunction with a crane to help set pieces that are too heavy for the crane to handle alone, and with too few pieces involved to warrant bringing in a higher capacity crane or even a second crane, the piece to be erected should be skidded or rolled into position. The pole is set up by the crane at one end of the piece, with the shoe of the pole on suitable blocking or support. The pole is guyed in place to adequate temporary anchorages, a set of load falls is reeved to its head to enable it to pick one end of the piece to be erected, and foot kickers are installed. The lead line from this lifting tackle can go to a hand crab, winch, or an air hoist clamped to the pole. Or one of the drums of the crane hoist or an auxiliary powered hoist can be used. The crane hooks onto the other end of the piece and the two rigs lift together.

A stronger pole can be used, with its lifting tackle reeved as before, but with sufficient parts in the load falls to pick the piece. The load line to this falls is actuated by one of the crane hoist drums or an auxiliary hoist drum. The pole alone is then used to set the piece.

Another method has the pole set up on one side of the piece, with the crane on the other side directly opposite. The crane lifting falls and the load falls on the pole are powered by one continuous wire rope from the crane drum, reeved through the two sets of load falls blocks. Enough parts must be reeved, depending on the available lead-line pull and the weight to be lifted. This permits the two sets of falls to act together in picking the piece, whereas with the pole at one end of a piece and the crane at the other, one set of falls may get ahead of the other. Using either method, once the piece has been erected, the crane moves the pole to a new position to erect the next piece to be handled by the two rigs. After all such pieces have been erected, the crane lowers the pole and loads it for shipment away after its load falls, guys, and kickers have been removed.

9.12   Basket-Pole and Tower Erection

The basket-pole, like the gin-pole, can be a spliced pipe or a steel-fabricated pole similar in shape and design to a derrick boom or mast, or a timber of sufficient strength and length can be used. The fittings for the top and bottom of the basket-pole differ from those for a gin-pole and the method of moving it vertically is entirely different.

For simplicity in explaining the use of this type of erecting rig, assume that a four-column tower is to be erected and that the tower is too high to be erected by any available crane or derrick. (Structures to be erected by a basket-pole are erected in a fashion similar to that for a four-column tower.) The pole is assembled and tripped upright, as described for the gin-pole, in approximately the center of the four footings for the tower columns. Frequently a crane is used in conjunction with a basket-pole, in which case the crane assembles the pole and sets it up in its first working position.

The hoist is located close by and anchored in place. The bottom fitting for the basket-pole differs from that for a gin-pole. There must be provision not only for securing the load-falls lead line from the hoist but there must also be at least four loops or eyes to which to fasten four supporting guys. The loops or eyes should also be large enough for the four jumping guys, or else four additional loops will be needed. In setting up the basket-pole only the four head wire-rope guys and the manila-rope falls from them are secured to four equally spaced anchorages, usually the four column footings.

To start erecting, the pole is leaned by slacking the proper hand falls at the ends of the wire-rope guys from the top of the pole until the head is over the first column location. The column is rolled or skidded to where the load falls can pick it and place it in position on its anchor bolts in the footing. Or, if a crane is being used in conjunction with the pole, the crane can set the pole in position and erect the first tier of columns and bracing. The column erected is temporarily guyed if necessary for safety. Without a crane to erect this tier, the pole is leaned by taking in or slacking off the proper hand falls on the four wire-rope guys until the head of the pole and the load falls are over the second column location. After this column is set, and also temporarily guyed if necessary, the pole is leaned halfway between the columns to erect the framing and bracing between these two columns.

Then the pole is leaned to erect the next column and framing, continuing on the fourth side. Now the pole is brought upright by means of the four hand falls. Four sets of jumping falls and guys are secured from slings placed on the top of each of the four columns to the fitting at the foot of the pole. The lower ends of the hand falls on the guys to the head of the pole are moved up one by one to be fastened to the tops of the four erected columns. Lashings or slings are used to hold them in place, and separate lashings or slings are used to hold the jumping falls in the same places.

By taking in on all four jumping guys equally and paying out the upper guy hand falls, the pole is raised until the top is high enough so that, when leaned, it will be able to erect the next section of columns (Fig. 9.12.1). At this time preferably fixed guys, or else shorter guys and hand falls, are secured between the foot of the pole and the tops of the four erected columns. The jumping falls can be removed or left in place until needed for the next jump.

By adjusting the upper hand falls, the pole can be leaned in various directions to erect the next tier of columns and framing on each of the four sides, and jumping when the steel in that tier has been erected, plumbed, fitted, and fastened. The procedure for jumping is similar to the one for the first jump, except that after the pole has been raised the upper ends of the lower supporting guys are moved up to the new tops of the columns.

Occasionally, if the weight is not too great and the pole, guys, and lead line are adequate, a side framing of two columns and the steel between is assembled on the ground and the connections permanently fastened so that it can be lifted as one piece. Then the side framing and columns on the opposite side are similarly assembled, fastened, and erected. Finally, these two sides are joined in place on the tower, with the pieces that belong between them. This eliminates some of the work aloft, and is safer and generally more economical and expeditious since the framing of the two sides can be proceeding while the pole is being jumped and the permanent fastening of the various members in the two assembled sides is done on the ground instead of in the air.

A small mobile crane can generally be used advantageously to unload and assemble this material, as well as having been used to set up the pole initially and possibly having erected the lowest portion of the tower steel. When the tower is completed, the pole is lowered to the ground by means of falls secured to the top of the tower. The dismantled sections of the pole are "snaked" out of the foot of the tower to be loaded and shipped away. The crane can aid materially in this last operation.

A hoist with enough drums can be used to power the falls on the four jumping guys, using wire-rope falls as guys instead of the manila hand falls on the ends of the shorter wire-rope guys. The extra cost of bringing in such a hoist is usually well worth while since it can expedite the jumping and eliminate the hazards of operating the jumping guys by hand.

Temporary or permanent guys must be installed as erection of the tower proceeds upward. The points where these are to be attached, as well as temporary or permanent anchorages for them, should be determined in advance. The tension to be achieved by tightening the turnbuckles or other devices installed with the guys must be determined. On the taller towers this must be checked frequently by accurate gauges or measuring devices so that equal and correct tension is secured in all guys at any one level. The tensions at the different levels are varied in accordance with stress requirements.

Another scheme for erecting a tower when neither a crane, derrick, nor basket-pole is to be used utilizes a Chicago boom on one leg of the tower. A seat supports the boom at each erecting level, and a fitting is provided for a boom falls between the head of the boom and the leg of the tower that supports the boom seat. Point lines from the head of the boom to either side permit the boom to be swung directly over a piece on the ground inside the area of the tower base. When the piece has been lifted high enough, it is swung over to its final location for setting in place.

When ready to jump such a Chicago boom, a second set of seat and boom-falls fittings must be secured to the section of column erected above. The boom is held temporarily to the tower leg acting as a mast, and the boom falls is moved up to the new fitting. By "going ahead" on the boom-falls lead line, after cutting loose the seat at the foot, the boom is jumped

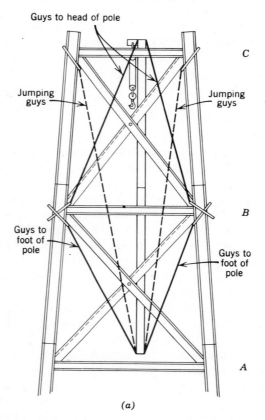

Guys to head of pole

Jumping guys

Jumping guys

Guys to foot of pole

Guys to foot of pole

C

B

A

(a)

FIG. 9.12.1. Jumping a basket-pole: (a) erection of steel to level **C** completed, jumping guys in place, pole ready to be jumped; (b) pole jumped to erect steel at next level above **C**.

up to its new level, secured to the new boom seat, and erection of the next tier of columns and bracing proceeds. After the erection has been completed, the boom is lowered inside the tower to the ground to be dismantled and shipped away. The tower must be adequately guyed during erection to resist the thrust at the foot of the boom and the horizontal forces at the boom-falls fitting above.

As an alternative, the boom seat can be secured on the outside of the tower. This is a more difficult method to use since most towers taper inward and the boom must reach out further to pick material from the ground.

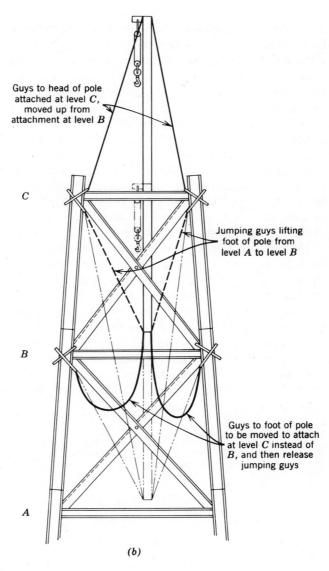

Guys to head of pole attached at level C, moved up from attachment at level B

C

Jumping guys lifting foot of pole from level A to level B

B

Guys to foot of pole to be moved to attach at level C instead of B, and then release jumping guys

A

(b)

FIG. 9.12.1 (*continued*).

9.13   Dutchman

A dutchman is used only for lightweight material. It can be assembled from timbers secured locally. It has a sill of about 4 × 4-in. timber, 6 to 8 ft long. A mast is fastened at its center at right angles to the sill (Fig. 9.13.1). The mast is a 4 × 4-in. timber or heavier, and generally about 10 to 20 ft long (as needed). A light scab secures the mast to the sill. Two braces, 2 × 4 in. or heavier (as needed), are secured at about 45° to 60° from the ends of the sill to the mast. The upper load-falls block is lashed or otherwise fastened to the head of the mast. Three manila falls, which act as guys, are also fastened to the top of the mast, one to each side and one directly to the rear. (Some erectors use only two rear guys spread apart instead of the safer three-guy method.) An additional single-part manila rope "lazy guy" is sometimes added toward the front to prevent the mast from going over backward in case the load is released suddenly and the mast should spring up and back.

The dutchman is used for handline work, and a crab can be secured to the mast at a convenient height for the men to operate the handles. The rig is set up by lifting it by hand into an upright position. The piece to be erected is moved on rollers or a dolly or carried by hand to a point under its location in the structure. The dutchman is then moved so it leans forward slightly with the load falls over the piece. The manila falls guys are tied to temporary anchorages and the piece is lifted into place by the load falls. Then the next piece is spotted under where it is to be erected, the guys are released, and the dutchman is lifted by hand to its new position or skidded or rolled on planks. The guys are again secured to new anchorages and the piece is erected.

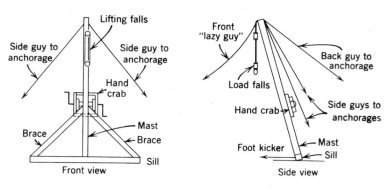

FIG. 9.13.1.   Dutchman.

Instead of assembling the dutchman from timbers secured locally, some erectors use permanent members, fabricating them in their toolhouse. Holes are provided in the mast, braces, and sill to bolt steel connection plates at the intersections. In addition, a head-iron is secured to the top of the mast with loops or eyes for the guys and for the upper load-falls block and with holes in the mast for bolting the hand crab in place. The time saved in assembling and later dismantling is usually worth more than the extra cost of shipping such a permanent set of members.

9.14   Gallows Frame

Basically, a gallows frame consists of two side frames, triangular or similar in shape, with a cross piece between the tops of these two frames, from which a hand or powered chain hoist or similar hoist is hung (Fig. 9.14.1). There must be provision for lateral movement of the hoist on the cross piece. It is most useful where a few pieces too heavy to be handled by hand have to be lifted from a delivering carrier with no crane or derrick available, or when it would be uneconomical to bring in powered equipment just to handle a few heavy pieces. A gin-pole can be set up to assemble the gallows frame.

By installing the gallows frame over a railroad track or a roadway, girders or other pieces in a railroad car or on a truck can be run under the frame. The chain hoist, or other form of hoist, is then used to pick one piece, whereupon the car or truck is pulled away and the piece is moved to one side and lowered onto a dolly or onto rollers; or even onto temporary blocking off to one side but within the area of the frames for storing. The remaining pieces in the car or on the truck are then unloaded in a similar fashion and the car or truck is released. Later, as a piece is

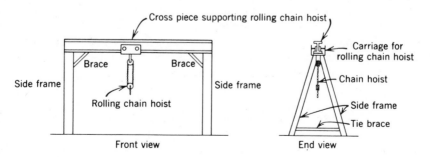

FIG. 9.14.1.   Gallows frame.

needed, it is picked off the blocking, rolled back into position by the hoist rolling on the cross piece, and reloaded on a truck or other carrier for delivery to the erecting rig. In some cases there may be tracks from this point to the erection site, in which case the pieces can be lifted out of storage and rolled over the track to be landed on a car for delivery to the erecting rig.

When a hoist with sufficient capacity for handling the heavy pieces would be too cumbersome or otherwise undesirable or unavailable, a jacking device can be used. This jacking device is mounted on rollers to travel back and forth on the gallows-frame cross piece. It is provided with pendants or slings to reach down for hooking onto the slings or hitches on the piece to be lifted. These pendants or slings, in turn, are secured to a horizontal structural member that is lifted by a pair of jacks at each end of the device.

As a substitute for a gallows frame, there is always the hand method using jacks and hand falls, if heavy pieces are delivered and there is no proper unloading equipment on hand. Blocking is laid on the ground up to the level of the car or truck platform or other form of delivering carrier. One end of the piece is jacked up high enough to install a roller and planking for the roller to ride on. Then the other end is jacked up and a roller and planking are inserted under it. Then, by means of manila rope falls, the piece is rolled off the carrier onto the blocking. The blocking should be built up in two piles, one for each end of the piece. After the piece has been rolled onto the blocking at each end, the piece is again jacked up, one end at a time, to remove the rollers and planking and then jacked down onto the blocking.

If no powered equipment is on hand, the next step to get the piece down to ground level, if it is to be rolled or skidded to the erection rig, requires jacking it down off the blocking. This is done safely by jacking one end at a time. An end is jacked up just enough to slide out the top layer of blocking timbers, and the jack is used to lower that end to the next lower layer of blocking timbers. This is then done at the other end. Repeatedly an end is raised slightly, blocking is removed, and the end is jacked down to the next lower layer of timbers. Care must be used to keep the piece steadily upright during each jacking and lowering step.

9.15   Derrick Boats and Barge-Mounted Rigs

Derrick boats have a single guyed mast to hold the boom, or a stiff A-frame mast with one stiff backleg to the center of the back of the boat,

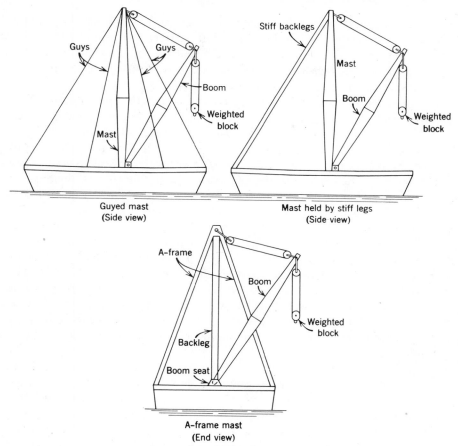

FIG. 9.15.1.   Derrick boats.

or a mast and two stiff legs to the two back corners of the deck (Fig. 9.15.1). An internal combustion or steam-powered hoist is used to operate the boom and load falls, and for slewing (swinging) the boom by means of swing lines between fixed connections on the two sides of the deck and attachments part way up or at the head of the boom. The hoist usually has spools on the ends of the rotating shafts for manila mooring lines or lines for moving the boat; or else separate hoists are mounted on the deck for this purpose.

Small derrick boats have only a limited capacity, whereas the larger ones can handle loads of 300 to 500 tons at a moderate radius over the side. Their capacity is limited not only by the strength of the derrick members and rigging, but also by the amount the boat is permitted to tip safely, which in turn depends on its buoyancy and the freeboard permitted.

A very high-capacity type of derrick boat uses a shear-leg mast consisting of two built-up front legs connected across the top of the two legs by a built-up strut, the three pieces being leaned forward over the front end of the deck, with two built-up backlegs from the intersection of the top of each front leg and the cross piece, to the rear corners of the boat. This type of derrick boat can usually handle loads of 800 to 1000 tons but must be maneuvered so that the head frame is over the piece on another barge, pick the piece, and then be maneuvered over the area where the piece is to be landed or erected. It has only a very special, limited use.

If a crawler crane is run onto a flat deck barge instead of a derrick boat it must be lashed down securely. Another alternative is to use one barge, or two barges lashed together, on which to set up an erection tower. A stiffleg derrick is set on top of this tower, with a boom that is long enough to reach the area over which the steel is to be erected, the tower being high enough for the derrick boom to clear the completed structure. The assembly of the derrick on top of the tower and the setting up of the tower itself can be done best at the erection site by an auxiliary crane or derrick handling the various pieces, or it can be done elsewhere and the barge towed to the site. The actual assembly of the derrick is done in a similar fashion to that described in Sec. 8.2.

When no auxiliary equipment is available, a gin-pole is used to set up the first tier of the tower. A second pole is then used as a "Chicago boom," on the inside of the portion of the tower erected at that time. It is supported on a seat secured to one of the tower legs. This boom or pole, in turn, erects the next tier of the tower and then picks the first boom bodily and sets it high enough on this newly erected steel, on a similar seat on another one of the columns, usually diagonally opposite the boom that is performing the operation. The boom in the new position proceeds to complete the next tier and then picks the first boom to set it in place above. This is repeated alternately until the tower has been completely erected, and then one of the booms is used to set the stiffleg derrick on top of the tower. Dismantling is done in similar fashion, each boom lowering the other as the tower is dismantled tier by tier.

To move the barges or boat from one point to another, tugs should be used if the equipment is difficult to move or if the tugs can be gainfully employed the balance of the time since they are usually quite expensive and ordinarily are charged for on a time basis. If tugs are not used mooring lines are used, being attached to some fixed, sturdy form of mooring post or to anchors in the water, ahead of the boat or barges. Then, by using the spools on the hoist or drums on an auxiliary hoist, these forward lines are taken in and lines in the opposite direction to similar mooring posts or anchors are slacked off. When the equipment has reached a new working position, the forward lines are made fast and the rear lines are moved up to help secure the boat in a reasonably fixed position.

Appendix *A*

Selected Reading

Periodicals

These magazines are recommended for their articles on the construction of various kinds of structures. They give information on the procedures used and the difficulties encountered, which will aid in planning for the best methods of erecting steelwork. They will also help to avoid details and processes found to be unsatisfactory. The articles and the advertisements give valuable information on new equipment that is available.

Civil Engineering (monthly), American Society of Civil Engineers, New York.

Construction Contracting (monthly), McGraw-Hill, New York.

Contractors and Engineers Magazine (monthly), Buttenheim, New York.

Engineering News-Record (weekly), McGraw-Hill, New York.

Welding Journal (monthly), American Welding Society, Miami, Fla.

Acier-Stahl-Steel (monthly), Brussels, Belgium.

Books and Catalogs

The following are valuable as references and source material on structural shapes, wire rope and slings, bolting, welding, rigging, and safety, and for understanding detail drawings.

ANSI A10.13 Safety Requirements for Steel Erection, American National Standards Institute, New York.

ANSI B30 Safety Standard for Cableways, Cranes, Derricks, Hoists, Hooks, Jacks, and Slings, American National Standards Institute, New York.

B30.1 Jacks

B30.3 *Hammerhead Tower Cranes*

B30.5 *Mobile and Locomotive Cranes*

B30.6 *Derricks*

B30.8 *Floating Cranes and Floating Derricks*

B30.9 *Slings*

B30.10 *Hooks*

B30.12 *Handling Loads Suspended from Rotocraft*

Data sheets on tools and equipment, National Safety Council, Chicago.

Handbook of Rigging for Construction and Industrial Operations, W. E. Rossnagel, McGraw-Hill, New York.

Hot Rolled Steel Shapes and Plates, United States Steel Corp., Pittsburgh, Pa.

Manual of Accident Prevention in Construction, The Associated General Contractors of America, Washington, D.C.

Manual of Steel Construction, American Institute of Steel Construction, New York.

Occupational Safety and Health Standards, Department of Labor, Occupational Safety and Health Administration, Washington, D.C.

Purple Strand Slings and Fittings, Bethlehem Steel Corp., Bethlehem, Pa.

Safety and Health Regulations for Construction, Department of Labor, Occupational Safety and Health Administration, Washington, D.C.

Safety Code-Erection Division, Bethlehem Steel Corp., Bethlehem, Pa.

Safety Maintenance Directory Combined with the Manual of Modern Safety Technique, Alfred M. Best, Morristown, N.J.

Safety Manual, Construction Department, American Bridge Division, United States Steel Corp., Pittsburgh, Pa.

Specifications for Structural Jobs Using ASTMA A-325 or A-490 Bolts Approved by Research Council on Riveted and Bolted Structural Joints, distributed by Industrial Fasteners Institute, Cleveland, Ohio.

Structural Drafting and the Design of Details, C. T. Bishop, Wiley, New York.

Structural Shapes, Bethlehem Steel Corp., Bethlehem, Pa.

Structural Welding Code (D1.1-79), American Welding Society, Miami, Fla.

Tiger Brand Wire Rope, United States Steel Corp., Pittsburgh, Pa.

Tiger Brand Wire Rope Slings, United States Steel Corp., Pittsburgh, Pa.

Wire Rope, Bethlehem Steel Corp., Bethlehem, Pa.

Appendix B

Suggested Erection Safety Code

Contents

297

Introduction

The erection of steel structures can be a hazardous occupation. The work requires steady nerves, healthy bodies, and the ability to carry on operations even under difficult circumstances. Each operation presents its own problems and generally no two jobs are exactly alike. Therefore, it is impossible to set up exact rules to control safety measures in erection work, and general rules in this code must be adapted to individual situations.

Good health is a primary necessity to an erection worker, and he should carefully regulate his manner of living to keep himself in a healthy condition. He may be exposed to all sorts of weather conditions, so the matter of proper clothing to afford sufficient protection without seriously hampering his movements should be given serious consideration by him.

Safety to the worker, as well as to others on the same operation, is not only dependent on the judgment of his supervisor but also on the individual himself. Proper safety devices and precautions must be used, but no matter how good the tools provided may be, or what kind of safety devices or methods are used, unless the men using them follow safe practices accidents will not be prevented.

As a guide to its field employees in the safe conduct of their work, the company has prepared this code, based on good practice in steel erection work, and to help prevent accidents. The code is based not only on the experience gained over many years of steel erection work by this company, but also on the records of other erection companies that have enforced safety over a long period with excellent results. The rules are mostly common-sense precautions that all skilled workers naturally follow because, in addition to preventing injuries, they are the best ways to do a job.

The safe way to do a job must always be found before going ahead. Some safety rules may have been inadvertently omitted from this code. In the event that no rule is found to cover a situation, remember *the safe way to work is the best way. Don't take a chance.*

General

The rules in this code will help new employees cope with the hazards of steel erection and will act as a reference for experienced men. They should be reviewed occasionally. The rules are intended to act as a guide for safe workmanship. If they appear to be annoying restrictions to your movements or methods in doing your work, stop and consider why the rules were written. They are for your own protection and to help you protect your fellow workers. There are good, sound reasons for every rule. To look on these rules as needless restrictions, to ignore their guidance, or to fail to learn them is to invite injury. Rules are responsibilities to be accepted and used. Ignorance of these rules is no excuse if you are injured.

Accident prevention must be on a cooperative basis. All employees must report promptly any unsafe conditions to the man in charge of their work. Suggestions and recommendations for improving the safety of a job should be reported by an employee to his supervisor. All supervisors are to cooperate fully in putting into effect all practical suggestions or recommendations that will aid in reducing the hazards of the job to a minimum. Suggestions and recommendations that cannot be acted on by the supervisor in charge are to be reported to the main office, where the necessary action will be taken.

The company conducts a safety program for the purpose of preventing accidents and injuries. Every employee is expected to participate in this program. Production with safety is a basic principle of the program. Your safety and welfare are of vital interest to your fellow workers and to management. With your cooperation we can eliminate accidents. Make your job a safe place to work.

1. General Rules

1.1. Always be careful; use good judgment and prudence in doing your work so as to protect yourself and others from injury, whether or not the acts necessary for safety are included in this safety code.

1.2. Report to your foreman or timekeeper when injured even slightly. Do not try to treat yourself.

1.3. Do not remove foreign bodies from your eye or allow anyone else to do so, except attendants at a dispensary, a doctor, or a first-aid attendant.

1.4. Notify the company through your foreman or timekeeper of any change in address. This is important in case of accident or sickness.

1.5. Do not engage in horseplay on the job. This is dangerous and is prohibited.

1.6. Always wear a safety hat even when you think there is no chance of falling objects or swinging cables, hooks, etc.

1.7. Always wear safe shoes, preferably with nonslip soles and without nails.

1.8. If your shoes are greasy, muddy, or otherwise slippery, clean them before you walk on steel or go up or down a ladder.

1.9. Look where you step, and be sure that what you step on is safe and secure. Do not step on the ends of loose planks. Avoid jumping from one level to another.

1.10. Never use planks to walk on, or for any other purpose, that are unsafe for the span. Never use timbers to pile loads on that are unsafe for the span or load.

1.11. Never leave nails or spikes protruding in planks, boards, or other timbers. Pull them out or clinch them into the wood.

1.12. Avoid wearing oily or paint-soaked clothing. If your clothing becomes oily or paint soaked, keep away from fires or operations where hot sparks can fly.

1.13. Always wear good gloves. Never wear torn gloves near any moving equipment.

1.14. Never permit your gloves to become saturated with anything that may cause them to become slippery.

1.15. When handling slings or other cables your leather-faced gloves should be in good condition to avoid serious cuts from broken wires.

1.16. Never wear torn or loose clothing that is likely to catch in any revolving or operating equipment. Avoid cuffs on trousers if your work is such that the cuffs can catch on anything and cause you to lose your balance.

1.17. Do not wear finger rings while at work.

1.18. Keep your hands away from moving fans or fan belts and any moving parts of equipment.

1.19. Keep your fingers and hands away from a connection being made where they can be caught or squeezed.

1.20. Keep out from under loads. Keep clear of moving loads. Stay in the clear in case a load must be landed quickly in an emergency.

1.21. Always stay clear of a swinging load, especially when connecting.

1.22. When unloading steel with a crane or derrick, see that men are out of the car or off the truck before the load is lifted.

1.23. Drinking by employees during working hours is absolutely prohibited. Employees under the influence of liquor must not start work. Violators will be dismissed.

1.24. Pick up your tools, bolts, pins, etc., so no one will slip on them. Keep small material in buckets or other containers to prevent their falling on men working below.

1.25. Keep the structure clear of loose tools and other loose material, especially when working over railroad tracks or highways in use, or over men working below.

1.26. Remove loose material from planks before moving them.

1.27. Avoid throwing or dropping bolts, washers, pins, or other tools. Use buckets or other containers and raise or lower by using a line.

1.28. Do not "ride the load."

1.29. All employees are prohibited from riding on the ball or empty hook.

1.30. Never slide down manila lines or wire-rope cables, guys, or falls.

1.31. Never cut loose a piece until it has been properly bolted at both ends. Do not rely entirely on a wrench or drift pin in the hole.

1.32. If it is necessary in erecting a beam, to use only one bolt at each end, be sure this bolt is pulled up tight with a wrench and is placed in such a way that the beam will not roll when walked on.

1.33. Avoid sitting or standing on a wrench that has been inserted in a bolt hole.

1.34. Never walk on a beam that can roll or vibrate excessively.

1.35. Never leave lines or cables hanging loose where they may foul passing men, cars, or moving objects.

1.36. Never touch a running rope.

1.37. Always place U-bolt wire-rope clips on wire rope in the proper direction.

1.38. If a wire-rope guy or lashing is to be used at or about its full safe working load, use at least the minimum number and the recommended spacing of proper-size wire-rope clips for the diameter of wire rope being clamped.

1.39. Never tighten a wire-rope clip so much that it cuts or unduly distorts the wire rope.

1.40. Retighten wire-rope clips after an hour's full running time and test frequently thereafter.

1.41. Never let the end of a running cable come so close to a sheave or stop that the clips are jammed together. If this occurs accidentally, stop and replace the clips in the proper position and spacing.

1.42. The proper-size thimble should be used on all beckets.

1.43. Avoid using slings or lashings on a hook in such a way that they tend to open up the hook. Place them well down in the base of the hook or use a shackle.

1.44. Avoid excessive bridling.

1.45. Never stand in the bight of a line or in front of a snatch or lead block.

1.46. Make sure gate or snatch blocks are properly closed before using.

1.47. Hooks should be moused where there is danger of their jumping out of a sling eye or shackle or other holding device.

1.48. Always use a tagline hook heavy enough so it will not straighten out when pulled.

1.49. Always use a tagline when lifting or lowering a load or a piece beyond a man's reach.

1.50. Never let spreader hooks swing where they can hit someone.

1.51. Use only ladders in good condition. Always tie or lash a ladder in place. Be sure the base is steady and secure. Provide a suitable platform at top and bottom of the ladder. Ladders should extend at least 3 ft above upper platform.

1.52. When going up or down a ladder always face the ladder.

1.53. Avoid carrying objects in your hands while using a ladder. Use a handline for raising and lowering tools or materials.

1.54. Ladders with tie-rods must be set up with the rods below the rungs.

1.55. Never slide down a ladder.

1.56. Always inspect scaffolds and floats before getting on them, and make sure that all lines are properly tied. Never jump down on a float or scaffold.

1.57. Never operate a machine with which you are unfamiliar until you have been properly instructed and have been authorized to do so.

1.58. Never place pneumatic tools in a fire to warm them or thaw them out.

1.59. Never fool with compressed air. Horseplay with compressed air is prohibited.

1.60. Place air hose and welding cables so others will not trip over them.

1.61. Examine all tools before using them. Do not work with burred, broomed, mushroomed, or otherwise defective tools.

1.62. Inspect tools frequently for defects. Do not use tools with split or loose handles.

1.63. When using a bar for prying, stand so that it cannot strike you or that you will not be thrown off balance if it slips.

1.64. Be sure your wrench, bull pin, or other equipment is secure in your belt holster so that it cannot fall out.

1.65. Do not use a wrench as a hammer.

1.66. Use properly fitted wrenches. Turn in wrenches that have worn or sprung jaws. Never use a pipe on the end of a wrench for additional leverage; use a longer handled wrench.

1.67. Always use properly fitted handles on jacks; never use a crow bar.

1.68. Safety goggles must be worn when grinding, cutting, caulking, chipping, drilling, reaming, or cutting concrete, or for any other operations where sparks, chips, etc., can fly into your eyes.

1.69. In wearing goggles, if you wear personal spectacles, secure the proper type goggles to fit over these; do not depend on the ordinary glass in the lenses of your spectacles.

1.70. Wear the proper shade goggles when working near the flash or arc of an electric welder. Wear the proper goggles, helmet, or shield when burning or welding.

1.71. Shield your eyes when passing an area where sparks, chips, dirt, etc., are flying.

1.72. Always wear protective goggles when pouring molten materials.

1.73. Never put wet lead or zinc in a pot with molten metal. It will cause an explosion.

1.74. Never pour molten material into a hole that contains water or is wet.

1.75. In storing a girder upright always shore in such a way that the shores cannot be accidentally removed or knocked out.

1.76. Always guy columns and trusses that may be unstable when free, before cutting loose.

1.77. Store material with sufficient clearance for safe passage around it.

1.78. Do not stand between loads or pieces being lifted.

1.79. If you have occasion to remove the cover from any opening, always guard that opening so no one will be endangered. Always replace the cover when you are finished.

1.80. Never oil, grease, remove guards, or attempt to repair machinery while it is in motion. See that all guards are replaced before restarting the machinery. Report to your foreman any guards missing from equipment to be used.

1.81. Never work on a crane runway unless there is a good crane-stop between you and an operating crane.

1.82. Load cars or trucks so nothing can fall off.

1.83. Make sure you understand your work and then proceed in the safest manner.

1.84. Always be sure guys are clear of passing cars or other moving vehicles.

1.85. Be sure the kickers at the bottom of any pole or mast are secure.

1.86. Never tie a guy to a railroad track in such a way that a car or truck can cut it off.

1.87. Always wear a proper respirator when working in a dust-laden atmosphere. Replace the filter regularly.

1.88. Beware of gas! Injurious gas is found in unexpected places. Sometimes you cannot see it or smell it. Keep away from such areas where there is a possibility of gas until you have received specific instructions from your foreman.

1.89. Use the greatest care in handling any acid. Use rubber gloves. Where the fumes are dangerous use a suitable respirator.

1.90. Never permit oxygen to contact oily clothes or rags.

1.91. Two men should carry pieces of material or equipment of unwieldy length or shape.

1.92. When two or more men are carrying long material together, they all should always carry the material on the same shoulder.

1.93. When carrying long pieces of material or equipment, watch the ends so that no one will be injured.

1.94. Be sure all men are clear before dragging or pulling any material ahead.

1.95. Men accustomed to wearing glasses are not permitted to go aloft without wearing glasses. Bifocals are not permitted aloft, but men accustomed to wearing them may go aloft when properly fitted with distance, single-vision glasses.

1.96. Report any suggestions for safer operations to your foreman.

1.97. Pile material safely. Use proper blocking. Never exceed a safe height of pile. Always land material so it cannot roll over.

1.98. Before a load is hoisted see that the load is free of any material that may loosen and fall out of the slings should the load foul or be jarred.

1.99. Never straighten a load by swinging it against a wall, column, or other object. Lower it and readjust.

1.100. Always be sure that cable spreaders or other hooks are properly in place before picking a load.

1.101. Do not walk on steel or planks that are icy. Do not walk on painted steel if the paint has not dried.

1.102. Planks covered with ice or snow must never be turned over since the ice will then be between the plank and the steel, and the plank will slide. Instead, clean off the planks.

1.103. In timber work, be sure any spuds or cleats used for temporary hand- or foot-holds are strong and secure.

1.104. Lift a load properly. Keep your back as upright as possible, set your stomach muscles, and allow your legs to take part of the load.

1.105. Cover exposed ends of reinforcing rods if they project where a man can fall on, or be caught by, the ends.

1.106. Wipe up spilled oil, grease, or paint promptly. These materials on wood or steel may cause a man to slip and fall.

1.107. Never turn on electricity, steam, air, or water, or set in motion any machinery without first carefully checking to see that no one may be injured or endangered by your act.

1.108. Do not allow sightseers or trespassers around your work where they would interfere or could be injured.

1.109. Report promptly any unsafe practices, conditions, or equipment to your foreman.

1.110. Never use other contractor's planks, slings, blocking, or any material or equipment unless specifically instructed to use them by the superintendent or foreman. Endeavor to use only materials and equipment furnished by this company.

1.111. Do not permit other contractors to use this company's tools, equipment, or any material that could break or cause an accident if used improperly.

2.   Manila Lines and Wire Rope

2.1. In tying lines to scaffolds, floats, or supporting beams, be sure your hitch is secure and safe before getting on the scaffolding.

2.2. Never jump onto a scaffold or float.

2.3. For taglines, use a minimum of ½-in. diameter new, or ¾-in. diameter used manila line. Always use a tagline hook heavy enough so it will not straighten out when pulled, preferably at least ⅜-in. diameter rod.

2.4. For ordinary hand falls use ¾-in. diameter manila line with a minimum of a 6-in. block.

2.5. For temporary derrick-guy falls, use four parts of 1-in. diameter manila line, with a minimum of a 10-in. block.

2.6. For floats or ship scaffolds use a minimum of 1-in. diameter manila lines.

2.7. For needle-beam scaffolds use a minimum of 1¼-in. diameter manila lines.

2.8. Always protect a running cable from chafing against abrading surfaces. Make sure all leads are perfect.

2.9. Never drag ropes or cables over rough or sharp surfaces.

2.10. Never slide down manila lines or wire-rope cables, guys, or falls.

2.11. Always place U-bolt clips on wire rope in the proper direction. See Table 3 in Appendix B.

2.12. If a wire rope is to be used at or about its full safe working capacity, install the proper number of wire-rope clips at the correct spacing. See Table 3 in Appendix B.

2.13. On beckets of running cable always use the minimum number and the recommended spacing of proper-size clips for the diameter of wire rope being clamped.

2.14. If you find any broken wires at a becket, cut off the end and refasten the becket.

2.15. Never tighten a wire-rope clip so much that it cuts or unduly distorts the wire rope.

2.16. Never let the end of a running cable come so close to a sheave or stop that the wire-rope clips are jammed together. If this occurs accidentally stop and replace the wire-rope clips in the proper position and spacing.

2.17. Retighten wire-rope clips after an hour's full running time and, when necessary, at all regular inspections.

2.18. Never permit a wire rope to kink.

2.19. Never use a hook that has spread.

2.20. Never use a hoisting spreader or a sorting spreader to choke or cradle a load. Always use slings, and be sure the load is properly choked or secured before hoisting.

2.21. When putting chokers on a bundle of planks, put the eyes over the edges of the bundle so the cable is completely in contact with the planks.

2.22. When using a sling or choker to make a heavy lift, use wood, pipe, or tire over the sharp edge of steel, and make sure it cannot fall off when the sling is released.

2.23. Be sure a choked sling is centered and tight before moving a load.

2.24. In selecting the proper size of wire-rope sling, consider the additional load resulting from the type of connection, such as pulling the web of a beam or girder into a knife connection.

2.25. Always try to lower a piece into a knife connection. If the piece should jam when pulled up into such a connection the sling may break if it is not stopped in time.

2.26. Use extra caution when lifting light pieces. Use a round turn on the choked sling or a shackle in a connection hole.

2.27. Always use separate choked slings when picking two pieces at the same time. Use slings of different lengths.

2.28. Never pick a load with sorting spreaders more than a few feet off the floor or working level.

2.29. Avoid excessive bridling.

2.30. Never use a block for wire rope that is designed for manila line.

2.31. The proper-size thimble should be used on all beckets.

2.32. Sheaves and blocks must be used only for the diameter wire rope for which they are grooved.

2.33. See that blocks are in good condition. Do not overload or damage them.

2.34. Keep all turns of wire rope tight and close together on the drum. Be sure the cable leads properly on the drum.

2.35. Do not take hold of a hoistline cable near a sheave block as your fingers may be drawn into the block.

2.36. Never overload a manila line, a wire rope, or a wire-rope sling.

2.37. Avoid kinking a manila line.

2.38. Always take the necessary precautions to prevent a turnbuckle from accidentally unscrewing when subjected to a load.

2.39. Never cut out part of the eye of a turnbuckle. Do not use a turnbuckle with such a damaged eye.

2.40. Before using a shackle to handle a load, make sure the shackle bolt or pin is secure. Do not use a spud wrench instead of the pin or bolt.

2.41. To remove wire rope from a reel, mount the reel securely but so it can turn freely, and unreel the rope.

2.42. Remove manila line from a new coil by pulling the inside end out. Never start on the outside end.

2.43. When loading manila line or wire rope for shipment, keep it away from storage batteries or anything containing acid that may break or spill.

2.44. If wet rope is used, tie it so as to provide for shrinkage when it dries.

3. Fitting, Reaming, and Bolting—Jacks

3.1. Always inspect scaffolds and floats before getting on them, and make sure that *all* lines are properly tied. Be sure your float or scaffold lines are tied to a secure support.

3.2. Make sure scaffolds or floats are in good condition before using. Make sure the ropes are long enough for tying properly.

3.3. When hanging a float, tilt it slightly in toward the work and secure it so it will not kick out.

3.4. When hanging or removing floats or scaffolds keep in a safe position, preferably astride the steel.

3.5. Be careful when carrying floats. Always be on the upwind or windward side so the wind will not cause the float to knock you over.

3.6. Never jump onto a float or scaffold.

3.7. Avoid throwing or dropping bolts, washers, pins, or other tools or small objects. Use bolt buckets or other approved containers, and raise or lower by using a line.

3.8. Do not leave your tools lying about for others to trip over.

3.9. Keep small tools, loose bolts, pins, washers, etc., in bolt buckets or other approved containers. Do not leave them lying around loose on scaffolds, floats, or tops of beams or girders, where they can be tripped on or knocked off onto men below.

3.10. Always have a good footing when reaming or drilling, especially when working on a scaffold, float, or top of a girder or beam.

3.11. Never use planks on a scaffold or across beams if the planks are unsafe for the weight to be put on them for that particular span.

3.12. Always be sure the boards in your float or other planks you use are in good condition and not cracked or split.

3.13. Never use less than three planks on a needle-beam scaffold.

3.14. Use drop bolts with nuts on the ends of loose scaffold planks on needle beams.

3.15. Never leave any loose lines hanging below your scaffold.

3.16. Do not use a wrench as a hammer.

3.17. Use properly fitted wrenches and do not work with wrenches that have worn or sprung jaws.

3.18. Avoid using erection bolts with worn or misshaped nuts or heads, from which wrenches are liable to slip.

3.19. Examine all tools before using them. Do not work with burred, broomed, mushroomed, or otherwise defective tools.

3.20. Inspect tools frequently for defects. Do not use tools with split or loose handles.

3.21. Before disconnecting a hose from a pneumatic machine, cut off the air supply and relieve the pressure in the hose.

3.22. Do not hold the hose and twirl the tool when connecting leader hose to an air tool, or when disconnecting the hose from the tool. Hold the tool and turn the leader hose to connect or disconnect them.

3.23. Do not pass air tools to another workman by using the hose as a line. If it is impossible to pass the tool from hand to hand, a line must be used.

3.24. Do not point an air hammer toward anyone. Remove snap and plunger when leaving job temporarily.

3.25. Do not leave an air hammer in such a position that the trigger might accidentally slip or be kicked.

3.26. Do not knock out a drift pin or bolt without making sure that the pin or bolt will not fly and strike someone.

3.27. Do not allow trucks to run over air hose. Lay it between two planks to protect it.

3.28. In using a riveting hammer, if a tapping sound is heard, return the hammer to the toolhouse for repairs.

3.29. Always wear proper safety goggles when reaming, drilling, or using any pneumatic equipment where the exhaust can blow something into your eyes.

3.30. When backing out rivets or knocking off heads, use a proper shield to keep the rivets or heads from flying. Never stand in front of a rivet

3.31. Keep your hands and clothing away from a rotating chuck or projections buster.

3.32. Keep your gloves away from an impact wrench universal joint so they cannot be caught.

3.33. Be sure your impact wrench socket is securely in place, with no projection that can catch your hands or clothing.

3.34. Familiarize yourself with the proper methods of using an impact wrench. Instructions on its proper use can be secured from your foreman.

3.35. Never guide a bit into place by taking hold of it with your hands.

3.36. Always stand in such a position that you will not be caught if an air machine fouls.

3.37. If any pneumatic tool you are using is not operating properly, return the tool to the toolhouse.

3.38. Use the proper-size jack when raising or lowering a load.

3.39. Make sure a jack is properly blocked so it will not slip or kick out.

3.40. When lifting with a jack use shims or blocking to follow up the load.

3.41. Never leave a load on a jack without blocking it securely.
on a pneumatic tool.

3.42. Remove a jack handle when not in use.

4. Signaling

4.1. Give all signals to the operator in accordance with the approved standard. See Table 12 in Appendix B.

4.2. Be sure the crane or hoist operator understands the signals being given.

4.3. Before giving the signal to pick a load be sure the load falls are plumb.

4.4. The foreman or signalman should notify the operator in advance if an extra-large or heavy load is to be handled.

4.5. The foreman or signalman, where necessary, shall walk ahead of a moving load and warn workmen and others to keep clear. He shall also see that the load is high enough to clear obstructions or is swung clear of same.

4.6. Do not give signals to the operator unless you have been authorized to do so by the foreman.

4.7. Signals must only be given by one designated man at a time.

5.   Cranes, Derricks, and Travelers

5.1. All employees are prohibited from riding on any loads being handled by a derrick, crane, or other lifting or lowering equipment.

5.2. Keep off cranes unless your work requires you to be there.

5.3. Be sure that material is properly balanced and secure before making a lift.

5.4. Keep clear of moving loads and keep others from getting under any load.

5.5. Avoid swinging loads over anyone.

5.6. In swinging a crane be sure there are no obstructions and that no man can be caught between the crane and a wall or pile of material.

5.7. When working overhead, do not drop material or tools unless absolutely necessary. In that case always make sure there is no one below and take steps to prevent anyone from walking under falling material.

5.8. Pile material safely; use proper blocking. Never exceed a safe height of pile. Always land material so it cannot roll over. Use shores when needed.

5.9. Do not allow an accumulation of oil or grease around derricks, platforms of cranes, or elsewhere where men are obliged to walk or work. This will avoid the danger of slipping or falling.

5.10. Engines, compressors, or other equipment must not be permitted to drip oil or grease on the structure.

5.11. Inspect your ropes, cables, and clips frequently. Do not overload or damage them.

5.12. Do not take hold of a hoisting cable near a sheave block as your fingers may be drawn into the block.

5.13. Keep your eyes on the entire load as it is being hoisted to see that nothing fouls.

5.14. Follow closely all operating instructions given for the crane or derrick on which you are working.

5.15. Cranes and hoists must be operated only by experienced operators.

5.16. Never try to pick a load beyond the safe working capacity of the equipment.

5.17. When truck-crane outriggers are required, the blocking must be properly placed to provide good bearing to suit ground conditions.

5.18. Never pick a load that will cause the rear end of the crane to raise off the ground.

5.19. Keep the boom low when traveling with an empty hook.

5.20. When working the auxiliary falls on the jib or when moving the crane any considerable distance, secure the main load block to prevent it from swinging and damaging the boom lacing.

5.21. The boom on truck and crawler cranes should be kept low when traveling on fill or on uncertain ground.

5.22. Make sure derrick anchors are properly and securely placed.

5.23. Never tie a guy to a railroad track in such a way that a car or truck can cut it off.

5.24. Always be sure guys are clear of passing cars or other moving vehicles.

5.25. Hoisting engines must be tied down, shored, or otherwise anchored.

5.26. Be sure the kickers at the bottom of any pole or mast are secure.

5.27. Make sure traveler tie-downs are properly and securely placed.

5.28. All working parts of a crane, hoist, derrick, or engine must be carefully inspected and lubricated regularly.

5.29. Gasoline engines must be filled only when stopped. Safety gasoline cans must be used when tank is not filled directly from the drums or other source of supply.

6.   Cars, Trucks, and Railroad Equipment

6.1. Keep off cars, trucks, cranes, and other conveyances unless your work requires you to be there.

6.2. The foreman in charge of work to be done in close or dangerous proximity to any track used for the passage of cars, engines, locomotives, cranes, trucks, or other moving machinery must take all necessary precautions for the safety of the men before the work is commenced.

6.3. Be careful when passing moving or standing cars or trucks, especially when they are being loaded or unloaded.

6.4. Keep clear of loads being picked when unloading cars or trucks.

6.5. Avoid crossing a track close to the end car of a train.

6.6. Look both ways before crossing a highway or railroad track.

6.7. Keep out from under and between cars.

6.8. Piled material must always be kept at least 6 ft from the nearest rail of a railroad track in service.

6.9. Do not walk on the track when there is any walkway available.

6.10. Keep off the track except when your work requires you to be there. Endeavor to lay no material on a track.

6.11. Load cars or trucks so nothing can fall off or roll over.

6.12. Railroad cars standing on an inclined track must have their brakes tightly set. A chock block must also be placed on the track to keep the car from rolling down the incline.

6.13. Endeavor to stop trucks to be unloaded on level ground. If this cannot be done, block truck wheels before loading or unloading and take necessary precautions to prevent the load from shifting or rolling over.

6.14. Block railroad-car wheels before loading or unloading to prevent movement of the car.

6.15. In removing a girder or truss from a car or truck containing one or more remaining girders or trusses in upright positions, always shore those left in the car or on the truck. Also block the car or truck when necessary to prevent tipping.

6.16. Always shore unloaded material that can fall over. Make sure the shores are secure and cannot be knocked out accidentally.

6.17. The driver of a truck must be out of the cab when the truck is being loaded or unloaded.

6.18. Never throw anything off a truck or out of a car if there is any chance of hitting anyone.

6.19. Avoid disturbing other material when picking a load or a piece from a car or off a truck.

6.20. Before moving a railroad car see that the hand brakes are in working order.

6.21. Never move a car by pushing it with a timber against the car or by using the boom, runner, or main falls of a crane or derrick.

6.22. If it is necessary to move a car with a crane, truck, or other piece of heavy equipment, pull the car with wire rope secured to the crane or truck frame and have a man on the brake to stop the car, especially if it is on a grade.

6.23. When moving a car, truck, or crane, make sure the track or roadway is clear of men and material. A positive warning or signalman must be used when the operator's view of the track or roadway is obscured.

6.24. Never move a railroad car against another stopped car if this might endanger men in either car.

6.25. When "spotting" cars do not block crossings or roadways.

6.26. At open ends of rails, rail stops or other chocks should be fastened securely.

6.27. Keep the structure clear of loose tools, small timbers, and other loose material, especially when working over railroad tracks or highways, etc., in use.

7. Crane and Hoist Operators

7.1. Machines must only be operated by experienced operators.

7.2. The operator is responsible for the proper upkeep, condition, and safe operation of his machine.

7.3. Familiarize yourself with your crane, hoist, or engine mechanism and its proper care. If adjustments or repairs that you cannot make are necessary, report them at once to the foreman.

7.4. Examine your crane, hoist, or engine for loose parts or defects each day it is used and after repairs have been made.

7.5. Keep all persons away from your equipment unless their work requires them to be there.

7.6. Keep your equipment clean. Clothing or any personal belongings, tools, extra fuses, oil can, waste, or any other necessary articles shall be stored in a tool box or other container and not left loose on or about the equipment.

7.7. The operator shall not eat or read while actually engaged in the operation of a crane, hoist, or engine, nor operate same when he is physically unfit.

7.8. Recognize signals only from the person supervising the lift or from the signalman assigned. Operating signals must follow an approved standard.

7.9. Proceed only on signal. If in doubt, have signal repeated.

7.10. When lowering a load, proceed carefully and make sure the load is under safe control.

7.11. When handling heavy loads, make sure your frictions are tight enough before picking the load, and be sure your brakes will hold the load to be picked.

7.12. When moving a crane, make sure the roadway is clear. When you cannot see the road, have someone who can see the road give you the necessary signals; don't rely on a whistle blast or warning gong or horn.

7.13. The operator must check conditions when working near electric wires, guy lines, or structures. Avoid contact of boom or cables with electric wires, guy lines, or structures. See Table 11 of Appendix B for minimum safe clearance from electric power lines.

7.14. Before moving a crane on which an empty sling is hanging, both ends of the sling shall be hooked to the block, hook, or shackle, high enough to clear all obstructions.

7.15. Never lift a load beyond the safe working capacity of your machine.

7.16. If the falls become twisted or entangled, stop all motion before clearing up the rope.

7.17. Shut down machinery before refueling.

8. Gasoline and Compressors

8.1. Never refuel, oil, grease, remove guards, or attempt to repair machinery while it is in motion. See that all guards are replaced before starting the machinery.

8.2. Always keep your hands and loose parts of your clothing away from compressor or engine fans or fan belts.

8.3. Hold a crank handle so it cannot hit you if the engine kicks back.

8.4. Never suck gasoline through a hose to start suction or for any other purpose.

8.5. Preferably use a gasoline pump for drawing gasoline from drums. Otherwise, *roll* the drum onto horses or other blocking after installing a tight spigot or valve.

8.6. Where necessary to use a can, always use a safety can for filling a fuel tank.

8.7. Never force gasoline out of a drum or other container by means of compressed air.

8.8. Never store an appreciable quantity of gasoline near an engine in actual operation.

8.9. Always store gasoline away from open fires, sparks, or any source of excessive heat.

8.10. Never permit a fire near an air receiver.

8.11. Never allow antifreeze, alcohol, or flammable liquids to enter an air receiver because of the danger of explosion.

8.12. Do not allow fumes from flammable liquids to enter the intake of a compressor.

8.13. When gasoline-driven engines, compressors, or other machines are operated in an enclosed space, even though large, always pipe the exhaust to the open air.

8.14. Use only a safety solvent to clean engines and motors. Never use gasoline or other flammable solvents.

8.15. Fuel system sump and settling tanks and sediment bulbs on all engines must be drained regularly of collected water or sediment.

8.16. In cold weather, use only approved type ether starting aids. Indiscriminate use of ether is dangerous.

9.   *Electrical Work and Equipment*

9.1. Do not try to do your own electrical wiring, fix electric lights, or make any other electrical repairs. This should be done only by the man selected by the foreman, or by a qualified electrician.

9.2. Be extra cautious when your work requires you to be near electric power wires or feeders. The absolute minimum clearances between men or equipment and high voltage lines as shown in Table 11 of Appendix B must be strictly observed.

9.3. Power higher than 440 V in insulated wires, particularly alternating current, and any power in bare wires, must be shut off while work is being done closer than permitted in Table 11 or where equipment can possibly come in contact with same. Assume that all wires, switches, etc., are alive or "hot" until you are sure they are not.

9.4. Do not use extension cords or electric hand tools with exposed wires or other defects.

9.5. Place extension cords or electric cables so that others will not trip over them. Feeder cables should be hung with insulators or laid clear of workmen or moving equipment.

9.6. Always ground conduits and the frames of all electrical equipment. Ground electric hand tools through a 3-wire feeder lead.

9.7. Ground all cranes working where their booms or cables can possibly contact any power lines.

10.   *Welding*

10.1. When actually welding with electric arc, always wear a helmet or shield with the proper colored glass, and safety glasses to prevent slag entering your eyes when cleaning the weld.

10.2. Never look at a welding arc with the naked eye.

10.3. When it becomes necessary to weld near another welder, wear proper goggles under your shield or helmet as an added protection to your eyes or have a protective shield set up between you and the other man's arc.

10.4. When working at any job near electric arc welding, always wear proper colored goggles or have a protective shield between you and the arc.

10.5. Your leather gloves should be worn while welding and your leather sleevelets when welding overhead.

10.6. Welding flashes can cause burns to the body. Be careful of holes in your clothes or exposed areas such as your legs between your socks and your trousers or overalls.

10.7. Do not leave welding wire or stub ends lying around. They can roll if stepped on and cause a man to fall.

10.8. Do not leave electrode holders and cable lying around. When you have finished welding wind up the cable on the machine or store it safely out of the way.

10.9. Do not lay down your electrode holder in such a way that a short circuit or contact can be made with anything metallic.

10.10. When welding under erectors or fitters or in other places where material may fall, always secure some form of overhead protection for yourself.

10.11. Shut off the machine when you are through welding.

10.12. Where welding is done inside any enclosed space, blowers should be used, an opening should be left to secure sufficient natural ventilation, or an airline respirator should be used.

10.13. Never inhale fumes from welding; they may be harmful.

10.14. Welding should not be completed on the outside of a closed container until an opening of some sort has been made in the container.

10.15. Do not attempt to weld or cut on a tank or enclosure, or smoke, or bring a light into same that has contained gasoline or any other flammable gas or liquid unless the space has been thoroughly purged and has been OK'd by your foreman.

10.16. Always warn a welder to move when it is necessary to pass over his work area with a load. A welder works behind a mask that limits his visibility.

10.17. At power-outlet connections only a reliable man should be permitted to throw switches, change fuses, etc. Fuses must always be installed on the cold side of a switch.

11.   Oxygen and Acetylene—Burning

11.1. Burners must be familiar with the instructions for the safe and proper handling of oxygen and acetylene, or other gases used, and the correct use of the torch before doing any burning. These instructions are issued by the manufacturer of oxygen, acetylene, and other burning gases and by the torch manufacturers.

11.2. Always use the proper goggles when burning or gas welding.

11.3. Your burning or welding gloves must be in good condition to avoid serious burns.

11.4. Do not wear ragged or oil-soaked clothes.

11.5. Never wear low shoes or trousers with cuffs unprotected by overalls when cutting or welding.

11.6. Do not burn or weld materials giving off smoke or fumes without the proper respirator or hood. See your foreman.

11.7. Always refer to oxygen and acetylene or other gases by their full name, not by the words "air" and "gas."

11.8. Always have a fire extinguisher available when burning over or near flammable material.

11.9. Never use a torch as a hammer or as a lever to move work.

11.10. Never swing the torch flame where it can burn another man or yourself, or where it can come in contact with flammable materials.

11.11. Avoid burning over other men or over flammable material until you have placed a good protection under your work.

11.12. When burning material be sure that you are standing clear of any pieces that could fall off during the burning, and warn other men to keep from under such falling pieces.

11.13. When burning any piece make sure you will not fall when the piece is finally cut off.

11.14. Do not burn material lying on a concrete floor. The concrete may explode from the heat and injure you.

11.15. Keep oxygen and acetylene cylinders as far as possible from the point where you are burning. Never burn directly over them.

11.16. Never lay welding or cutting work on the cylinders.

11.17. Never use oxygen, acetylene, or other gas cylinders for any purpose other than for supplying oxygen, acetylene, etc. to a torch or lance.

11.18. Never use a cylinder as a roller.

11.19. Never drop cylinders or allow them to be bumped.

11.20. Never use a sling when moving cylinders with a crane, derrick, or other rig. Use a suitable cradle or box.

11.21. Store oxygen and acetylene cylinders separately—never together. In summer keep cylinders out of the sun as much as possible.

11.22. Keep sparks, flames, and oil away from cylinders at all times. Keep cylinders away from open fires, lights, or excessive heat, and away from electric wires and electrical equipment.

11.23. Always leave the cylinder valve caps screwed in place when the cylinders are not in use.

11.24. Cylinders should be stored and used in an upright position and must be secured to prevent their falling.

11.25. When it is absolutely necessary to lay cylinders down, as on the derrick working floor, they should be blocked to prevent rolling. The valve end of the acetylene cylinder must be at least 6 inches higher than the bottom.

11.26. Use the pressures for oxygen and acetylene and other gases as recommended by the torch manufacturer.

11.27. Never use gases from the cylinders without the use of proper pressure-reducing regulator valves and pressure gauges.

11.28. Before attaching the pressure regulators to cylinders, "crack" the cylinder valve by opening each valve for an instant to blow dirt out of the nozzle. Do not stand in front of valves when opening them.

11.29. Always release regulator-adjusting screws before opening cylinder valves. Open cylinder valves slowly.

11.30. Never remove the bushing from an acetylene cylinder in order to open the valve.

11.31. Always set regulating pressures with the torch needle valves open. Correct adjustments can be made only when the gases are flowing.

11.32. Always close cylinder valves and disconnect pressure regulators from cylinders when making moves with the burning outfit.

11.33. Release all pressures from regulators when closing down for the day.

11.34. Close all cylinder valves and replace all cylinder caps when the cylinders are empty.

11.35. Always turn off cylinder valves when you discontinue burning for any appreciable time.

11.36. Always leave the acetylene key with the acetylene cylinder when the valve is turned on.

11.37. Never tamper with regulating valves.

11.38. Tops of cylinders must be free at all times so that there will be no obstruction to the quick closing of valves.

11.39. Do not use leaky or defective regulators, torches, or hose. The glass of regulators should be protected against breaking.

11.40. If torches, hose, or regulators leak, report same to your foreman immediately. This is especially important if the leak is in the acetylene gauges or in the cylinder.

11.41. Do not try to fix faulty gauges, tips, or torches. Turn them in.

11.42. Never look for a leak with a match or other light. Use soapy water.

11.43. If you discover a leaking cylinder in an enclosed space, take it to the open air at once. Keep it away from all fires and open lights. Notify your foreman at once.

11.44. If a cylinder catches fire and you cannot extinguish it promptly, play a stream of water on the cylinder to keep it cool.

11.45. Never attempt to extinguish fire in an oxygen hose by doubling and pinching the hose. If you do this the fire will follow the hose to that point and can cause a severe burn. Always cut it off at the cylinder.

11.46. Always light the torch with a spark lighter, never with a cigarette or match.

11.47. Never light a cigarette with the torch flame.

11.48. Turn off your torch when not in use.

11.49. Keep the burning hose out of the path of other men. If it is stepped on a flashback may occur in the hose, igniting and burning it.

11.50. Use only brass or bronze splice connections to repair broken hose.

11.51. Do not attempt to weld or cut on a tank or vessel, or smoke, or bring a light into same if it has contained gasoline or other flammable gas or liquid unless the space has been thoroughly purged and has been OK'd by your foreman.

11.52. When burning inside of tanks, barges, or any enclosed space, be very careful to close the acetylene regulator when torch is not in use. If acetylene escapes from your torch, an explosion may result.

11.53. In confined areas be sure you have proper ventilation before using the burning torch.

11.54. Do not leave burning equipment lying around on the ground or steelwork when the job is finished. Coil your hose and store it clear of other workmen or falling objects.

11.55. Do not hold slack in hose between your legs while burning.

11.56. Use no oil! Never permit oil or grease to come in contact with oxygen. Never permit oxygen to contact oily clothes or rags.

11.57. Keep cylinders, valves, regulators, hose, fittings, clothes, and gloves free of oil and grease. Your burning equipment does not require lubrication.

11.58. Never interchange oxygen or acetylene regulators, hose, or other pieces of apparatus, and never use them with any other gas. Oxygen hose and regulator fittings will have right-hand threads and the hose will be either black or green. Acetylene hose and regulator fittings will have left-hand threads and the hose will be red. See that all fittings are tight.

How to Lift Correctly

You may think you are strong but be careful what you lift. Even if your back muscles were made of steel, there would be a limit to the strain they can stand.

Size up the load before you lift it. Get help if necessary.

Use your legs to lift and save your back.

It's not *what* you lift but *how* you lift it!

Keep your shoulders back and spine straight. Take a deep breath as you lift. Divide the load equally between your hands so that the weight is evenly supported by your entire frame.

Bend your knees to put the strain on your leg muscles instead of your abdomen. Lift slowly and safely by straightening your knees and standing erect. The strong leg muscles will then do the work, the leg bones will be the levers.

Crouch with your feet close to the object. Bend ankles and keep your weight on the balls of your feet. Keep your feet rather close together, between 8 and 12 in. apart. Place your hands in a firm, balanced lifting position.

Do not try to lift a greater load than you can carry safely. If an object is too heavy to carry safely alone, get someone to help you or use powered equipment.

TABLE 1. Strength of 3-Strand Manila Rope and Tackle

Diameter of Rope (in.)	Minimum Size of Block (in.)	Maximum Weight per 100 ft (lb)	Safe Working Load: Lead Line Pull (lb)	Safe Working Loads (lb)					
				1-Part Falls – 1 Single Block	2-Part Falls – 2 Single Blocks	3-Part Falls – 1 Single and 1 Double Block	4-Part Falls – 2 Double Blocks	5-Part Falls – 1 Double and 1 Triple Block	6-Part Falls – 2 Triple Blocks
½	4	7.5	660	535	1020	1,450	1,840		
¾	8	16.7	1350	1090	2080	2,960	3,760	4,480	5,120
1	10	27.0	2250	1820	3460	4,940	6,270	7,460	8,540
1¼	12	41.8	3375	2730	5190	7,410	9,400	11,190	12,800
1½	14	60.0	4625	3750	7120	10,150	12,880	15,340	17,550

The above table is for new manila rope.

For rope in fair condition but in no place shaved or chafed to a smaller size reduce the above values 50 per cent.

Tackle values are based on use of one (1) snatch block, snatching lead line to engine or winch spool.

Always use the fewest snatch blocks possible. If more than one snatch block must be used, add one extra part for each additional snatch block used, in addition to the number of parts shown above for the weight to be lifted.

TABLE 2. Strength of Wire Rope, Improved Plow Steel

(6 x 19 type, fiber core)

Diameter of Rope (in.)	Minimum Size of Sheave (in.)	Safe Working Load— Running Rope (lb)	Safe Working Load— Eye and Eye Slings Straight (lb)	Choked (lb)	Approximate Weight per Foot of Rope (lb)
⅜	9	3,050	2,450	1,850	0.23
½	12	5,350	4,300	3,200	0.40
⅝	14	8,350	6,700	5,000	0.63
¾	16	11,900	9,500	7,150	0.90
⅞	18	16,100	12,900	9,650	1.23
1	20	20,900	16,700	12,550	1.60
1⅛	24	26,300	21,000	15,800	2.03
1¼	28	32,300	25,800	19,400	2.50
1⅜	32	38,800	31,100	23,300	3.03
1½	36	46,000	36,800	27,600	3.60
1⅝	40	53,500	42,800	32,100	4.23
1¾	44	62,000	49,600	37,200	4.90
1⅞	48	70,500	56,400	42,300	5.63
2	52	80,000	64,000	48,000	6.40

The above table is for new wire rope or rope as good as new.

For rope in good condition but not new, not excessively worn or flattened, which has few or no broken wires reduce the above values 25 per cent.

Strength of Wire-Rope Connections

Properly applied socket	100% of rope
Eye splice with thimble or spool	85% of rope
Eye splice in sling	80% of rope
Cable clips properly installed	80% of rope

TABLE 2 (continued). Safe Working Loads—Wire Rope Falls

Figures given are based on 10,000 lb lead line pull at hoist.
For other than 10,000 lb adjust values accordingly.

Number of Parts in Falls	Bronze Bushed Sheaves in good condition, Safe Load (lb)	Ball Bearing Sheaves in good condition, Safe Load (lb)
1	8,300	9,120
2	16,100	18,000
3	23,450	26,500
4	30,350	34,900

TABLE 2 (*continued*). Safe Working Loads—Wire Rope Falls

Number of Parts in Falls	Bronze Bushed Sheaves in good condition, Safe Load (lb)	Ball Bearing Sheaves in good condition, Safe Load (lb)
5	36,850	43,000
6	42,950	50,800
7	48,650	58,400
8	54,050	65,800
9	59,100	72,950
10	63,850	79,900
11	68,350	86,600
12	72,550	93,150

The above values are based on two (2) snatches between the hoist and the falls. For each additional snatch block or sheave 6 per cent additional lead-line pull is required for bronze bushed sheaves and 3 per cent additional lead-line pull is required for ball bearing sheaves, or reduce the above values accordingly.

TABLE 3. Wire Rope Clips

How to attach U-bolt type wire rope clips

U-bolts of *all* clips must only be placed on dead end of rope.

Never stagger U-bolt-type clips.

Never put U-bolt of clip on live portion of rope.

Diameter of Rope (in.)	U-Bolt-Type Clips, Number of Clips for Improved Plow Steel Rope	Fist-Grip-Type Clips, Number of Clips for Improved Plow Steel Rope	Minimum Spacing Between Clips (in.)
3/8	3	3	3
1/2	4	3	3 1/2
5/8	4	3	4 1/4
3/4	5	4	5
7/8	5	4	5 3/4
1	6	5	6 1/2
1 1/8	6	5	7 1/4
1 1/4	7	6	8
1 3/8	7	6	8 3/4
1 1/2	8	6	9 1/2

TABLE 4. Additional Stress Due To Bridling

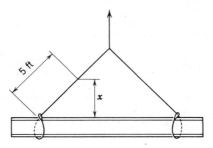

Lay off 5 ft along one leg from point where sling is choked at top of beam. Measure vertical distance X from that point to the top of the beam.

The table below gives the remaining strength in each leg of the spreader.

Vertical Distance X to point along sling 5 ft from top of beam	Reduction of Strength of Wire Rope, etc. due to Bridling (percentage)	Strength of Wire Rope Remaining (percentage)
5 ft. – 0 in.	0	100
4 ft. – 9 in.	5	95
4 ft. – 6 in.	10	90
4 ft. – 3 in.	15	85
4 ft. – 0 in.	20	80
3 ft. – 6 in.	30	70
3 ft. – 0 in.	40	60
2 ft. – 6 in.	50	50

TABLE 5. Strength of Plank, Dressed Plank

(¼ in. under sizes shown)

Size (in.)	Safe Loads (lb) Span (ft)							
	4	6	8	10	12	14	16	18
1 x 6	54	33						
1 x 8	72	43						
1 x 10	90	55						
2 x 8	400	260	190					
2 x 10	500	325	235	180				
2 x 12	600	390	285	220	165			
3 x 10	1245	820	600	465	370	305		
3 x 12	1495	980	720	555	445	370	300	
3 x 14	1740	1145	840	660	520	430	350	290

TABLE 5 (*continued*). Rough Planks

(sizes as shown)

Size (in.)	Safe Loads (lb) Span (ft)							
	4	6	8	10	12	14	16	18
1 x 6	97	61						
1 x 8	129	82						
1 x 10	161	102						
2 x 8	525	340	250					
2 x 10	655	430	310	240				
2 x 12	785	515	370	285	230			
3 x 10	1435	975	715	555	450	365		
3 x 12	1780	1170	860	670	540	440	370	
3 x 14	2080	1360	1000	780	630	510	430	360

The loads given are in pounds concentrated at center of span and are in addition to the weight of the plank itself.

The above loads are for fir or spruce planks in first-class condition. For yellow pine planks in first-class condition add 10 per cent to the above allowable loads.

Safe loads given are based on dressed planks ¼ in. thinner than sizes shown, and for rough planks of the full sizes shown.

Planks should not be used on any spans where no load is given.

TABLE 6. Safe Loads on Dressed Timber Used as Beams

Safe Loads (lb)

Size (in.)	Position: Horizontal Dimension (in.)	Span (ft)									
		4	6	8	10	12	14	16	18	20	24
2 x 4	4	200	130	90							
4 x 4	4	990	650	480	380	310	260	220			
4 x 6	6	1,530	1,010	750	590	480	400	340			
6 x 6	6	3,440	2,290	1,700	1,340	1,110	930	800	690	610	
6 x 8	8	4,700	3,120	2,320	1,830	1,510	1,270	1,090	950	830	
6 x 8	6	4,900	4,260	3,180	2,520	2,080	1,760	1,520	1,330	1,170	940
8 x 8	8	6,690	6,660	4,330	3,440	2,840	2,400	2,070	1,810	1,600	1280
8 x 12	8	10,250	10,200	10,160	8,150	6,740	5,730	4,970	4,370	3,890	3160
8 x 14	8	12,050	12,000	11,940	11,260	9,320	7,940	6,890	6,070	5,410	4410
8 x 16	8	13,820	13,760	13,690	13,630	12,310	10,500	9,130	8,040	7,180	5870
10 x 10	10	10,740	10,690	10,640	7,020	5,800	4,930	4,270	3,740	3,320	2680
10 x 12	10	12,960	12,900	12,840	10,320	8,540	7,260	6,300	5,540	4,930	4000
10 x 14	10	15,220	15,150	15,080	14,280	11,810	10,050	8,730	7,690	6,860	5590
12 x 12	12	15,690	15,620	15,550	12,490	10,340	8,790	7,630	6,710	5,970	4840
12 x 14	12	18,430	18,340	18,260	17,240	14,300	12,170	10,570	9,310	8,300	6760
12 x 16	12	21,170	21,070	20,970	20,880	18,870	16,090	14,000	12,330	11,020	9000
14 x 14	14	21,630	21,520	21,420	21,320	16,780	14,290	12,410	10,930	9,750	7940

The loads given are in pounds concentrated at the center of the span and are in addition to the weight of the timber itself.
The above allowable loads are for fir or spruce timber in first-class condition.
For yellow pine timber in first-class condition add 10 per cent to the above allowable loads.
Inspect timber carefully for cross-grain. If badly cross-grained reduce above safe loads accordingly.
All loads are based on surfaced timber.

TABLE 7. Safe Loads on Timber Columns, Posts, or Braces

Size (in.)	Safe Loads (tons) — Length of Column (ft): 8	10	12	14	16	18	20	25	30	35	40	45	50	60
	B													
4 x 4	4.7	4.0												
	A													
6 x 6	13.4	12.2	11.0	9.8	8.6	7.5	6.3							
8 x 8	27.3	25.6	24.0	22.4	20.8	19.2	17.5	13.5	9.4					
10 x 10	45.9	43.9	41.9	39.8	37.7	35.7	33.6	28.5	23.4	18.2	13.1			
12 x 12	69.4	67.0	64.5	61.9	59.4	57.0	54.5	48.3	42.1	35.9	29.7	23.4	17.2	
14 x 14	97.7	94.8	91.8	89.0	86.0	83.1	80.2	72.9	65.5	58.3	51.1	43.7	36.5	21.9

The above loads are based on yellow pine or fir timber in first-class condition.

Loads below line *A* can only be permitted if the end grain of column rests on a steel beam or slab.

Loads between lines *A* and *B* can only be permitted if the load is transferred in and out of column by bearing on oak or similar pedestal or corbel, to distribute the load over a greater area of yellow pine cap or sill.

Permissible pressure across the grain of yellow pine is 400 lb per square inch and for oak is 750 lb per square inch.

Inspect timber carefully for cross-grain or other defects. If badly cross-grained or defective reduce above safe loads accordingly.

TABLE 8. Strength of Shackles

Diameter of Pin (in.)	Diameter of Shank (in.)	Width between Eyes		Safe Working Load (tons)
		Nominal (in.)	Plus or Minus (in.)	
¾	⅝	1 1/16	1/16	2.2
⅞	¾	1 ¼	1/16	3.2
1	⅞	1 7/16	1/16	4.3
1 ⅛	1	1 11/16	1/16	5.7
1 ¼	1 ⅛	1 ⅞	1/16	6.7
1 ⅜	1 ¼	2	1/16	8.3
1 ½	1 ⅜	2 ¼	⅛	10.0
1 ⅝	1 ½	2 ⅜	⅛	12.0
1 ¾	1 ⅝	2 ½	⅛	14.0
2	1 ¾	2 ¾	⅛	16.0
2 ¼	2	3 ¼	⅛	21.0
2 ½	2 ¼	3 ¾	¼	27.0
2 ¾	2 ½	4 ⅛	¼	34.0
3	2 ¾	4 ½	¼	40.0
3 ¼	3	5	¼	48.0

Size of shackle is to be identified by diameter of pin but diameter of shank of the shackle must be checked.

All shackle pins must be straight and undamaged.

If the width between eyes exceeds the maximum given above, the safe load must be materially reduced.

TABLE 9. Strength of Sling Hooks

Inside Diameter of Eye (in.)	Throat Opening (in.)	Rated Capacity (tons)
¾	1	0.50
⅞	1 1/16	0.60
1	1 ⅛	0.78
1 ⅛	1 ¼	1.5
1 ¼	1 ⅜	1.8
1 ⅜	1 ½	2.2
1 ½	1 11/16	2.6

TABLE 9 (continued). Strength of Sling Hooks

Inside Diameter of Eye (in.)	Throat Opening (in.)	Rated Capacity (tons)
1 ⅝	1 ⅞	3.4
1 ¾	2 ¹⁄₁₆	4.5
2	2 ¼	5.0
2 ⅜	2 ½	6.8
2 ¾	3	8.4
3 ⅛	3 ⅜	9.9
3 ¼	3 ⅝	13.0
3 ½	4	17.0
4	4 ½	22.0
4 ½	5	30.0

The capacity of a hook can be found by measuring the diameter of the hole in the eye of the hook.

If the throat opening of any hook exceeds the dimension given for the corresponding diameter of eye, the hook has probably been overstrained and must not be used until the manufacturer has checked it. (Some manufacturers may deviate from the standard dimensions shown above.)

TABLE 10. Strength of Turnbuckles

Drop-Forged Turnbuckles (Eye and Eye or Eye and Clevis) Diameter of Threaded Portion of Turnbuckle (in.)	Safe Load (tons)	Wire Rope to be Used (6x19 type – I.P.S., fiber core)	
		Maximum Diameter to be Used (in.)	Safe Load – Guys, etc. (tons)
¾	2.2	⅜	2.1
⅞	3.1	½	3.6
1	4.1	⅝	5.5
1 ¼	6.6	¾	7.9
1 ½	9.7	⅞	10.7
1 ¾	13.1	1	14.0
2	17.2	1 ⅛	17.6
2 ¼	22.5	1 ¼	21.6
2 ½	28.0	1 ⅜	26.1
2 ¾	36.7	1 ⅝	36.0
3	41.0	1 ¾	41.3

TABLE 11. Minimum Electrical Clearances

Voltage	Minimum Clearance (ft)
Up to 50 kV	10
50 kV to 200 kV	15
200 kV to 350 kV	20
350 kV to 500 kV	25
500 kV to 650 kV	30
650 kV to 800 kV	35
800 kV to 950 kV	40
950 kV to 1100 kV	45

Equipment must not be operated closer to high voltage lines than the distance listed above *unless* the lines have been grounded and the grounding device can be checked visibly.

Properly insulated 440-V ac feeders to equipment are excluded from the above restrictions.

The clearances given apply in any direction, vertical or horizontal.

TABLE 12. Standard Signals

1. *Hoist.* Forearm vertical, forefinger pointing up. Move hand back and forth repeatedly in a small horizontal circular motion.
2. *Lower.* Arm extended downward, forefinger pointing down. Move hand in small horizontal circle.
3. *Stop.* Arm extended, palm down. Move arm back and forth.
4. *Swing.* Arm extended, palm down, forefinger indicating direction of swing.
5. *Boom up.* Arm extended, fingers closed, thumb pointing up, jerk hand up and down repeatedly a short distance.
6. *Boom down.* Arm extended, fingers closed, thumb pointing down, jerk hand down and up repeatedly a short distance.
7. *Travel.* Arm extended forward, hand open, making pushing motion in direction of travel.
8. *Make movement slowly.* With one hand make signal for desired operation; hold the other hand open close to the hand giving the signal.
9. *Emergency stop.* Both arms extended, hands open, palms down, move hands quickly back and forth repeatedly.
10. *Raise boom and lower load* (Keeping load at same elevation). Give *boom up* signal and repeatedly open and close fingers.

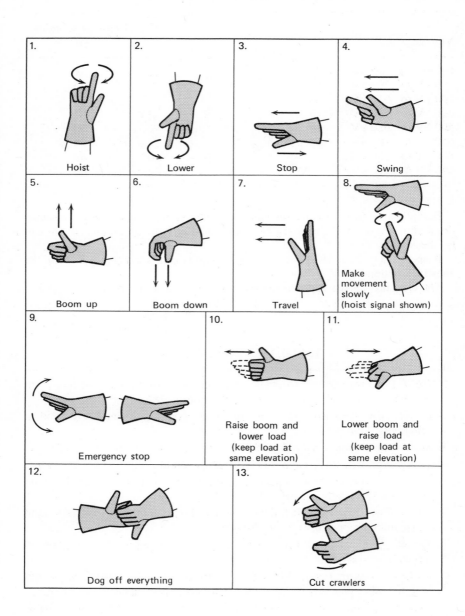

1.	2.	3.	4.
Hoist	Lower	Stop	Swing

5.	6.	7.	8.
Boom up	Boom down	Travel	Make movement slowly (hoist signal shown)

9.	10.	11.
Emergency stop	Raise boom and lower load (keep load at same elevation)	Lower boom and raise load (keep load at same elevation)

12.	13.
Dog off everything	Cut crawlers

330

11. *Lower boom and raise load* (Keeping load at same elevation). Give *boom down* signal and repeatedly open and close fingers.

12. *Dog off everything.* Clasp fingers of one hand with the fingers of the other, palms facing.

13. *Cut crawlers.* Forearms horizontal, fists closed, bring one hand back and the other forward, to indicate direction of rotation desired. Repeat until no longer needed.

In giving hand signals, where both a runner and a main load are being used, before giving the signal to hoist or lower the load:

Tap hand on head if the signal applies to the main load.

Tap elbow with the other hand if the signal applies to the runner (whip).

Bell Signals

When operating: 1 bell means *stop*.

When stopped: 1 bell means *hoist*.

When stopped: 2 bells means *lower*.

When stopping temporarily: 3 or 4 bells alternately on boom and load bells means *dog it off* or *stopping for some time*.

When dogged off as above, before starting: 3 or 4 bells alternately on boom and load bells means *get ready to start work again*.

Appendix C

Typical Forms

1. Erection Cost Report (Tier Building—Welded Construction)
2. Progress and Production Report (Tier Building—Welded Construction)
3. Progress and Production Report (Tier Building—Bolted Construction)
4. Report of Contracts in Course of Erection
5. Report of Distribution of Men
6. Report of Distribution of Men (Alternate)
7. Running Record of Labor Costs and Production
8. Report of Accident Investigation

Contract No. _____

Report No. _____ Purchaser _____ At _____ 19 ____

DAILY PAYROLL

	No.	Hours	Rate	Amount
Foreman				
Timekeepers				
Pushers				
Engineers				
Ironworkers				
Helpers				
Apprentices				
Laborers				
Watchmen				
Total for Day				
Previously Reported				
Total to Date				

DAILY WELD REPORT

Gang	Diam.	Description	Size of Bead	Length Feet	Type of Weld	Labor Cost
1						
2						
3						
4						
5						
6						
7						
8						
9						
10						
11						
12						
13						
Total						

Labor: Weld pusher, compressor engineer, etc.

Average cost per foot to date

Time of Monthly Men			Time of Cranes		
Current	Previous	Total	Current	Previous	Total

TYPICAL ERECTION COST REPORT FORM
(TIER BUILDING – WELDED CONSTRUCTION)
DAILY OR WEEKLY

| Description | DISTRIBUTION OF COST | | | | | | | WORK DONE TODAY |
| | For Day | | Previously Reported | | Total to Date | | | (Give briefly, but clearly, a statement of the work done on this contract.) |
	Quantity	Cost	Quantity	Cost	Quantity	%	Cost	
Grillage or slabs								
Derrick, pole, crane								
Jump or moving derricks								
Steel unloaded (tons)								
Steel distributed (tons)								
Erection—Derrick or crane								
Erection—Hand								
Erection—Misc.								
Relay hoisting								
Plumbing								
Permanent bolting								
Fitting up								
Lintel adjusting								
Plank								
Drill (Dia.)								
Old structure								Weather
Welding								Date to finish
Air plant								Enclosures (list of papers sent in envelope with this report)
Small tools, office, etc.								
Repairs to crane								
Shop errors								
Drawing room errors								
Watching								
Fore'n, timek, & civil eng.								
Lost time								
Total payroll								

Timekeeper _____

Foreman _____

335

PROGRESS AND PRODUCTION REPORT

Purchaser _____

Address _____

Contract No. _____

Date _____

Floor	Tier	Steel Rec'd.	Steel Erected	Derrick Jumped	Welding	Col. Splices	Paint	Forms	Arches

	Today	Previous Total	Total to Date
Number of welder (days)			
Number of feet welded			
Average feet per man—day			

Derrick or Crane	No.	Total		Previous		Total to Date		Tons Erected			Pieces Erected			Average Tons Erected			Average Pcs. Erected			Ave. Pcs.	Per Ton
		Hours Work	Days Work	Hours Work	Days Work	Hours Work	Days Work	Today	Prev.	Total	Today	Prev.	Total	Today	Prev.	Total	Today	Prev.	Total		

336

	Raising Gang	Filling In	Fitting	Bolting	Plumb.	Burning	Tools	Fitting for Weld.	Welding	Plank	Non-prod.	Elec.	Total
Foreman													
Timekeepers													
Pushers													
Engineers													
Ironworkers													
Helpers													
Total													
Ave. gang													

Per man day

Per man day

Per man day

Contract No. _____

Date _____

Purchaser _____

Report No. _____

Job Location _____

TYPICAL
PROGRESS AND PRODUCTION REPORT FORM
(WELDED CONSTRUCTION – TIER BUILDING)

337

PROGRESS AND PRODUCTION REPORT

Installment	Steel Received at Yard	Steel Unloaded at Site	Steel Erected	Bolt	Weld	Foundations	Paint

High Strength Bolts	Today	Previous Total	Total to Date
Number of men (8 hr. days)			
Bolts installed (dia.)			
Avg. bolts per man (8 hr. days)			

Machine Bolts	Today	Previous Total	Total to Date
Number of men (8 hr. days)			
Bolts installed (dia.)			
Avg. bolts per man (8 hr. days)			

Welding	Today	Previous Total	Total to Date
Number of men (8 hr. days)			
Lin. ft. welded (Size—type)			
Avg. ft. per man (8 hr. days)			

	Today	Previous Total	Total to Date
Number of men (8 hr. days)			
Units installed			
Avg. units per man (8 hr. days)			

Erection	Today			Previous		Total to Date		Tons Erected			Pieces Erected			Average Tons Erected			Average Pieces Erected			Average Pieces
Derrick, Crane or Traveler	No.	Hours Work	Days Work	Hours Work	Days Work	Hours Work	Days Work	Today	Prev.	Total	Today	Prev.	Total	Today	Prev.	Total	Today	Prev.	Total	Per Ton

338

Hand erect'n					

DESCRIPTION OF WORK DONE

Weather _____ Temperature _____

Date to complete erection _____ Date to complete contract _____

_____

_____

_____

_____

Contract No. _____ Purchaser _____

Date _____ Job Location _____ Bldg. No. _____

TYPICAL
PROGRESS AND PRODUCTION REPORT FORM
(BOLTED CONSTRUCTION)

339

Show Percentage Until Item Completed
Then Show Date of Completion.
Under "Ship" Also Show Date Shop
Will Start Shipping.
– Indicates None Required
• Indicates Final Weight.
X Working Overtime per Contract Obligations.
XX Working Overtime with Office Approval.

REPORT OF CONTRACTS IN COURSE OF ERECTION

Week Ending

Contract	Purchaser	Description	Tons	Erection Dates					Superintendent	Ship	False Work	Un–load	Erect	Bolt	Misc.	Paint–ing
				Contr. Req.		Act. Sched.		Foun–dation								
				Start	Comp.	Start	Comp.									

TYPICAL FORM FOR
REPORT OF CONTRACTS IN
COURSE OF ERECTION

DAILY DISTRIBUTION

Work Done	Position:	Supt.		Fld. Engr., Tkpr., Etc.		Foremen		Engrs.		Oilers		Iron—workers		Apprentices		Total Men	Total Hours	Gang Hrs.	Derk. or Cr. No.	Derk. or Cr. No.	Derk. or Cr. No.			Daily Costs
	Rate:																		Hours	Hours	Hours			
	Men	Hrs.	Men	Hrs.	Men	Hrs.	Men	Hrs.	Men	Hrs.	Men	Hrs.	Men	Hrs.									

342

Contract:																			
Date:																			
Report No.																			
Total Hours																			
Painters																			
Trucks—Haul (a)																			
Trucks—Transp. (a)																			

(a) Show total number of trucks and hours worked.

TYPICAL FORM FOR
REPORT OF DISTRIBUTION OF MEN

DISTRIBUTION OF MEN

Rate: _____

Supt.	Field Engrs. Timekeeper Clerks	Foremen	Engineers	Ironworkers	Apprentices	Oilers				Total
Totals										

344

DESCRIPTION OF WORK DONE

Weather _____

Temperature _____

Date to Complete Erection _____

Date to Complete Contract _____

	Tons Unloaded
Today	
Previous	
Total to date	

Superintendent _____

Field Supervisor _____

TYPICAL FORM FOR
REPORT OF DISTRIBUTION OF MEN

345

To be completed and sent to the Office by the Foreman or Timekeeper of the injured employee *on the day of the accident*

Date of Report _____

Name of injured man _____

Date and hour of accident _____ _____ A.M. _____ P.M.

Place where accident occurred _____

Name of foreman _____

Man's occupation at time of accident _____

Equipment involved in accident _____

How accident occurred _____

_____

(if insufficient space, continue on other side)

Witnesses _____

Nature of injury and part of body injured _____

Did man return to work after treatment? _____

If not, give probable length of time of disability _____

Name of doctors and/or hospital _____

_____

Unsafe Acts Causing Accident

_____   1. Failure to obey instructions or rules
_____   2. Lack of knowledge or skill
_____   3. Unsafe stepping, tripping, jumping, slipping
_____   4. Hit self on material or by tools
_____   5. Insufficient observance of hazardous conditions
_____   6. Caught by or between material
_____   7. Unsafe movement about hazard: torch, welding, etc. (specify)
_____   8. Unsafe handling of load, material, equipment (specify)
_____   9. Nonuse of proper equipment, tools
_____   10. Using defective tools, equipment
_____   11. Improper use of tools or equipment
_____   12. Unsafe position of man, hands, legs, etc.
_____   13. Improper position for lifting
_____   14. Insufficient help for lifting, handling, etc.
_____   15. Not wearing eye protection
_____   16. Flash from welding arc
_____   17. Insufficient personal protection (gloves, goggles, clothing, etc.)
_____   18. Stepping on nails
_____   19. Unsafe hold on material, equipment, tools
_____   20. Failure to warn others
_____   21. Any other acts not listed above (state)
_____   22. No personal act

Suggested Form for
Report of Accident Investigation

Unsafe Conditions Causing Accident

_____	1.	Defective air hose
_____	2.	″ air tools
_____	3.	″ equipment (specify)
_____	4.	″ jacks
_____	5.	″ tools
_____	6.	″ wire rope
_____	7.	Equipment failure
_____	8.	Timber breaking
_____	9.	Unsafe piling, positioning of material
_____	10.	Poor housekeeping
_____	11.	Slippery material
_____	12.	Short fall of material
_____	13.	Small pieces falling (long fall)
_____	14.	Material swinging
_____	15.	Material rolling
_____	16.	Insufficient protection from welding
_____	17.	″ ″ ″ burning
_____	18.	″ ″ ″ other (specify)
_____	19.	Lack of proper guards (mechanical)
_____	20.	″ ″ ″ ″ (electrical)
_____	21.	″ ″ ″ ″ (other—state)
_____	22.	Any other cause or condition not listed above (specify)
_____	23.	No physical or mechanical cause or condition

Measures to be taken to prevent similar accidents: _____

_____

_____

_____

Are there any tools, equipment, or conditions that could cause the same type of accident? (describe) _____

_____

Additional space for description of accident: _____

_____

_____

Comments: _____

_____

Reviewed by: _____ Prepared by: _____

(signature)

Suggested Form for
Report of Accident Investigation (*continued*)

Appendix *D*

Examples of Completed Reports

The cost and production figures are based on actual projects, but some of the figures, quantities, and locations have been changed as well as the names of persons. The wage rates have been revised to reflect present-day rates, but these will probably be obsolete as soon as this book is published. The dates have been revised and brought up-to-date, but the time intervals of the operations have been maintained.

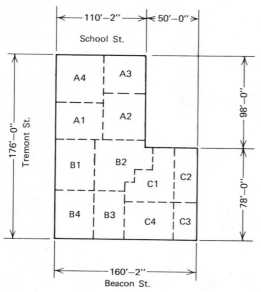

Purchaser: XYZ Corporation
Structure: 20-Story Building
Location: Beacon Street, Boston, Mass.

Contract: B-4244
Prepared by: W.G.R.
Date: 5/21/79

	Estimated	Actual
Tons	3,990	3,928
Bolts	70,000	76,079
Foundations Compl.		2/2/79
Work Started		2/25/79
Start Erecting		3/4/79
Start H.S. Bolting		3/10/79
Complete Erecting		5/19/79
Complete Bolting		5/21/79

Tier	Floor	Ship Orig.	Ship Curr.	Ship Actual	Erection Orig.	Erection Curr.	A	B	C	H.S. Bolts Date Compl.	H.S. Bolts Total & per Tier	Total Tons & Tons per Tier
11	BHR & MR	4/30	4/28	4/21	5/12	5/15	5/19	5/19		5/21	76,079	3,928
											2,887	86
10	R & 18	4/23	4/23	4/18	5/5	5/12	5/12	5/12		5/18	73,192	3,842
											4,591	190
9	17 & 16	4/20	4/20	4/16	4/30	5/8	5/8	5/8		5/13	68,601	3,652
											6,970	292
8	15 & 14	4/13	4/13	4/10	4/24	4/30	4/30	4/30		5/8	61,631	3,360
											7,197	300
7	13 & 12	4/7	4/7	4/3	4/20	4/27	4/27	4/24		5/1	54,434	3,060
											7,238	298
6	11 & 10	4/2	4/2	3/28	4/13	4/22	4/22	4/21	4/21	4/27	47,196	2,762
											7,317	342
5	9 & 8	3/27	3/27	3/25	4/8	4/17	4/17	4/17	4/14	4/23	39,879	2,420
											7,190	389
4	7 & 6	3/24	3/24	3/20	4/3	4/13	4/13	4/6	4/6	4/20	32,687	2,031
											7,502	384
3	5 & 4	3/19	3/19	3/16	3/31	4/3	4/3	4/2	4/1	4/14	25,187	1,647
											7,750	396
2	3 & 2	3/16	3/16	3/11	3/26	3/31	3/31	3/30	3/26	4/2	17,437	1,251
											8,600	505
1	1 & C	3/9	2/26	3/6	3/20	3/26	3/25	3/18	3/18	3/26	8,837	746
												655
Grillage		3/2	2/20	2/19	3/13	3/5	3/5	3/5	3/3			91

Example of Weekly Progress Report
(Derrick Erection)

350

Week Ending	Tons Shipped To Date	Tons Unloaded To Date	Tons Erected		Tons per Gang-Day To Date	Pieces per Gang-Day To Date	High-Strength Bolts Installed		Bolts per Man-Day
			This Week	To Date			This Week	To Date	
3/5	571	308	121	121	34	28			
3/12	1,251	501	221	342	41	61	1,050	1,050	86
3/19	1,752	746	188	530	40	63	3,520	4,570	88
3/26	2,442	1,251	496	1,026	43	71	5,020	9,590	90
4/2	2,911	1,837	580	1,606	44	73	8,920	18,510	91
4/9	3,096	2,259	459	2,065	42	71	7,475	25,985	93
4/16	3,614	2,613	346	2,411	43	73	4,715	30,700	94
4/23	3,928	3,060	504	2,915	44	75	9,475	40,175	93
4/30		3,540	445	3,360	44	78	11,395	51,570	93
5/7		3,652	264	3,624	44	79	6,880	58,450	94
5/14		3,928	304	3,928	42	81	9,851	68,301	95
5/21							7,778	76,079	94

Example of Weekly Progress Report
(Derrick Erection) (*continued*)

Purchaser:	TAW Airlines, Inc.	Contract:	B-4250
Structure:	Small Airport Hangar	Prepared by:	W.G.R.
Location:	Philipsburg, N.J.	Date:	4/25/79

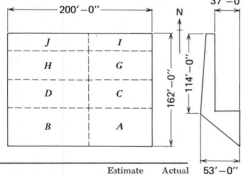

	Estimate	Actual
Tons	890	903
Machine Bolts	4,500	210
High-Strength Bolts	13,100	19,200

Start job	2/13/79
Start erection	2/22/79
Start bolting	2/27/79
Complete erection	3/27/79
Complete bolting	4/10/79
Complete job	4/17/79

Week Ending	Tons Shipped to Date	Tons Unloaded to Date	Tons Erected		Tons per Gang-Day	Pieces per Gang-Day	H.S. Bolts		Bolts per Man-Day
			This Week	To Date			This Week	To Date	
2/25	354	354	75	75	24	4			
3/4	683	683	310	385	34	27	3,730	3,730	127
3/11	867	867	173	558	34	45	2,700	6,430	124
3/18	867	867	85	643	33	37	2,025	8,455	125
3/25	903	903	195	838	33	50	2,655	11,100	124
4/1			65	903	34	50	2,280	13,390	122
4/8							3,195	16,585	120
4/15							2,615	19,200	120

Example of Weekly Progress Report
(Crane Erection)

Purchaser: ABC Co., Inc.
Structure: Mill Building
Location: Maspeth Avenue, Brooklyn, N.Y.

Contract: B-4245
Prepared by: W.G.R.
Date: 5/13/79

Grid diagram (540'-0" × 360'-0"):

A1	B1	C1	D1	G1	H1
A2	B2	C2	D2	G2	H2
A3	B3	C3	D3	G3	H3
A4	B4	C4	D4	G4	H4

(N, Main Street, Maspeth Avenue)

	Estimate	Actual
Tons	6,000	5,977
Bolts	150,000	147,570

	Schedule	Actual
Work Started	1/3	1/3
Start Erection	1/8	1/9
Start H.S. Bolting	1/10	1/12
Complete Erection	3/5	3/10
Complete Bolting	4/1	4/7

Week Ending	Tons Shipped To Date	Tons Unloaded To Date	Tons Erected This Week	Tons Erected To Date	Tons per Gang-Day	Pieces per Gang-Day	High-Strength Bolts This Week	High-Strength Bolts To Date	Per Man-Day
1/14	1,155	770	41	41	20	52	300	300	50
1/21	1,624	1,107	235	276	60	156	1,300	1,600	60
1/28	2,144	1,282	521	797	65	169	2,660	4,260	75
2/4	2,932	1,679	858	1,655	107	278	6,050	10,310	89
2/11	3,627	2,638	1,162	2,817	148	384	16,900	27,210	96
2/18	5,579	3,482	612	3,429	77	200	19,975	47,185	99
2/25	5,977	4,214	1,404	4,833	175	455	15,480	62,665	98
3/4		5,010	1,015	5,848	127	330	20,400	83,065	100
3/11		5,977	129	5,977	65	169	20,200	103,265	101
3/18							22,300	125,565	100
3/25							10,500	136,065	103
4/1							5,500	141,565	104
4/8							6,035	147,570	103

Example of Weekly Progress Report
(Crane Erection)

Install-ment	Ship Orig. Sched.	Ship Revised Sched.	Ship Actual	Erect Orig. Sched.	Erect Revised Sched.	Erect Actual	High-Strength Bolts Orig. Sched.	High-Strength Bolts Actual
A 1-2	1/2	1/2	1/2	1/10	1/10	1/18	1/24	2/4
A 3-4	1/2	1/2	1/3	1/10	1/10	1/23	1/24	2/11
B 1-2	1/3	1/5	1/6	1/12	1/13	1/27	1/27	2/14
B 3-4	1/4	1/5	1/7	1/12	1/13	2/1	1/27	2/17
C 1-2	1/5	1/9	1/10	1/13	1/16	2/7	1/30	2/25
C 3-4	1/6	1/9	1/11	1/13	1/16	2/10	1/30	3/2
D 1-2	1/9	1/11	1/13	1/16	1/18	2/16	2/1	3/7
D 3-4	1/10	1/13	1/20	1/16	1/18	2/24	2/1	3/10
G 1-2	1/15	1/20	1/27	1/22	1/27	2/27	2/8	3/15
G 3-4	1/16	1/26	2/3	1/22	2/2	3/2	2/8	3/17
H 1-2	1/18	2/2	2/10	1/25	2/9	3/8	2/15	3/24
H 3-4	1/20	2/10	2/17	1/25	2/17	3/11	2/15	4/7

Example of Weekly Progress Report
(Crane Erection) (*continued*)

	Tons	Pieces	Cost		Cost @ $1.00 Rate	
			Per Ton	Per Piece	Per Total Tons	Total
Grillage & Slabs		215		2.20	.04	475
Derricks		9		340.00	.23	3,060
Relay Steel	5,856		.55		.24	3,220
Jump Derricks		76		40.00	.23	3,040
Plank					.16	2,190
Derrick Erection	13,300	23,610	1.45	.82	1.45	19,280
Hand Erection					.03	400
Plumbing					.21	2,760
Machine Bolts		72,690		.04	.22	2,920
High-Strength Bolts		242,250		.092	1.68	22,290
Sag & Bracing Rods		160		1.58	.02	250
Align Lintels		478'		.22	.01	110
Miscellaneous Erection					.05	730
Small Tools & Office					.24	3,200
Maintenance Engineer					.06	800
Nonproductive Overtime					.02	300
Lost Time					.37	4,900
Supt., Tkpr., Fld. Engr.					.38	5,120
Totals	13,300	23,610	5.64	3.18	5.64	75,045

(By keeping costs at $1.00 hourly rate, estimates can be prepared for any hourly rate.)

Production

	Tons	Pieces	Gang-Days	Man-Days	Tons per Gang-Day	Pieces per Gang-Day	Pieces per Man-Day
Derrick Erection	13,300	23,610	320		41.5	73.8	
Machine Bolts		72,960		365			200
High-Strength Bolts		242,250		2,842			85

Description

This is a building of 41 stories and roof, with dimensions of 200' x 315' at the base, and a tower 90' x 210' above the 11th floor.

Six derricks were used to the 11th floor and three were then used to erect the tower.

Two of the six derricks used to erect the lower portion of the building were left in place at the 11th floor for use as relay derricks, only one being used at a time for relaying. One relay gang was able to supply three erecting derrick gangs with steel and also help carry planks.

This building is of generally light framing except for 35 girders at the 2nd floor used to carry the building over the lobby. The heaviest girder weighed 25 tons. The heaviest column pick was 12 tons.

The manpower obtained for this job was better than average.

Example of Labor Costs and Production
(Derrick Erection)

Contract B-4244
Structure: 20-Story Office Building
Estimated Tons: 3990
Actual Tons: 3928
Superintendent: A. Cameron
Field Engineer: B. Capel

Foremen: Derrick Erection: C. Cecil
D. Coker
E. Harper
Hand Erection: F. Harris
Miscellaneous: G. Hudson

Purchaser: XYZ Corporation
Location: Beacon Street, Boston, Mass.
Estimated Bolts: 70,000
Actual Bolts: 76,069
Equipment: 3 20-ton Derricks
1 Truck Crane
1 300 cu. ft. Compressor
Labor Rate: Estimated: $11.50
Actual: $12.00
Operating Engineers: H. Lilly
J. Smith
M. Monk

Description

This contract was for a 20-story office building with tower for machinery and water tank making a total of 22 floors.

Labor was excellent. Little other work was going on when this contract was started, so there was a choice of the better men available.

Our hauling contractor did an excellent job of hauling steel to our derricks from the railroad yard.

We took steel from both Beacon Street and Tremont Street. Both streets were only 25 feet wide but, fortunately, were routed for one-way traffic. Both streets were heavily traveled throughout the day. The police cooperated to help move the continuous lines of traffic from two lanes to one in order to bypass our equipment or trucks hauling steel.

The weather for this time of year was average. A total of 8.5 lost time days was less than anticipated.

There were no serious accidents.

Job Started: 2/25/79
Completed: 5/28/79

Labor Costs

					Labor						
			@ $12.00 Rate				@ $1.00 Rate				
	Quantity		Unit Costs		Per Total Tons	Total Costs	Unit Costs		Per Total Tons	Total Costs	
	Tons	Pieces	Per Ton	Per Piece			Per Ton	Per Piece			
Grillage	90	80	30.78	34.63	.71	2,770	2.55	2.88	.06	230	
Set up & Dismantle Dks.		3		3,440.00	2.63	10,330		287.00	.22	860	
Jump Derricks		25		356.00	2.27	8,910		29.70	.19	743	
Unload Steel	3,838	7,620	3.64	1.83	3.56	13,980	.30	.15	.30	1,165	
Distribute & Erect	3,838	7,620	14.20	7.15	13.88	54,504	1.18	.60	1.16	4,542	
Plank					2.52	9,900			.21	825	
Adjust Lintels		4,235'		1.54	1.66	6,540		.13	.14	545	
Plumbing					4.12	16,200			.34	1,350	
Fitting-Up					2.59	10,164			.21	847	
Machine Bolts		25,160		.54	3.48	13,680		.05	.29	1,140	
High-Strength Bolts		76,079		1.08	20.92	82,200		.09	1.74	6,850	
Small Tools & Office					2.40	9,420			.20	785	
Paid Holidays					1.02	4,008			.09	334	
Lost Time					2.34	9,180			.19	765	
Supt., Tkpr., Fld. Engr.					3.93	15,420			.33	1,285	
Totals	3,928	7,700	68.02	34.70	68.02	267,206	5.67	2.89	5.67	22,266	

Example of Final Costs and Production Report
(Derrick Erection)

354

Total Costs

	Original Estimate		Actual Costs		Summary	
	Unit	Total	Unit	Total	Profit	Loss
Total Labor	88.00	350,000	68.03	267,206	82,794	
Field Expense	8.80	35,000	4.97	19,525	15,475	
Transport Equipment	3.00	11,970	2.74	10,750	1,220	
Taxes	5.35	21,350	4.15	16,300	5,050	
General Overhead	10.50	41,900	8.16	32,060	9,840	
Comp. Insurance	17.74	70,770	13.75	84,030	16,740	
Public Liab. Insurance	8.87	35,380	6.88	27,015	8,365	
Tools Charge	13.15	52,500	10.20	40,080	12,420	
Hauling Steel	9.70	38,700	9.54	37,480	1,220	
Demurrage			.43	1,675		1,675
Surveying	3.25	12,960	3.52	13,844		884
Engineering	.50	2,000	.48	1,900	100	
Welfare, etc.	44.00	175,560	34.01	132,600	42,960	
Safety Incentive			2.11	8,280		8,280
Totals	212.50	848,090	168.97	662,745	196,184	10,839
Profit					185,345	

Production

	Tons	Pieces	Gang-Days	Man-Days	Tons per Gang-Day	Pieces per Gang-Day	Tons per Man-Day	Pieces per Man-Day
Erection	3,928	7,700	98		40	79		
High-Strength Bolts		76,079		809				94
Machine Bolts		25,160		148				170
Adjust Lintels		4,235'		76				56'

Example of Final Costs and Production Report
(Derrick Erection) (*continued*)

Contract B-4245
Structure: Mill Building
Purchaser: ABC Co., Inc.
Location: Brooklyn, N.Y.
Estimated Tons: 6000
Actual Tons: 5977
Superintendent: L. Gibbs
Field Engineers: W. Brown
                   J. Wallace
General Foreman: M. Jones
Foremen: Crane Erection: J. Scott
                   R. Able
                   C. Brown
                   E. Cross
                   G. George
                   H. James
         Hand Erection: L. Frank
         Bolting: M. Bliss
                   P. Smith
         Miscellaneous: R. Watts
Operating Engineers: J. Cusack
                   R. Gillespie
                   W. Novick
                   F. Connors
                   D. Austin
                   W. Moore
                   L. Beedle
                   E. Masters

Estimated Bolts: 150,000
Actual Bolts: 147,570
Equipment: 2 30-ton Truck Cranes
             4 50-ton Crawler Cranes
             4 300 cu. ft. Compressors
             2 Air Hoists
             3 Trucks
Labor Rate: Estimated: $11.50
             Actual: $12.00
Job Started: 1/3/79
Completed: 4/7/79

Description

This contract consisted of truss-type construction with an intermediate floor at about 25 feet above the ground elevation. All the field connections were bolted, using high-strength bolts in most of the connections.

The inspection on this project was performed by a local inspection company, rather than by the customer. Their inspection force was very unreasonable and had no concern for the schedule which was important to the customer. The two most unreasonable demands of their inspection was the plumbing of the steel and their adherence to the specified torque on the high-strength bolts rather than to an established torque tension ratio arrived at by using our calibrated wrench.

The foundation contractor was slow in finishing and releasing areas for our use, and we were tight up against him for better than half of the contract, which made it necessary to use a yard for nearly all of the steel.

The supply of men was plentiful, but the quality was poor because of the availability of lighter, easier work for the cost plus fixed fee contractors in the area.

There was only one serious accident and no fatals on this contract.

Labor Costs

			Labor							
			@ $12.00 Rate				@ $1.00 Rate			
	Quantity		Unit Costs		Per Total Tons	Total Costs	Unit Costs		Per Total Tons	Total Costs
	Tons	Pieces	Per Ton	Per Piece			Per Ton	Per Piece		
Unload at Yard	5,977	15,370	7.82	3.04	7.82	46,740	.65	.25	.65	3,895
Relay from Yard	335	510	5.73	3.76	.32	1,920	.48	.31	.03	160
Base Plates	102	525	228.00	44.34	3.89	23,280	19.02	3.70	.32	1,940
Cranes		6		1,320.00	1.33	7,920		110.00	.11	660
Unload at Site	5,977	15,370	5.65	2.20	5.65	33,780	.47	.18	.47	2,815

Example of Final Costs and Production Report
(Crane Erection)

356

Distribute	5,977	15,370	5.33	2.07	5.33	31,860	.44	.17	.44	2,655
Erect	5,810	12,238	17.74	8.42	17.25	103,080	1.48	.70	1.44	8,590
Hand Erection	66	2,610	302.00	7.63	3.33	19,920	25.15	.64	.28	1,660
Fit-Up		147,600		.14	3.55	21,240		.01	.30	1,770
Machine Bolts		20,800		.67	2.32	13,860		.06	.19	1,155
High-Strength Bolts		147,600		.87	21.56	128,880		.07	1.80	10,740
Inspect H.S. Bolts					1.39	8,280			.12	690
Plumbing					4.10	24,480			.34	2,040
Sag Rods		2,320		6.62	2.57	15,360		.55	.21	1,280
Crane Rails		10,190'		2.33	3.97	23,700		.19	.33	1,975
Tools & Office					2.44	14,580			.20	1,215
Nonprod. Overtime					17.67	105,600			1.47	8,800
Reporting Time					1.76	10,500			.15	875
Lost Time					6.04	36,120			.50	3,010
Salaries					5.68	33,960			.47	2,830
Totals	5,977	15,370	117.96	45.87	117.96	705,060	9.83	3.82	9.83	58,755

Total Costs

	Original Estimate		Actual Costs		Summary	
	Unit	Total	Unit	Total	Profit	Loss
Total Labor	113.00	678,000	117.96	705,060		27,060
Field Expense	11.30	67,800	9.95	59,490	8,310	
Transport Men	1.50	9,000	.34	2,000	7,000	
Transport Equipment	9.90	59,400	3.38	20,200	39,200	
Taxes	7.35	44,100	7.67	45,830		1,730
General Overhead	13.56	81,360	14.15	84,600		3,240
Comp. Insurance	17.57	105,420	18.34	109,640		4,220
Public Liab. Ins.	8.79	52,740	9.17	54,820		2,080
Tools Charge	16.95	101,700	17.69	105,760		4,060
Hauling Steel	6.00	36,000	4.79	28,600	7,400	
Demurrage			1.00	6,000		6,000
Surveying	5.00	30,000	3.35	20,000	10,000	
Engineering	.40	2,400	.34	2,000	400	
Welfare, etc.	54.80	328,800	57.21	341,950		13,150
Safety Incentive			.84	5,000		5,000
Totals	266.12	1,596,720	266.18	1,590,950	72,310	66,540
Profit					5,770	

Production

	Tons	Pieces	Gang-Days	Man-Days	Tons per Gang-Day	Pieces per Gang-Day	Tons per Man-Day	Pieces per Man-Day
Crane Erection	5,810	12,238	179	1,073	32.4	68.3	3.3	7.0
Hand Erection	66	2,610		208			0.3	12.5
Passing	332	510	3	20	66	102	16.6	25.5
Rods		2,320		160				14.5
Machine Bolts		20,800		144				144
High-Strength Bolts		147,600		1,342				109

Example of Final Costs and Production Report
(Crane Erection) (*continued*)

Contract B-4250
Structure: Small Airport Hangar
Purchaser: TAW Airlines, Inc.
Location: Philipsburg, N.J.
Estimated Tons: 890
Actual Tons: 903
Superintendent: L. Gibbs
Field Engineer: W. Brown
Timekeeper: A. Smith
Foremen: Crane Erection: J. Scott
         R. Able
   Hand Erection: L. Frank
   Bolting: M. Bliss
   Miscellaneous: R. Watts
Operating Engineers: J. Cusack
         R. Gillespie
         W. Novick
         E. Masters

Estimated Machine Bolts: 4,500
Actual Machine Bolts: 210
Estimated High-Strength Bolts: 13,100
Actual High-Strength Bolts: 19,200
Equipment: 2 50-ton Crawler Cranes
     1 20-ton Truck Crane
     1 350 cu. ft. Compressor
Labor Rate: Estimated: $11.50
      Actual: $12.00
Job Started: 2/13/79
Job Completed: 4/17/79

Description

This project was started in February. High winds, cold weather, and snow seriously hampered the start of operations. All of the material was trucked directly from the fabricating shop to the job site. The General Contractor graded the apron area in front of the hangar and covered this with stone subgrade, which made it an excellent place to unload and distribute our steel-work. The hangar area was extremely muddy and great difficulty was encountered in maneuvering the crane in the work area.

Considerable care was given to the erection and adjustment of the four trusses so that the ends of these cantilevered trusses would be at their proper elevation. Final results indicated an excellent job was done in the field.

Good men were available and the General Contractor was cooperative. There were no serious accidents.

Labor Costs

| | Quantity | | @ $12.00 Rate | | | | @ $1.00 Rate | | | |
| | | | Unit Costs | | Per Total Tons | Total Costs | Unit Costs | | Per Total Tons | Total Costs |
	Tons	Pieces	Per Ton	Per Piece			Per Ton	Per Piece		
Slabs		50		31.20	1.73	1,560		2.60	.14	130
Cranes		2		660.00	1.46	1,320		55.00	.12	110
Passing Steel	250	370	4.80	3.24	1.33	1,200	.40	.27	.11	100
Assembling Trusses		8		195.00	1.73	1,560		16.25	.14	130
Unld, distr, erect	840	1,730	25.43	12.35	23.65	21,360	2.12	1.03	1.97	1,780
Joists	60		100.00		6.64	6,000	8.33		.55	500
Hand Erection	3	50	160.00	9.60	.53	480	13.33	.80	.05	40
Miscel. Erection					3.72	3,360			.31	280
Plumbing					.27	240			.02	20
Machine Bolts		210		.57	.13	120		.05	.01	10
High-Strength Bolts		19,200		.81	17.14	15,480		.07	1.43	1,290
Sag Rods		170		2.82	.53	480		.24	.05	40
Adjust—Shim Trusses		4		360.00	1.59	1,440		30.00	.13	120
Adjust Door Guides					.80	720			.07	60
Small Tools & Office					2.39	2,160			.20	180
Lost Time					7.97	7,200			.66	600
Salaries					12.62	11,400			1.05	950
Totals	903	1,830	84.25	41.57	84.25	76,080	7.02	3.46	7.02	6,340

Example of Final Costs and Production Report
(Crane Erection)

Total Costs

	Original Estimate		Actual Costs		Summary	
	Unit	Total	Unit	Total	Profit	Loss
Total Labor	105.00	93,450	84.25	76,080	17,370	
Field Expense	10.50	9,350	15.39	13,900		4,550
Transport Men	1.00	890	.66	600	290	
Transport Equipment	6.80	6,050	2.49	2,250	3,800	
Taxes	6.41	5,700	5.14	4,640	1,060	
General Overhead	15.75	14,020	12.64	11,410	2,610	
Comp. Insurance	16.28	14,480	13.06	11,790	2,700	
Public Liab. Insur.	8.14	7,245	6.52	5,890	1,355	
Tools Charge	8.39	7,470	6.73	6,080	1,390	
Crane Rental	8.00	7,120	5.73	5,700	1,950	
Falsework			3.67	3,310		3,310
Engineering	5.50	4,900	3.48	3,140	1,760	
Welfare, etc.	26.25	23,360	21.06	19,020	4,340	
Miscellaneous			1.71	1,550		1,550
Safety Incentive			1.33	1,200		1,200
Totals	218.02	194,045	183.86	166,030	38,625	10,610
Profit					28,015	

Production

	Tons	Pieces	Gang-Days	Man-Days	Tons per Gang-Day	Pieces per Gang-Day	Tons per Man-Day	Pieces per Man-Day
Erection	840	1,730	28	222.5	30	62		
High-Strength Bolts		19,200		161				119
Machine Bolts		210		1.25				168
Sag Rods		170		5.0				34

Example of Final Costs and Production Report
(Crane Erection) (*continued*)

Contract: B-4247; 34-Story Office Building; Four Derricks to Eighteenth Floor, One Above

	Tons	Pieces	Cost Per Ton	Cost Per Piece	Cost @ $1.00 Rate Per Total Tons	Cost @ $1.00 Rate Total
Grillage & Slabs	166	95		3.89	.06	370
Derricks		4		375.00	.23	1,500
Relay Steel	1,570	3,800	.57	.24	.14	900
Jump Derricks		34		56.00	.29	1,900
" " with outriggers		7		110.00	.12	770
Derrick Erection	6,390	12,200	1.87	.98	1.82	11,950
Move Derrick					.02	160
Plumbing					.20	1,350
Machine Bolts		41,700		.047	.30	1,960
High-Strength Bolts		96,780		.087	1.29	8,470
Adjust Lintels		6,390'		.15	.15	960
Small Tools & Office					.16	1,020
Maintenance Engineer					.05	300
Nonproductive Overtime					.01	70
Lost Time					.18	1,200
Supt., Tkpr., Fld. Engr.					.45	2,960
Totals	6,390	12,200	5.47	2.94	5.47	35,380

(By keeping costs at $1.00 hourly rate, estimates can be prepared for any hourly rate.)

Production

	Tons	Pieces	Gang-Days	Man-Days	Tons per Gang-Day	Pieces per Gang-Day	Pieces per Man-Day
Derrick Erection	6,390	12,200	173		37	71	
Machine Bolts		41,700		245			170
High-Strength Bolts		96,780		1,059			91.4

Description

This is a building of 34 stories, roof, and two machine levels, with dimensions of 150' x 220' at the base, and a tower 80' x 95' above the 18th floor.

Four derricks were used to erect to the 18th floor and one was used to erect the tower.

One of the four derricks used to erect the lower portion of the building was left in place at the 18th floor for use as a relay derrick, only one being needed to erect the tower.

This building is of generally normal beam and column construction. The manpower for this contract was average since there was no scarcity of work at the time the building was being erected.

Example of Labor Costs and Production
(Derrick Erection)

Purchaser: XYZ Corporation
Structure: 20-Story Office Building
Location: Beacon Street, Boston, Mass.

Contract: B-4244
Prepared by: W.G.R.
Date: 6/5/79

Installment	Pieces per Install-ment	Tons per Install-ment	No. of Dks., Cranes or Travelers	Pcs. per Dk., Cr. or Trav.	Tons per Dk., Cr. or Trav.	Est. Days	Estimated Date To Ship	Arrive Via	At Site	Erect
Grillage	1380	90	1 Cr.	80	90	3	3/2	Truck	3/10	3/13
Tier Floor										
1 Lobby						4	3/9			
& 1	1300	650	3 Dks.	430	220	4	3/9	Rail	3/16	3/20
2 2 & 3	1000	500	″ ″	330	170	3	3/16	″	3/23	3/26
3 4 & 5	800	400	″ ″	260	135	3	3/19	″	3/26	3/31
4 6 & 7	800	400	″ ″	270	130	3	3/24	″	3/31	4/3
5 8 & 9	700	350	″ ″	240	120	3	3/27	″	4/6	4/8
6 10 & 11	800	400	″ ″	260	140	3	4/2	″	4/8	4/13
7 12 & 13	600	300	2 Dks.	300	150	4	4/7	″	4/15	4/20
8 14 & 15	600	300	″ ″	300	150	4	4/13	″	4/20	4/24
9 16 & 17	600	300	″ ″	300	150	4	4/20	″	4/29	4/30
10 18 & R	300	150	″ ″	150	75	3	4/23	″	4/30	5/5
11 M.R. & B.H.R.	200	100	″ ″	100	50	4	4/30	″	5/6	5/12

(This schedule is sent to the Fabricating Shop to determine if shipments can be made on the dates requested. If not, a revised schedule must then be made.)

Example of Preliminary Schedule

A. Preparation.
  1. Study: plans, specifications, contract requirements, estimate.
  2. Visit site.
    a. Locate curbs, streets, poles, wires, obstructions.
    b. Note adjacent buildings.
    c. Check traffic requirements for steel delivery.
  3. Note heavy pieces and their locations.
B. Erection Scheme.
  1. Set up a general method of erection.
    a. Decide on the number, type, and approximate location of derricks.
    b. Establish shipping installments and tier divisions and send them to the Drafting Department.
      (1) Balance work to be done by the various derricks.
      (2) If possible, have each installment large enough to make a minimum 40,000-lb carload shipment.
      (3) In general, use four divisions per guy derrick for the usual 100' × 100' area covered by a derrick.
      (4) Letter shipping installments using letters that will not be misread under difficult conditions. Avoid using both E and F.
  2. Work out a Preliminary Schedule, sending shipping dates to the fabricator with request to make sure those dates can be met.
  3. Send the drafting room a letter together with "1-b," noting any changes from the Erection Department Standards (drafting) Requirements, and calling attention to any special details or connections required. Advise that the derrick reactions will be furnished later.
  4. Locate derricks.
    a. Check the reaches and drift for setting critical pieces and for unloading from the street.
    b. Check the capacity for unloading and setting heavy pieces.
    c. Check the guy distances.
    d. Consider the beam layout in derrick panel.
      (1) Keep the derrick out of elevator shafts and stairways.
      (2) Keep the leads 1 foot or more from all beams.
      (3) Allow room for jumping derricks. Jumping beams on upper floor when pulled over to column must clear one-half of the footblock when the mast is raised.
      (4) The derrick should be located far enough from upper header beam to be able to reach over it and erect filling-in beams on the far side. If not possible, rope off this panel (for safety) until after the derrick is jumped, in preference to overhauling beams for erection after jumping.
    e. Compute stresses.
      (1) Use uniform floor load of 1½ tier of steel plus plank.
      (2) Use the concentrated load of maximum derrick reaction including lead line pulls.

362

(3) Using (1) and (2) above, compute stresses in the beams support-ing the derrick, reducing allowable stresses for unsupported length of any beam. Reduce allowable stresses drastically for un-supported beams perpendicular to the direction of the boom when unloading steel. For this condition, the foot guys are not very effective since any rolling of the beam will loosen them.

(4) Decide whether the beams in the structure will support the derrick.
   (a) If overstressed, would shore to floor below be satisfactory?
   (b) Or would a second tier of jumping beams reduce the stresses to allowable limits?
   (c) Or should the second tier of beams be connected directly to the columns?

(5) Give drafting room the derrick reactions for which they should provide connections.

f. Layout guys.
   (1) Try to have three back guys when derrick is unloading steel.
   (2) Balance the guy layout since extremely short guys on one side with long guys opposite make it hard to turn derricks.
   (3) Check on upper floors and design outriggers if required because of setbacks.
   (4) Make certain guys as laid out yield required capacity for unload-ing and setting heavy pieces.
   (5) Check that the floor system of the building will transmit the hori-zontal guy reaction back to the footblock.
   (6) Check that the foundations are heavy enough to resist the guy pull.

g. Make a drawing incorporating the above and showing:
   (1) Unloading areas required.
   (2) Guy anchor locations.
   (3) Hoist locations.
   (4) Derrick locations.
   (5) Maximum picks.

h. Send a copy of "g" to customer for his information and for his use in setting guy anchors. If rock anchors are required, insist that the erector's men drive them into holes drilled by the customer.

C. Operation.
1. Give a letter of instruction to the superintendent, calling attention to any unusual conditions.
2. Watch that the derrick floor is not overloaded and that the connections are properly fitted up prior to unloading steel or operating the derrick.
3. Check that the superintendent is following the scheme as laid out.
4. Schedule the deliveries of steel and equipment.

Checklist for Engineer on Tier Building
(Derrick Erection)

Building: Veterans Hospital, Bronx, N.Y.

To Superintendent J. Jones:

Our estimate on the above work is based on the following production:

Unloading steel at yard and storing	67 tons per yard gang-day
Unloading, distributing, and setting slabs to line and grade	15 tons, 47 pieces per gang-day
Crane in and out	2 yard gang-days per crane
Guy derricks up and down (includes unloading and loading out)	5 derrick gang-days per derrick
Unload, distribute, and erect steel on site, excluding slabs but including jumping derrick and planking derrick floors	25 tons, 59 pieces per gang-day
Store for erection by others	67 tons per derrick gang-day
Handling plank in and out, and planking extra floor for bolters	46 derrick gang-days
Fitting-up for erection	530 holes per man-day
Machine bolts	160 per man-day
High-strength bolts	100 per man-day
Plumbing	197 man-days
Tools and office	156 man-days

A yard gang is to consist of a foreman, operating engineer, oiler, and three ironworkers.

A derrick gang is to consist of a foreman (pusher), operating engineer, six ironworkers, and a half-day for an apprentice.

High-strength bolting production is based on using one foreman (pusher), one compressor operator, 30 ironworkers, and two apprentices.

In order to meet our estimate on this contract, it will be necessary for you to equal or better the above production.

> A. B. Frank
> Manager, Erection Dept.

Example of Letter for Expected Production
(To Superintendent of Job)

Appendix *E*

Rivets and Riveting

Contents

Introduction

Although today welds and high-strength bolts are being used extensively instead of rivets as fasteners for structural steel members, rivets are sometimes required for special situations.

Accordingly, this Appendix has been prepared for those erectors with little or no experience with riveting, which has become a vanishing art. It should be used in conjunction with the other sections of the book.

1. Estimating

1.1 Preparation

Records should be kept of production and the cost of riveting from actual field-work. When feasible, the production in rivets per gang-day for ordinary carbon rivets should be kept separately from production in high-strength alloy rivets; they should be grouped by different diameters. Long, straight or tapered rivets should be grouped separately from short lengths. Otherwise an average diameter for all the rivets on the job, or all in any particular type, can be calculated and will then be usable in estimating a new job with approximately the same average diameters.

Additional information should be developed in the field for use in the office by the estimator, covering operations other than riveting but involving rivets, such as removing rivets in old structures being altered by cutting out or burning.

1.2 Estimate

In preparing the actual estimate, it is advisable to try to determine the number and type of rivets to be driven as shown on the design drawings. The estimate must then be based on the probable number of rivets that will be driven per gang per day, and the number of gangs needed to complete riveting within the contract time. This will determine whether a separate foreman will be needed just to handle the riveting gangs or whether one foreman will be able to cover both the riveting and the bolting gangs.

A riveting gang normally consists of four ironworkers with the time of a foreman or pusher, a compressor operator, and an apprentice divided among several gangs.

1.3 Forms

In preparing a standard form for estimating, if riveting will be involved in work the estimator will be bidding, spaces should be included for rivets: ordinary carbon, high-strength alloy; by diameters; and for normal lengths and excessive lengths.

There should also be a space for fitting-up for riveting as a reminder to include the cost. Costs should be included for setting up and dismantling the air plant.

When estimating the time for completing a contract being bid, care must be taken to allow for overlapping some of the erection time with the riveting time, since some riveting will be going on while erection continues, and some time will be needed to complete the riveting after the actual setting of the steel has been completed.

1.4 Labor

Estimates of production and costs will depend on the number of rivets driven per gang-day; on the types of men available; on the diameter, length, and type of rivets; on the type and locations of connections; the need for scaffolds or floats; and the need for extra gangs to swing scaffolds in advance. Time depends on the number of men available, whether in organized gangs, and how experienced. (Because rivets are used less and less today in favor of high-strength bolts and welds, riveters are becoming scarce.) These are all costs in addition to those involved in the balance of the work other than riveting.

2. Office Procedure

2.1 Checklist

Checklists for making sure all possible operations involved in erecting the structure are taken care of should include the preparation of rivet lists indicating which points are to be riveted rather than bolted, and quantities broken down into categories by tiers or areas and totals.

2.2 Reports and Records

When a job is completed, the field labor costs by type of rivets, etc., should be secured and recorded in such a way that they can be used for subsequent estimates as well as for comparison with production on other jobs.

The cost and production report form should have spaces for rivets by totals and by rivets per gang-day, breaking down the production into types where radically different production may be expected. Costs should be kept of fit-up for riveting, air plant, special riveting scaffolds, cutting out of improperly driven rivets, and spotting of driven rivets if painting is included in the erector's contract.

2.3 Instructions to the Field

A copy of the shipping schedule must be given to the field, and it must take into consideration when rivets for the various tiers or areas will be needed, because they are often shipped separately from the main structural steel and must be scheduled to arrive in ample time but not so far ahead that there will not be enough space to store them until needed.

The rivet lists for the use of the field should be included with the erection drawings sent to the job so that the superintendent has enough information as to what will be needed for each tier or area of the structure.

2.4 Field Supervisory Personnel

Once riveting is started, the superintendent, the resident engineer or field engineer, and the office representative should all watch the riveting progress to see that it conforms to the schedule because falling behind may impede the progress of the contractors who follow the steel erector with their operations, as well as increase the cost over that which had been estimated.

3. Riveting Tools and Equipment

3.1 Toolhouse Personnel

In the toolhouse, facilities should be provided for servicing riveting tools and equipment. The toolhouse superintendent should have men available who can repair and service the rivet hammers, rivet dollies, etc., as well as the compressors.

The superintendent should have had experience in the field as a superintendent or foreman on riveted work and preferably have actually performed the various phases of riveting so that he will know how the tools and equipment will be used and how to service them.

3.2 Tool List

A standard job tool list should include all the various tools, etc., that might be needed, so that the list will remind the person preparing the list of tools to be shipped and he will not forget any items that might otherwise be involved. This can cause delay on the job since such items will then be sent after they are needed, aside from the extra shipping costs.

A handy list can be set up for the needs of one riveting gang, and one for several gangs. One gang may require a certain size manifold and valves at the compressor, while more than one gang might need not two manifolds, but simply one with more outlets.

3.3 Tools Particularly Applicable to Riveting

Backing-out punch. Hand (Fig. E3.3.1). After rivets that must be removed have one head cut off by a buster, cutter, or burning torch, the backing-out punch is used to drive the beheaded rivet out of its hole.

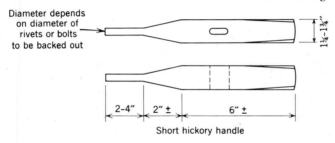

Diameter depends
on diameter of
rivets or bolts
to be backed out

2-4"   2" ±   6" ±

Short hickory handle

FIG. E3.3.1. Hand backing-out punch (known as "B. and O.").

Buster. Hand (Fig. E3.3.2). Useful for cutting off the heads of rivets to be removed.

Dolly, riveting. Jam (Fig. E3.3.3); striking. These are usually pneumatically operated. The jam dolly has a compressed-air-actuated plunger acting against a rivet set at one end; and that is placed against the manufactured head of a hot rivet placed in its hole for driving. At the other end of the dolly a screw-end spud or a provision for a long pipe with a spud at its other end permits jamming the dolly against adjacent steel. With the compressed air applied, the rivet set is held firmly against the rivet being driven. The striking dolly is similar except that, instead of a constant pressure against the rivet set, the plunger is actuated to strike repeatedly against the end of the rivet set in similar fashion to the rivet

1¼-1¾"

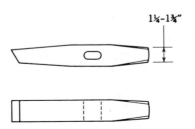

FIG. E3.3.2. Hand buster.

Air inlet connection

Throttle

Feed screw

Buck-up snap

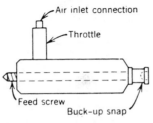

FIG. E3.3.3. Jam dolly (holder-on).

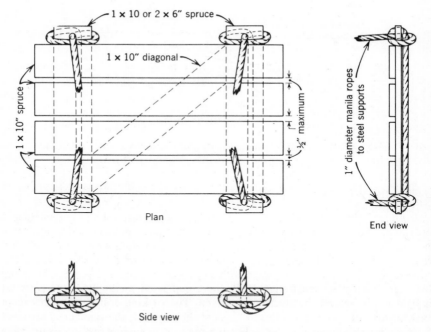

FIG. E3.3.4. Plank float; 1 × 10s preferably bolted to supports with carriage or elevator bolts, or nailed with nails cinched underneath.

hammer being used to form the new head. It is most useful for large-diameter and long-shank rivets. The jam dolly is also known as a "holder-on."

Floats. Plank (Fig. E3.3.4); plywood (Fig. E3.3.5). Some erectors prefer plywood floats, while others like plank floats with spaces between the boards. Both types are used for hanging from the structure to permit riveters to work at points not readily accessible from the steel itself.

Hammer, powered. Rivet: regular, close-quarter, electric, pneumatic (Fig. E3.3.6); spare plungers. The close-quarter rivet hammer usually has an inverted handle to reduce its overall length.

Rivet buster (see also *Buster, hand*). Pneumatic; backing-out punch; chisel; chisel retainer; lock spring; rubber bumper; sleeve: upper, lower; bits; diamond points, etc.; "hell-dog." The ordinary pneumatic rivet buster is small enough for one man to handle in knocking off the heads and driving out the shanks of reasonable-size rivets. The "hell-dog" is a long, heavy, powerful buster requiring two or three men to handle and operate, and is used on larger diameter rivet heads.

Rivet catching can. (Fig. E3.3.7).

Rivet dolly. Banjo (Fig. E3.3.8); club; heel (Fig. E3.3.9); hook (Fig.

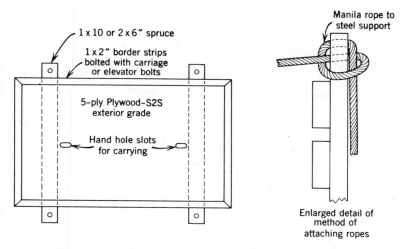

FIG. E3.3.5. Plywood float. Ropes attached as for plank float.

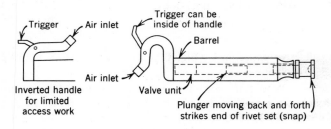

FIG. E3.3.6. Pneumatic rivet hammer.

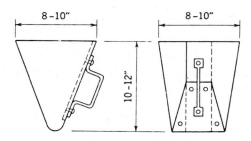

FIG. E3.3.7. Catching can.

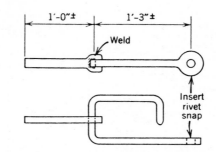

FIG. E3.3.8. Banjo dolly (number 9 bar).

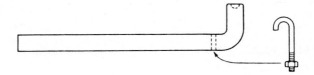

FIG. E3.3.9. Heel dolly.   **FIG. E3.3.10.** Hold-up or dolly hook.

E3.3.10); offset or goose-neck (Fig. E3.3.11); spring (Fig. E3.3.12); straight (Fig. E3.3.13); chain.

Rivet forge. With fan; extra fan; extra tuyere iron. The tuyere iron is a plate with holes to permit the air to pass through and provide the draft to keep the fire going. It is laid in the bottom of the forge pan over the opening through which the fan drives the air for a draft. Some forges are oil-fired, the oil being forced in under pressure.

Electric heating of rivets is usually unsatisfactory for any appreciable quantity of rivets and is not satisfactory for use in the field. It can be used in the tool-house for heating one or two rivets occasionally, eliminating the need to start and later quench a forge fire which is uneconomical and inefficient for only a few rivets to be driven.

Rivet heater's tongs. (Fig. E3.3.14); shovel; pick-up tongs (Fig. E3.3.15).

Rivet sets (*snaps*). Cup (Fig. E3.3.16a); flat or flush (Fig. E3.3.16b); emery stone (shaped); gages.

4. Safety

4.1 Aids to Safety

Girders, extra heavy pieces, and awkward pieces may need special hitches, which can be shop-assembled and bolted, riveted, or welded to the piece to be left permanently in place. With a number of such members involved, one hitch

FIG. E3.3.11.   Offset dolly (gooseneck dolly).

FIG. E3.3.12.   Spring dolly.

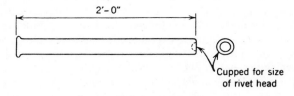

FIG. E3.3.13.   Straight dolly.

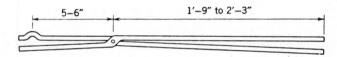

FIG. E3.3.14.   Heater's tongs.

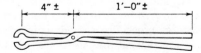

FIG. E3.3.15.   Pick-up tongs.

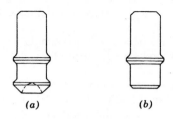

(a)   *(b)*

FIG. E3.3.16.   Rivet sets (snaps): **(a)** cupped for button-head rivets; **(b)** flush for countersunk rivets.

373

is often provided to be field-bolted in turn as each piece is erected. When the hitch must be removed, if riveted, it will probably be burned off.

When riveters can work from a planked floor or a planked area, the work can be done safely and usually more expeditiously. When this is not feasible or is inexpedient, scaffolds or floats should be used. When several men will work in the same area without planking, needle-beam scaffolds are best, especially if one hanging of the two needle beams will serve several points on which the work is to be done. When only one or two men will work at a point, floats or ship scaffolds, or even a boatswain's chair, can probably serve. Care must be used to tie them so that they cannot be pushed out and drop a man. Manila lines for needle beams and for floats should have sufficient extra strength and size so that if a strand is accidentally burned or otherwise damaged, the remaining strength will be sufficient for the load on the float or scaffold.

Material must not be dumped overboard when a scaffold is to be moved. Rivet heaters should have safe containers or buckets for burned rivets or hot rivets left over at the end of the day. Water should be on hand in case a hot rivet is misthrown and not caught, as well as to help the heater quench his forge fire at quitting time.

4.2 Protection to Be Provided

A fire-protection procedure should be set up if there is to be any riveting or any operation that could start a fire. A dropped hot rivet must be checked immediately as it can start a blaze after smouldering unnoticed in flammable material. The fire-protection program should provide a plan for quenching fires, for locating dropped hot rivets, and for placing protective coverings over flammable material that might be endangered by a misthrown hot rivet.

Part of the fire-protection procedure should include assigning someone to check at quitting time to see that all fires in forges are quenched, that all hot rivets have been located and are safely contained against starting a fire. The fire watch should be maintained where advisable for at least an hour after all work involving a fire hazard ceases. The fire watcher should be equipped with a portable extinguisher and with means of communication in case he cannot control an incipient fire.

For maximum safety, men handling forges should be taught the safe methods of using such equipment, and how to leave them at lunchtime and at quitting time.

If there is a hazard of red-lead vapors on lead-painted steel to be riveted or in cutting out rivets in old steel that has been painted with lead paint, respirators should be provided. On new steel, the fabricator should have been cautioned to use nontoxic paint on connections to be riveted. On such connections, lacquer is best on faying surfaces with red oxide or similar paint on the outer surfaces.

5. Erection Scheme

5.1 Study of Contract Drawings

Where one heavy piece can be erected with the equipment planned for use at the site, instead of assembling, fitting, and riveting various parts together before erecting, it is a real advantage to have the fabricator do the assembling and riveting before shipping, provided the size and weight will permit delivery by available carriers. Often such assembly at the site may require falsework with additional expense, time, and danger.

On some tight column splices for which no fillers are provided, it is advisable to have the shop omit the upper one or two rows of rivets so that the splices can be sprung slightly to enter the upper column easily. Similarly, on a girder or truss splice, if the first row of rivets is left for the field to complete, it will usually aid the erector materially, even though he must then do some work originally intended for the shop. If the erector performs some of the work the shop should have done, this will usually be balanced by the fabricator against additional work requested by the erector, such as furnishing lifting hitches or erection seats.

5.2 Installments

In arranging for installments of the structural steel, a study should be made of the rivet installments. By dividing the rivets into larger installments than for the structural steel, but satisfactory for the job requirements, the rivet shipments can cover greater areas and can then be shipped as truckloads or carloads, unless the fabricator is willing to ship the rivets for a particular erection installment with the steel for that installment.

At this time it is also expedient, if no standard has already been established between the erector and the fabricator, to notify his drafting room what percentage of excess rivets should be added to the required list of rivets to be furnished. This is to take care of losses or burned rivets, since most contracts for fabricated structural steel include an excess for which the customer is expected to pay. This percentage can vary from 2 to 5 percent, depending upon sizes, diameters, and type, but must be kept below that for which the fabricator will be paid by the customer.

When connections are to be made with rivets, this will be shown on small sheets usually termed rivet lists. The lists should give the location of every connection, such as beam numbers to column numbers, or beam number to another beam number in the case of a filling-in beam framing into a header. They will give the size of the rivets for that connection, with the length, diameter, and type such as H.S. (for high-strength) rivets, tapered rivets, etc.

5.3 Erection-Scheme Drawings

If trusses or girders are to be assembled at the site before raising them as completed members, the assembly areas should be marked on the erection-scheme drawing so that no other steel is unloaded there that may foul the area. The amount of blocking on which to assemble the individual pieces, if assembled flat, should be given and should be enough to permit men to work safely in installing the rivets while the pieces are still on the blocking. A fairly level, unobstructed area should be used, with timber blocking laid so that the truss or girder will be level, and the pieces can be readily assembled. The blocking should be high enough for the riveters to work safely on the connections on the top, bottom, and sides as the assembly has been laid out.

6. Starting the Job

6.1 Personnel

For his rivet foreman, the superintendent should try to use a man who has actually driven rivets and preferably, one who is able to heat. He should be able to instruct inexperienced men in the work they are to do, especially the safe tying of hitches for their floats or scaffolds. He should be able to know if the rivets driven are satisfactory, and if the number driven are up to the ability of the gang to produce. He should know enough about compressors to recognize malfunctioning and secure the proper repair man; know how to clean and lubricate riveting hammers, impact wrenches, reaming machines, jam dollies, etc.; and how to make minor repairs. (Major repairs should be made by the erector's toolhouse force or by the manufacturer of the equipment.) He should be familiar with the proper care of air hose or pipe for the compressed-air supply, and methods of coupling and the use of necessary valves and all the equipment involved.

On a good-sized job it will be advisable to have foremen in charge of the raising gangs, plumbing and fitting gangs, handline or detail gangs, bolting crews, and the riveting gangs. On a small job a foreman can usually take care of the plumbing and fitting gangs as well as the bolters and riveters.

A riveting gang normally consists of four men: a heater, catcher, bucker-up, and driver. The men often change places and duties during the day. An apprentice or helper feeds rivets and coal, if used, to the heater, and a compressor operator takes care of the compressor if pneumatic riveting hammers and/or impact wrenches or reaming machines are used.

If the work is to be done by union workmen, the local rules in some localities require additional men in the gangs, the minimum then being clearly stated in the union agreement. For example, an apprentice ironworker feeding rivets, or the rivet foreman, can take care of fueling the compressor and starting and stopping it. Under some agreements, a "compressor engineer" is required to start and stop the machine and see that it runs properly. Similarly, with an electrically operated compressor, some union agreements require an electrician to throw the switch to start the motor, disconnect it at lunchtime and at quitting time, and take care of minor electrical troubles.

6.2 Starting Work

When the preliminary work has been completed, rivet lists should be given to the rivet foreman so that he can sort rivets in advance of need and plan on how the correct sizes, lengths, types, etc., will be marked at the connections to guide the men doing the actual work.

6.3 Fitting

Sometimes the steel has been fabricated a little long or a little short. In that case, the erection bolts are removed and the holes reamed, using the plumbing guys to pull in the columns if long, or wedges if short. The holes are then reamed, and oversized rivets will be needed to fill the holes. In this case, additional larger diameter rivets will be needed and should be ordered in ample time to drive rivets in those holes so that if those connections are needed for a derrick or climbing crane to jump, there will be no unnecessary delay. Normally the reaming is done by the fitting gang, but if there are not too many holes involved, the riveting gang can take care of this. Occasionally the riveting gang will perform fitting operations as part of their regular work. In that case, they will usually be equipped with a pneumatically or electrically powered impact wrench and a powered reaming tool, in addition to their normal riveting tools.

In fitting, whether by a fitting gang or by a riveting gang, a pattern of holes is selected, starting near the center of a connection (if of many holes) and working out to the edges, tightening scattered erection bolts sufficient in number to bring the various faces of the different plies of steel into close contact. Drift pins are driven through the remaining open holes, and rivets are then installed in those holes. The fitting-up bolts are next removed, drift pins driven through those holes, and the remaining rivets driven.

6.4 Work under Way

The foreman should check to see that his gangs are getting an adequate supply of compressed air. If the length of the supply line between the compressor and the work is great, an air receiver should be set up close to the work. This will help to keep the pressure up to that which is required for good workmanship. The size of the compressor and of the receiver depend on the sizes and number of riveting hammers and any other pneumatic tools being used.

The hammer sizes depend on the diameter of the rivets involved. A good foreman will arrange to mark the steel at each point before the gang starts there, giving the type of rivet to be used if several different types are to be driven on the job. The diameters of the rivets to be driven are self-evident from the diameter of the holes, and the lengths will be determined by the heater, allowing enough stock to drive a good head.

6.5 Riveting

The four-man riveting gang will usually set up a platform of eight or ten 2 × 12-in. planks for the heater before starting to fit up or drive. This will support his forge, hold kegs for the size rivets to be driven, for buckets of coal (if the forge is coal-fired—some forges are oil fired, and occasionally an electrically operated machine is used to heat a few rivets), and a bucket of water in which to place hot rivets not needed or to extinguish his fire at quitting time. The heater will be equipped with a long-handled heater's tongs to arrange the rivets in his fire so that they will be heated cherry red in the order and sequence needed. He will use it to grab a hot rivet and toss it to the catcher who will be close to the point to be driven. The catcher, in turn, uses a catching can for the purpose; he will use a short-handled pick-up tongs to grab the hot rivet he has caught and place it in the hole to be driven. The bucker-up then pushes the rivet with a dolly, against the manufactured head, and holds it against the driver forming a new head on the shank of the hot rivet sticking through the hole on the other side of the connection. When feasible, the bucker-up should use a jam or striking dolly instead of a hand dolly. As soon as the new head has been formed, the bucker-up moves the hand dolly, or the jam or striking dolly with its pipe and snap, in line with the next hot rivet the catcher will have placed in another hole while the first rivet was being driven.

The work of driving is more arduous than bucking up; during the day the catcher will relieve the driver, who then catches, or all three men will take turns driving. Heating is a highly skilled operation and the heater is thus really a specialist and will stay at the operation while the others change about.

For good production, the four men should be an organized gang, that is, they should have worked together previously. Then each man knows what to do and when to do it, and good production will be attained by a smoothly operating group. Too often if the heater is sick or fails to report for work for any other reason, the other three men will go home rather than use a substitute heater, and the job is short one riveting gang for that day. If one of the other men is out, the rivet foreman will often fill in, although this prevents him from doing what he is hired to do, namely, to supervise all the riveting gangs.

Riveting gangs usually work from floats or needle-beam scaffolds. Only one pair of needle beams is generally needed at a point to support the planks laid across them if they are slung in such a way that there is support on both sides of the connection for the men to work on each side. With floats, two are usually required for safe operations, one hung on one side and the other on the opposite side of the connection. A line should be tied between the two floats so that as a bucker-up pushes against the dolly held on the rivet head and the driver forms the new head, they will not push their respective floats away from each other and permit the men to fall or cause loose tools on the floats to slide off.

A good foreman will check his gangs to convince himself of their ability to hang their needle-beam scaffolds or their floats safely, using proper knots, and he will see that the men inspect their ropes frequently. The rope lines must be adequate for the loads, with some extra reserve strength in case a misthrown rivet burns part of the rope or the rope is cut or damaged inadvertently while the men are on the scaffold or float.

The heater should be checked to make sure he knows the proper additional stock to add to the grip to make a good head on the rivet and to make sure that he knows how to heat his rivets cherry red (too much more would burn them). The rest of the gang should be checked to be sure they can catch the rivets safely and that they are driving tight rivets with properly formed heads, with no collars and no excessive scale; that they are fitting up the points satisfactorily before driving, are keeping the floats annd scaffolds clear of as much loose material as possible, and know how to move them from point to point safely. There should always be floor planks or other protection under the riveting gangs, not only for their own protection in case a man should fall, but also to protect men working under them from hot sparks or a dropped rivet.

Good, clean heating coal for the forge, if coal-fired, should be on hand and distributed to the heaters. Hammers should be serviced ahead of time and checked to be sure that the screens are clean, the hammers properly lubricated, the correct size snaps on hand, and that the gangs know how to secure the snaps to the hammer. The forges should be working properly with hand fans or an air-driven draft correctly connected to the compressed-air supply. Half-inch tail hoses should be readied and enough ¾-in. main hose should be on hand to reach all parts of the work from the receiver or directly from the compressor.

On a tall building where the compressor must be left below, 2-in. diameter air hose or pipe is used between the compressor and the receiver on the floor where the work is being done. Additional sections of hose or pipe are added as the receiver is moved up successively to each new floor. The compressor itself should be located where it is protected from falling objects or burning (or cutting) slag; or else some form of protection should be provided over it.

7. Safety Code

In addition to the General Rules given in Appendix B Suggested Safety Code, under Section 3. Fitting, Reaming, and Bolting—Jacks, ADD: Use care in throwing rivets to avoid hitting others.

Rivet heaters must wear safety goggles or spectacles for protection against dirt and sparks from the forge.

Index

Page numbers in italics refer to illustrations.

381